CLASSIFICATION AND ORDINATION

Advances in vegetation science 2

Edited by

EDDY VAN DER MAAREL

Dr. W. JUNK bv PUBLISHERS THE HAGUE – BOSTON – LONDON 1980

Classification and ordination

Symposium on advances
in vegetation science,
Nijmegen, The Netherlands, May 1979

Edited by

EDDY VAN DER MAAREL

Dr. W. JUNK bv PUBLISHERS THE HAGUE – BOSTON – LONDON 1980

Distributors:
for the United States and Canada

Kluwer Boston, Inc.
190 Old Derby Street
Hingham, MA 02043
USA

for all other countries

Kluwer Academic Publishers Group
Distribution Center
P.O. Box 322
3300 AH Dordrecht
The Netherlands

Library of Congress Cataloging in Publication Data CIP

Symposium on Advances in Vegetation Sciences, Nijmegen,
The Netherlands, 1979.
Classification and ordination.

(Advances in vegetation science; v. 2)
'Reprinted from Vegetatio, vol. 42, 1980.'
1. Vegetation classification – Congresses.
I. Maarel, E. van der. II. Vegetatio. III. Title.
IV. Series.
QK911. S94 1979 580'. 12 80–23092

ISBN-13: 978-94-009-9199-6 e-ISBN-13: 978-94-009-9197-2
DOI: 10.1007/978-94-009-9197-2

Reprinted from *Vegetatio*, Vol. 42, 1980

CONTENTS

INTRODUCTION

Eddy VAN DER MAAREL

This volume is the first of two volumes covering the Symposium 'Advances in vegetation science', which was held at Nijmegen, The Netherlands, from 15–19 May 1979. This symposium was organized on behalf of the Working Group for Data-Processing of the International Society for Vegetation Science. After this group held its final meeting two years earlier it decided to continue its activities, but within a wider scope. Most members of the Group felt that the original aim, i.e. the introduction of data-processing and multivariate methods for use in the systematic description of plant communities, was more or less fulfilled. The book Data-Processing in Phytosociology, largely based on papers in Vegetatio, edited by E. van der Maarel, L. Orlóci & S. Pignatti, and to be published by Dr. W. Junk Publishers, may be considered as a comprehensive report on the Working Group's activities.

The wider scope I mentioned can be described as 'theoretical vegetation ecology' or 'descriptive and experimental analysis of vegetation'. The further development of multivariate methods is certainly a matter of interest for the extended Working Group, but those methods should also be applied to other than synsystematical purposes, for instance to site – time series. In relation to this development general models for vegetation analysis and for vegetation succession should be elaborated. Besides general aspects of diversity and structural – functional properties and population dynamics are worth while studying.

Indeed, this all would imply a new organisation of vegetation ecologists. I hope the Nijmegen Symposium will contribute to this new development.

As I indicated already multivariate methods will remain important for the period to come. Unlike some colleagues who are satisfied with the present state of the art and now wish to achieve results of large scale data treatments I believe that we have not reached that stage. To mention one example: we do have a rather complete series of computer programmes for the rapid clustering and ordination of very large sets of relevés and for (subsequent) table rearrangement (this volume as well as the book Data-Processing in Phytosociology contain various new programs). What we do not have is a manual in which the apparently successful methods are compared and applied to some data-sets. H. Lieth, editor-in-chief of a new Junk series 'Tasks for vegetation science' already suggested to produce such a manual in this series.

The present volume contains the texts of the lectures and most of the poster demonstrations of the first three sessions of the Symposium, dealing with classification and ordination, as well as some manuscripts which were submitted by the time of the Symposium and fitted very well in the Volume. (Such a rather effective procedure can easily be developed if Symposium Proceedings are published as special volumes of Vegetatio!).

The first session presided by L. Orlóci, started with a broad introductory survey by P. Greig-Smith and was continued by M.P. Austin and I.C. Prentice. Session 2, presided by P. Greig-Smith included lectures by L. Orlóci, R. del Moral, M.B. Dale, R.C. Jancey and D. Lausi. Session 3 largely contained examples of applications and comparisons of methods, by M.A. Ayyad, S. Jensén, V. Komárková, O. Wildi, E. Feoli, P. van Hecke, J. Lepart and R.S. Clymo. The contribution by Ayyad and an inserted lecture by H. Lieth will be published in current issues of Vegetatio. (Herewith we have a second advantage of incorporating Symposium Proceedings in Vegetatio: there are always contributions which do not fit in the Symposium theme. By exluding them from the proper proceedings those become more homogeneous and by publishing them in the same way we do justice to the contributors!).

Poster demonstrations were given by most of the speakers mentioned above, and in this way we would compensate for the short speaking time (15–20 min.) most of them got.

Besides a poster demonstration by G. Wiegleb and a discussion contribution by E. van der Maarel were incorporated.

The organisation of the volume is as follows:

A. *General*: Orlóci's opening remarks and the lectures by Greig-Smith and Austin.

B. *Ordination, theory and methods*: Symposium contributions by Orlóci, Prentice, Feoli & Feoli Chiapella, van der Maarel, del Moral and Wiegleb, and inserted manuscripts by Hill & Gauch and Carleton. The contribution by Lepart & Debussche on ecological variables fits in this section.

C. *Classification, theory and methods*: Symposium contributions by Dale, Jancey, Feoli & Lausi, Jensén & van der Maarel and an inserted manuscript by Gauch.

D. *Classification and ordination: examples and comparisons*: Symposium contributions by Clymo, Komárková, van Hecke et al. and Peet. Wildi's paper (dealing with a new package of programs) joins this series.

E. *Epilogue*: Pignatti's closing contribution on the epistomological basis of vegetation science forms a logical end of our Symposium.

Looking back at the Symposium and reading through the manuscripts I came to the conclusion that this volume reasonably reflects the variety and quality we can find in the theories and methods of numerical vegetation science. I am sure that most of the participants feel the same way. I hope many readers will join this conclusion! What the reader cannot detect is the cooperative and stimulating atmosphere during the many discussions, both plenary and private in the poster hall. We can only hope, that more of such symposia, with short but representative lectures and intensive personal contact will follow!

Finally I wish to remark that this symposium was sponsored by the Faculty of Science of the University of Nijmegen. I thank the Director of the Faculty, Dr. C.J.M. Aarts, for his support and cooperation and I am especially grateful for the considerable help which was given by Peter Toll, L.L.D. of the Faculty's Directorate, both in the organization of the symposium and in taking care of the symposium guests! From our Department Rita Dubbers, Henk Butteling and Dr. Marinus Werger deserve being acknowledged for their help.

PREFACE TO 'CLASSIFICATION AND ORDINATION'

László H. ORLÓCI

The Symposium comes at a time when our outlook about the nature of the relevant statistical universe in plant ecology is changing, and at a time when we have a strongly felt need for a new methodology to deal with the new realities in data analysis.

As I see them, these realities may present problems on three separate counts:

1. The assumption of a non-linear species response, and non-linear covariation among species is pivotal in current thinking. Yet, the methodologies in such important fields as the multivariate analysis of variance, discriminant analysis, canonical analysis, component analysis, etc., cannot handle anything but linear covariation.
2. The heuristic approaches in data analysis are closely in line with current preferences for flexibility in method selection and maximum local relevance of the results. Yet what is being offered by way of the prevailing formal methodologies is characterized by adherence to rigid axiomatic formulations, stress on generality, arbitrary confinement of the practice to the simple case of linear data structures, and all these in the interest of elegant, axiomatic developments.
3. The incorporation of the idea of probability in the methodologies to render them statistical is highly desired. Yet, not much has been done to develop a new statistics which can handle non-linear covariation.

My feeling is that at this stage in the evolution of data analysis, ecologists should not feel bound by the orthodox statistical tradition in which data analysis degenerated into a rigid exercise constrained by formality and hosts of restrictive assumptions. They should rather view and practise data analysis as an opportunity in experimentation, to see ideas and techniques evolve, and results emerge which can illuminate as many aspects of the information in the data as reasonable possible.

THE DEVELOPMENT OF NUMERICAL CLASSIFICATION AND ORDINATION

P. GREIG-SMITH

School of Plant Biology, University College of North Wales, Bangor LL57 2UW, United Kingdom

The invitation to open this session set me thinking about the development of numerical procedures of classification and ordination. Their technical development has been reviewed by various authors from various viewpoints, e.g. Cormack (1971), Orlóci (1975, 1978), Dale (1975), Goodall (1970), Greig-Smith (1954, 1964, 1980), Whittaker (1967, 1973). I do not intend to discuss this in more than broad terms, but there is another aspect that has received less attention. This concerns the influences and constraints which have affected the development of numerical methods and their acceptance by phytosociologists. Their acceptance is particularly important; numerical methods are tools and unless they are used in the investigation of real ecological problems we are wasting our time in developing them.

What I have to say represents a personal view, but will, I hope, be of some interest. My excuse for attempting a broad survey of this kind is that I have been closely involved with numerical classification and ordination throughout their development. This has given me the opportunity to look back at the misconceptions and the failures to recognise what now seems obvious, that occur in all scientific development, but which are not so often talked about.

Numerical analysis of plant communities as we now understand it originated approximately 30 years ago, though interest in certain community attributes e.g. species/area relationships and in numerical approaches to the distribution of individual species had developed earlier and was sufficient to justify review articles in 1936 and 1948 (Ashby 1936, 1948). At this time classification was a long established approach, though there was controversy about the most appropriate system to use and the importance of classification to a broader understanding of vegetation. Although Ramensky had developed a technique of ordination (see Sobolev & Utekhin 1973) this was little known outside Russia and it came as a new approach to most phytosociologists. Ordination was thus linked with numerical approaches from the start and this, together with its association with the continuum concept of vegetation, influenced its reception.

The attitude of most ecologists to mathematics at that time can fairly be described as one of suspicion. This is curious because there was at the same time considerable respect for quantification, so that much time was sometimes devoted to obtaining quantitative data in the field, data largely wasted because no further analysis was made. It is interesting that Tansley could write in 1923 'in proportion to the advance of a branch of science its methods become more quantitative. This is true of biology in general and of ecology in particular as of other branches of science.' With characteristic percipience he went on to warn against gathering quantitative data for their own sake, but this warning was not infrequently ignored. The first edition of Fisher's *Statistical Methods for Research Workers*, which was to have such a profound impact on biology in general, appeared in 1925. One wonders whether this may paradoxically have delayed the development of numerical methods of phytosociology. For most biologists the kind of statistical analysis developed by Fisher, with its emphasis on fit to hypothesis, probability and tests of significance, became the only kind of mathematics that was relevant; it did not prove helpful in dealing with plant communities.

The early advocates of ordination techniques were all supporters of the interpretation of variation in vegetational composition as a continuum e.g. Ramensky, Curtis, Goodall, Whittaker. Acceptance of the continuum view was undoubtedly a powerful stimulus to the development of techniques of ordination, but it is now generally accepted that the choice between classification and ordination depends on the objective of data analysis and the structure of the data set being examined, rather than on preconceptions about the nature of vegetation. Presentation of ordination techniques in the context of continuum undoubtedly led to their being ignored by many ecologists. Misunderstanding was increased by the chance that the first practical technique (Curtis & McIntosh 1951) was illustrated by a set of data in which the first and only axis extracted was a successional one and the method was regarded as aiming at elucidating succession rather than of more general application.

The initial techniques were mostly crude and some could scarcely be regarded as numerical. There was little to attract the interest of mathematicians. Later, as techniques developed, mathematicians were to look at them critically and helpfully, as in Cormack's (1971) review of classification, but in the early stages numerical methods were regarded as irrelevant by most ecologists and as unworthy of notice by mathematicians.

Because of the influence of ordination techniques on classificatory techniques, it is convenient to consider ordination first. Dale (1975) has pointed out that there were three major sources of methods of ordination. One, direct gradient analysis, was dependent on recognition of environmental gradients and sought to relate vegetational data to them (Whittaker, 1952, 1956). The other two both sought patterns in the vegetational data and only after these had been identified was their relation to environmental gradients established. The first of these accepted, though not always explicitly, that a set of data could be considered in relation to as many independent axes as there were species present and argued that the dimensionality of the set could be reduced without serious loss of information if species occurrences are correlated (Goodall 1954, Bray & Curtis 1957). The second, path-seeking or more recently (Noy-Meir 1974) 'catenation', aimed to order stands so that species had a unimodal distribution along the axis (Curtis & McIntosh 1951). Reduction in dimensionality emphasises the overall relations between stands, catenation emphasises the relations between a stand and those most similar to it. Though the distinction between these approaches is evident in retrospect, it was not, I think, generally realised at the time. Even those most interested regarded Bray & Curtis' method at the time as a promising extension of that of Curtis & McIntosh to more than one axis, rather than a fundamentally different approach.

The first approaches were in terms of a single axis only. It is true that Whittaker (1956, 1960), by examining moisture gradients at different altitudes, produced what was in effect a two-dimensional ordination and, by comparing moisture and altitudinal effects on two contrasting soil types, was even able to include two different levels on a third axis, but the technique was essentially one of examination of single environmental gradients. There are two considerations of interest. These very simple ordinations were readily understood, even by those with little or no feel for a numerical approach, provided they were prepared to accept the context of continuum in which they were presented. On the other hand, the emphasis on single recognisable environmental gradients has had an influence on later developments which has not been wholly beneficial.

At this stage the limitations of available techniques were that indirect gradient analysis allowed for the derivation of a single axis only and was difficult to apply unless a single gradient was overriding, as in a successional situation, and direct gradient analysis depended on prior recognition of the most important environmental gradients; it was no accident that Whittaker's method was developed in a region of marked topographic variation. It was not surprising that attention was then concentrated on indirect methods that allowed more flexible derivation of several axes.

There were two independent approaches to multiaxis ordination, both dimension-reducing. Goodall (1954) applied a form of principal component analysis and Bray & Curtis (1957) used an essentially geometric approach to reduce the dimensionality of a matrix of interstand dissimilarities or distances. There is an interesting contrast in the subsequent developments from these two pioneer papers. Principal component analysis was not followed up as a practical tool until considerably later, though Dagnelie (1960) discussed the use of both it and factor analysis. Bray & Curtis' technique was relatively soon being used and appears as an important tool in Curtis' *The Vegetation of Wisconsin* (Curtis 1959). Although principal components analysis is more difficult to understand than the very simple Bray & Curtis technique, the delay in following up Goodall's suggestion resulted primarily from the computational load involved. Not until electronic computers of sufficient speed and capacity had been developed did it become feasible to use principal component analysis for any considerable body of data. This is not the only case where the application of analytical techniques has been delayed by the unavailability of adequate computing facilities, rapid though the development of the latter has been. By contrast, Bray & Curtis' technique is feasible 'by hand' i.e. with only a simple desk calculator; Ashton (1964) analysed a very considerable body of data from a species-rich tropical forest by hand. That it was developed in the very active group led by the late J.T. Curtis at the University of Wisconsin also contributed to its rapid adoption.

The simple ordination of Bray & Curtis was criticised on various grounds and modifications were proposed (e.g. Orlóci 1966, 1974, Austin & Orlóci 1966, Swan, Dix &

Wehrhahn 1969). Swan, Dix & Wehrhahn, noting that the use of the two stands farthest apart from one another in the unreduced species space as the endpoints of the first axis might not give the most efficient ordination, proposed to try all possible pairs of stands as endpoints and accept that pair giving the most efficient analysis, i.e. that retaining the maximum total interstand distance along the axis. To do this is computationally demanding and calls for consideration whether, given that the computational load is comparable, the technique has advantages on other grounds. This demonstrates a general risk, often ignored, that modification of a simple technique may eliminate one of its major advantages, its speed and simplicity.

Principal component and factor analyses have provoked more discussion and, I think, have been more misunderstood than any other numerical technique. There has been misunderstanding of the relation between principal component analysis and factor analysis (e.g. Greig-Smith 1964). They are deceptively similar in form, but involve different assumptions. Principal components analysis is essentially a different presentation of the data without any reduction in dimensionality; the initial variates, normally species, are transformed to an equal number of orthogonal variates, or components. Its value lies in the derivation of components in order of decreasing 'variance accounted for' and we may choose to ignore all but the first few components and still retain a large proportion of the information in the data. Factor analysis involves an assumption about the number of independent factors needed to account for the observed correlations between the occurrences of species. As Dale (1975) has commented, factor analysis appears attractive as an ordination technique, but there are formidable difficulties in practice (see also Williams 1976) and it has not been widely used.

Principal components analysis goes back to a paper of Karl Pearson in 1901 with the austere title 'On lines and planes of closest fit to a system of points in space'. Factor analysis was developed and both it and principal components analysis were initially mainly used in psychology, in attempts to define a limited number of independent factors of human ability from the results of psychological tests. This early association with psychology constrained the development of the use of principal component analysis in vegetational ordination. Principal component analysis involves two distinct stages: an initial transformation of the data and the extraction of the eigen values and eigen vectors of the matrix of cross-products between the transformed data. The transformation may involve either or both centering and some form of standardisation. Psychologists, for reasons which need not concern us, had necessarily to centre their data and to standardise by standard deviate i.e. to operate on a matrix of correlation coefficients. When principal component analysis was introduced into vegetational analysis, it was accepted uncritically that the correlation coefficient was the appropriate similarity measure to use. In many data sets all species are measured on the same scale and covariance was therefore sometimes used instead, but it remained for Noy-Meir (1973a, Noy-Meir, Walker & Williams 1975) to clarify the situation and to point out the advantages of not centering the data if they are markedly heterogeneous, as extensive field data often are, and to emphasise that different standardisations represent different weightings. Data standardisation is essentially a matter for biological, not mathematical, decision and depends on the answers to such questions as 'Are rare and common species to be given equal weight?' 'Are differences in standing crop to be ignored?'

It was early recognised that principal component analysis could operate on either cross-products between species or cross-products between stands, 'R' and 'Q' techniques, to give different ordinations, often both interpretable in ecological terms, and there was argument about which was more appropriate. This again resulted from the failure to recognise the two stage nature of principal component analysis. A correlation coefficient between species implies centering by species, a correlation coefficient between stands centering by stands and it is not surprising that they give different ordinations. If the same centering is used, R and Q analyses give the same ordination. Independent recognition of this by Gower (1966) and Orlóci (1967) allowed the useful economy in computing of choosing an R or Q analysis according to whether fewer species or stands are involved.

Misunderstanding of the use of principal components analysis in vegetational analyses was not confined to ecologists. Statisticians, the mathematicians most concerned, viewed principal components analysis as usefully applicable only if each of the variables is normally distributed, which is certainly not true of most real vegetational data. This discouraging judgement, which has probably deterred many ecologists, was presumably based on the mistaken idea that knowledge of the number of 'significant' components is of prime importance. Certainly, non-normal distribution prevents the valid application of significance tests, but these are irrelevant when the objective is data exploration rather than the testing of hypotheses (see below).

Principal components analysis has severe limitations,

which were soon recognised, as an ordination technique. The underlying model assumes linearity of response to each component and additivity of response to different components. These are clearly unrealistic assumptions in relation to the control of species performance by the environment and non-linearity especially has been much discussed (see Austin 1976). There is abundant evidence from experimental work on the response of species to the levels of environmental factors and from direct gradient analyses that response curves are not only not linear, but they are not even monotonic except over narrow ranges. Typically they are unimodal, but may be bimodal as a result of competitive effects (Ellenberg 1953). The result is that if a single gradient with species showing bell-shaped response curves along it is ordinated by principal component analysis, the gradient is not recovered by a single axis but requires two or more dimensions to display it and may be infolded, making interpretation difficult (Swan 1970, Noy-Meir & Austin 1970).

Attempts over the last few years to develop more satisfactory techniques of ordination raise several interesting questions. Do the effects of non-linearity of response curves on the resulting ordination matter? If the objective is to examine individual species response curves, they clearly do. If, however, the objective is to explore the data in order to erect hypotheses about the control of composition of the vegetation by environment, the answer is less certain. A considerable amount of non-linearity in the pattern of a gradient of composition on the ordination will still permit recognition of correlation with values of environmental factors, the basis of hypothesis generation. To assess this we must turn to cases of the use of ordination as a tool in a real situation. Hall & Swaine (1976) examined a very extensive set of data from forests in Ghana by reciprocal averaging ordination (Hill 1973) and found it a fruitful approach. Reciprocal averaging, which can be regarded as a particular form of non-centered principal components analysis, is admittedly less vulnerable to the effects of non-linearity but still shows them. Greig-Smith, Austin & Whitmore (1967) used a conventional centered principal components analysis on data from rain forest in the Solomon Islands and found it profitable.

Procedures have been suggested for ordinating stands in such a way that the individual species values give the best fit to smooth response curves (Gauch, Chase & Whittaker 1974, Ihm & van Groenewoud 1975). These raise the problem of the appropriate form of response curve to use. It has commonly been assumed that response curves are Gaussian in form. Even if they are symmetrical, and Austin (1976) has argued convincingly that this assumption is unjustified, there seems to be a fundamental misunderstanding here. The Gaussian curve reflects the influence of effectively random deviations, due to numerous minor influencing factors, on the probability of observing a particular value of a variable in any one observation. This seems irrelevant to the response of a species to an ordered environmental gradient, though it may give an approximation to a symmetrical response curve, the exact form of which we do not know. Are attempts to ordinate data by fitting to response curves chasing a 'will o' the wisp' ('Irrlicht')?

It is true that if simulated data constructed from a series of Gaussian curves are analysed in this way, an efficient retrieval of the gradient is achieved, but this begs the question how the technique will perform with real data and opens up the whole problem of assessing the efficiency of ordination methods. The earlier approach was to calculate the percentage of variation in the original data accounted for by the ordination, but this is in terms of variation fed into the analysis; the choice of similarity or distance measure determines what information is used. The alternative of testing methods on simulated data has little relevance until we know how to simulate realistic data.

It is worth emphasising that a dichotomy has developed in the objective of ordination, a dichotomy between data exploration as a basis for generating hypotheses about the relation between composition of the vegetation and its environment on the one hand, and elucidation of the relationship of individual species to environmental gradients on the other. The former appears to be dominant in most practical applications of ordination but can, so far, only be assessed empirically by the degree to which it is found to be helpful. The latter has figured prominently in recent methodological studies, perhaps as a result of the chance that many of the earlier procedures were concentrated on single axes, but has played much less part in practical application. I wonder how useful this methodological concentration on the narrower objective will prove to have been; I suspect that problems of species response are better tackled more directly.

Unlike numerical methods of ordination, numerical methods of classification developed against a background of a range of well-established non-numerical methods. It is interesting to consider the interaction between established views of vegetational classification and the development of numerical methods.

Three objectives can be identified in the classification of vegetation, though they are often not explicitly stated and more than one objective may be covered by one procedure. 1) Classification has one very practical function, as a basis of inventory and mapping, either as an objective in itself or as a basis of management. This is present in all the traditional systems, and at its most empirical represents a convenient partition of a range of variation which may or may not be continuous. 2) Classification may aim to identify 'real' entities with clear discontinuities between them, the antithesis of the concept of vegetation as a continuum. It is not always clear whether there is an element of this objective in a particular classificatory system or not. 3) Classification may be a tool in the exploration of correlations between vegetation and environment.

In addition to the general aversion to numerical methods, already referred to, other considerations contributed to a reluctance to accept numerical classification.

Though in practice most non-numerical systems based their classification on detailed recording of a limited number of stands, they aimed to produce a generally valid system into which further stands could be placed, i.e. a system comparable to a taxonomic treatment. The earliest numerical procedures, in contrast, were presented in the context of the examination of the relationships of a particular set of stands, with the implication that a different set from the same range of vegetation could give a different classification; the emphasis was almost entirely on the third objective.

Numerical and non-numerical approaches have in common the aim of producing final groups which are as homogeneous in composition as possible, but there were deeply entrenched convictions about the kind of species that would provide the most efficient criteria for doing so, e.g. dominant species, constant species, species of a particular life form. The distinctive contribution of numerical methods is to allow the data themselves to indicate the most efficient criteria; this came as a novel and unfamiliar idea.

As with the early development of ordination, the history of numerical classification shows ideas running ahead of computational facilities. The strategy of producing a classification may be either divisive or agglomerative. The whole set of data may be successively divided into subsets on an appropriate criterion to produce a hierarchy (divisive strategy) or individual stands may be grouped on an appropriate criterion and the resultant groups in turn grouped until all stands are finally fused into a single group, building a hierarchy from the bottom (agglomerative strategy). Both approaches are used in non-numerical systems, e.g. classification by dominant species is essentially divisive, the Braun-Blanquet system is agglomerative. Further, a strategy may be *monothetic*, based on a single criterion at each stage, i.e. the presence or absence of a single species, or *polythetic*, using many species as the criteria at each stage, i.e. assessment of overall similarity between stands. Again, both strategies are found in non-numerical systems; classification by dominants is monothetic, the Braun-Blanquet system is polythetic.

The first numerical method to be used at all widely was divisive (Williams & Lambert 1959, 1960, following on a suggestion of Goodall 1953) although Sørensen (1948) had earlier proposed an agglomerative method. It is perhaps no more than chance that the former was produced by workers trained in the Anglo-American tradition, but Sørensen came from the Scandinavian agglomerative tradition.

In principle, divisive-monothetic classification is straightforward. The data are divided on the presence or absence of each species in turn and that division is accepted which gives the minimum residual heterogeneity, measured in some appropriate way, within the two resulting subgroups. To do this, however, initially involved an unacceptable amount of computation. This led Williams & Lambert (1959) to suggest that division on that species which had the greatest amount of association with other species would tend to give the greatest reduction in heterogeneity. The resulting *association-analysis* was widely used, but with the increasing speed and capacity of computers, it became feasible to try division on each species in turn and a variety of methods resulted, differing only in the measure of heterogeneity used.

Monothetic procedures have the disadvantage that they ignore much of the information in the data. A divisive-polythetic strategy is not possible non-numerically in most circumstances, but the numerical approach is again straightforward in principle (Edwards & Cavalli-Sforza 1965); all possible divisions of the data into two are examined and that one is accepted which gives the maximum reduction in heterogeneity. With increasing number of stands, the number of possible divisions ($2^{n-1}-1$) increases rapidly and the method is still not possible. The impossibility of this direct approach in practice led to various forms of 'directed search' which aimed to eliminate the less efficient divisions without having to test them (Macnaughton-Smith et al. 1964, Gower 1967, Lambert 1972). An alternative approach which has also produced a variety of methods is to start with the first axis of an ordination of the data and accept the most efficient split of that axis as the criterion for subdivision (Lambert 1972, Lambert et al. 1973, Noy-Meir 1973b, Hill. Bunce & Shaw 1975).

Agglomerative classification presents two problems, the choice of similarity or distance measure, and the strategy of fusion. Both have been the subject of misunderstanding, at least by potential users. Just as the preliminary data transformation in principal component analysis was confused with the analysis itself, the distinctive part played by data transformation in agglomerative classification has been misunderstood. Some measures involve no transformation, unless this is done as a separate preliminary operation e.g. Euclidean distance, others a readily recognised standardisation e.g. the correlation coefficient, and others, and this is where misunderstanding has been most evident, a standardisation differing for each comparison e.g. Sørensen's coefficient, standardised by the sum of the two stands being compared. Different standardisations give markedly different hierarchies (Austin & Greig-Smith 1968) because different aspects of species representation are emphasised; a conscious decision on standardisation is necessary, but this has often not been realised.

Choice of fusion strategy determines the way the distance between a group and a single stand, or between two groups, is measured. To take two contrasting strategies only, a stand may be regarded as having a distance from a group equal to its distance from the nearest member of that group (nearest-neighbour or single-link sorting). Alternatively, the stand may be regarded as having a distance from the group equal to its distance from the member of the group furthest away from it (furthest neighbour or complete-linkage sorting). There are a number of other possible strategies but only nearest-neighbour sorting is free from ambiguity if there is more than one case of the shortest observed difference at any stage. Unfortunately, nearest-neighbour sorting produces very strongly 'chained' hierarchies; once a group is formed further stands tend to be added to it rather than form new groups. Chained hierarchies are almost useless ecologically, either for producing a general purpose classification or for examining correlation with environment. There has been an interesting controversy over the importance of ambiguity. Sibson (1971) has argued that classification must be unambiguous and hence only nearest-neighbour sorting should be used, a view vigorously opposed by Williams et al. (1971), who take the more pragmatic view that a classification must be useful.

There is, I think, more to this disagreement that the contrast between the views of theorists and those who deal with real data. The numerical classification of vegetation has many apparent similarities with numerical taxonomy and each has influenced the development of the other, but there are important differences in assumptions and objectives. Taxonomy, in most cases, deals with what are believed to be real entities, having discontinuities between them, however difficult these may be to identify. Any ambiguity in procedures is therefore disturbing. Paradoxically, ambiguity is not a real problem in numerical taxonomy because operations start some way up a hierarchy – with 'orthodox taxonomic units' rather than individuals – and some procedures have been used successfully that are theoretically capable of giving rise to ambiguities.

Though the relation between taxonomy and phylogeny is a matter for argument, most taxonomists do appear to accept that taxonomic arrangement reflects phylogenetic relationship. *Degree* of similarity between groups is then more than a tool in constructing a classification; it is of interest in itself. This is not true of a vegetation classification, where we are interested only in erecting useful categories (for inventory, mapping, etc.) or in elucidating correlation between vegetation and environment as a means of generating hypotheses about the factors determining the composition of vegetation.

Interaction with numerical taxonomy has perhaps also influenced the relative attention paid to divisive and agglomerative strategies. At least until the opening up of divisive-polythetic classification, agglomerative strategies were more likely to give efficient classifications. Because it starts with orthodox taxonomic units, taxonomy normally deals with a relatively small number of individuals in any one analysis and the computational load of agglomerative strategy is not an obstacle. Ecologically the situation is different; in any real data set there are liable to be a large number of individuals and not until a relatively long way up the hierarchy are results likely to be of interest. Divisive strategy, which can be stopped when the appropriate level is reached, is attractive. Much effort has been put into developing agglomerative techniques which then proved unattractive to users because they involved so much unrewarding computation with large data sets.

As with ordination techniques, assessment of the efficiency of techniques of numerical classification presents problems. Blackith & Rayment (1971) have put it well '. . . there are no objective criteria against which the classifications can be judged. There is, therefore, a tendency for multivariate techniques to be condemned when they disagree with conventional methods, and regarded as superfluous when they agree.' Again, we can only judge by results, not by whether they reproduce our preconceptions, but by whether they are useful in practice or are fruitful of hypotheses. As experience accumulates, we are likely to be able to make a more informed guess as to which techniques are likely to be satisfactory in a given situation.

I have ranged rather erratically over the development of numerical methods. What is their future? They have three principal advantages. They can disclose features which are not revealed by non-numerical methods because relationships are too complex to analyse subjectively. They are particularly useful in little-known or very complex vegetation, such as tropical rain forest. They allow more efficient use of a scarce resource, the skill to interpret the complexity of vegetation in the field. Much, though not all, of the sorting of information that numerical methods achieve can be done by someone with the necessary aptitudes and experience, but such people are better employed in the ultimate interpretation.

There are certain dangers. Are we perhaps too concerned with refinements of methodology? I reiterate my belief that numerical methods are only worth developing if they are to be used on real data in attempts to answer real questions. There are limitations to real data, limitations not only of accuracy of quantitative measures, but also of the reliability of human observation. Hall & Okali (1978) have made revealing observations on the degree to which data from secondary forest in Nigeria are affected by season and by the experience of observers. It is clear that we must expect a considerable degree of inaccuracy in field data from all but very simple vegetation. Are such data adequate input for very refined methods of analysis?

Related to practical use too is the danger of what may be termed the 'black box syndrome'. With increasingly complex methods, and the increasing availability of computer programmes for these methods, it becomes all too easy for the user to take a programme and use it without understanding what it does. Association-analysis, one of the earliest classificatory methods, has been widely used. It is revealing to examine the user literature and see how frequently the method has been misunderstood in important respects although it is a very straightforward one and was clearly explained when it was introduced (Williams & Lambert 1959, 1960). With more complicated methods and the sophistication of modern computers, the risk of misunderstanding and consequent misuse is greater.

Summary

The paper reviews the constraints and influences which have affected the development of numerical classification and ordination of vegetation.

Initial development of ordination techniques and their reception by ecologists was hindered by the mistaken idea that ordination involved acceptance of variation in vegetation as a continuum, as well as by a general suspicion of mathematical approaches.

Three distinct approaches to ordination, largely unrecognised at the time, are apparent in earlier work: direct gradient analysis, reduction in dimensionality and path-seeking (catenation) (Dale 1975).

Modifications of simple initial techniques made them more efficient at the cost of increased computation. Acceptance of heavier computation as computers increased in capacity and speed turned attention to principal component analysis and the superficially similar factor analysis. These have been widely misunderstood largely because they were initially applied in the same way as in the analysis of psychological data, in which different constraints and objectives apply. The initial failure to recognise that principal component analysis involves a preliminary data transformation, the form of which depends on answers to biological, not mathematical, questions, was particularly unfortunate.

Principal component analysis has limitations as a technique of ordination resulting from its assumptions of linearity and additivity of plant responses. Attempts to devise more effective techniques raise questions about the practical importance of non-linearity if the objective is data-exploration rather than elucidating the nature of species-response curves and about the adequacy of using simulated data as test data when we do not know how to simulate realistic data.

Data-exploration has been more prominent in practical uses of ordination but many methodological developments have concentrated rather on species-response curves.

Numerical classification also met obstacles to its acceptance additional to a general aversion to numerical techniques. The first numerical techniques were presented in the context of the relationships of a particular set of data, rather than of a generally valid system, which was the more familiar concept in non-numerical classification.

Both numerical and non-numerical classification aim to produce as homogeneous groups as possible. The distinctive contribution of numerical methods is to allow the data to indicate the most efficient criteria of classification; this was an unfamiliar idea.

The strategy of classification may be either divisive or agglomerative and either monothetic or polythetic. Choice of strategy in earlier work was not only constrained by computational limitation but may also have been influenced by an author's previous experience of non-numerical classification. As with ordination, the distinction between preliminary data transformation and subsequent analysis was at first not appreciated.

Numerical classification has been influenced by parallel numerical developments in formal taxonomy. Because objectives and assumptions are not always the same, this influence has not been altogether helpful.

The limitations of real data suggest that developments of technique are at risk of becoming too concerned with refinements of methodology. Increasingly complex methods and increasing availability of programmes for such methods carry the risk that they may be used without adequate understanding of what they do.

References

Ashby, E. 1936. Statistical ecology. Bot. Rev. 2: 221–35.

Ashby, E. 1948. Statistical ecology. II. A reassessment. Bot. Rev. 14: 222–34.

Ashton, P.S. 1964. Ecological studies in the mixed dipterocarp forest of Brunei State. Oxf. For. Mem. 25.

Austin, M.P. 1976. On non-linear species response models in ordination. Vegetatio 33: 33–41.

Austin, M.P. & P. Greig-Smith. 1968. The application of quantitative methods to vegetation survey. II. Some methodological problems of data from rain forest. J. Ecol. 56: 827–44.

Austin, M.P. & L. Orloci. 1966. Geometric models in ecology. II. An evaluation of some ordination techniques. J. Ecol. 54: 217–27.

Bray, J.R. & J.T. Curtis. 1957. An ordination of the upland forest communities of southern Wisconsin. Ecol. Monogr. 27: 325–49.

Blackith, R.E. & R.A. Reyment. 1971. Multivariate Morphometrics. Academic Press, London and New York.

Cormack, R.M. 1971. A review of classification. Jl R. statist. Soc. A. 134: 321–67.

Curtis, J.T. 1959. The Vegetation of Wisconsin. Univ. of Wisconsin Press, Madison.

Curtis, J.T. & R.P. McIntosh. 1951. An upland forest continuum in the prairie-forest border region of Wisconsin. Ecology 32: 476–96.

Dale, M.B. 1975. On objectives of methods of ordination. Vegetatio 30: 15–32.

Dagnelie, P. 1960. Contribution à l'étude des communautés végétales par l'analyse factorielle. Bull. Serv. Carte phytogéogr. Sér. B 5: 7–71, 93–105.

Edwards, A.W.F. & L.L. Cavalli-Sforza. 1965. A method for cluster analysis. Biometrics 21: 39–63.

Ellenberg, H. 1953. Physiologisches und ökologisches Verhalten derselben Pflanzenarten. Ber. dt. bot. Ges. 65: 350–61.

Fisher, R.A. 1925. Statistical Methods for Research Workers. Oliver and Boyd, Edinburgh.

Gauch, H.G., G.B. Chase. & R.H. Whittaker. 1974. Ordination of vegetation samples by Gaussian species distribution. Ecology 55: 1382–90.

Goodall, D.W. 1953. Objective methods for the classification of vegetation. I. The use of positive interspecific correlation. Aust. J. Bot. 1: 39–63.

Goodall, D.W. 1954. Objective methods for the classification of vegetation. III. An essay in the use of factor analysis. Aust. J. Bot. 2: 304–24.

Goodall, D.W. 1970. Statistical plant ecology. Ann. Rev. Ecol. Syst. 1: 99–124.

Gower, J.C. 1966. Some distance properties of latent root and vector methods used in multivariate analysis. Biometrika 53: 325–38.

Gower, J.C. 1967. Multivariate analysis and multidimensional geometry. Statistician 17: 13–28.

Greig-Smith, P. 1957. Quantitative Plant Ecology. Butterworth, London.

Greig-Smith, P. 1964. Quantitative Plant Ecology, 2nd edn. Butterworth, London.

Greig-Smith, P. 1980. Quantitative Plant Ecology, 3rd edn. In preparation.

Greig-Smith, P., M.P. Austin & T.C. Whitmore. 1967. The application of quantitative methods to vegetation survey. I. Association-analysis and principal component ordination of rain forest. J. Ecol. 55: 483–503.

Hall, John B. & D.U.U. Okali. 1978. Observer-bias in a floristic survey of complex tropical vegetation. J. Ecol. 66: 241–9.

Hall, J.B. & M.D. Swaine. 1976. Classification and ecology of closed-canopy forest in Ghana. J. Ecol. 64: 913–51.

Hill, M.O. 1973. Reciprocal averaging: an eigenvector method of ordination. J. Ecol. 61: 237–49.

Hill, M.O., R.G.H. Bunce & M.W. Shaw. 1975. Indicator species analysis, a divisive polythetic method of classification, and its application to a survey of native pinewoods in Scotland. J. Ecol. 63: 597–613.

Ihm, P. & H. van Groenewoud. 1975. A multivariate ordering of vegetation data based on Gaussian type gradient response curves. J. Ecol. 63: 767–77.

Lambert, J.M. 1972. Theoretical models for large-scale vegetation survey. Mathematical Models in Ecology (ed. by J.N.R. Jeffers), pp. 87–109. Blackwell, Oxford.

Lambert, J.M., S.E. Meacock, J. Barrs & P.F.M. Smartt. 1973. AXOR and MONIT: two new polythetic-divisive strategies for hierarchical classification. Taxon 22: 173–6.

Macnaughton-Smith, P., W.T. Williams. M.B. Dale. & L.G. Mockett. 1964. Dissimilarity analysis: a new technique of hierarchical subdivision. Nature, Lond. 202: 1034–5.

Noy-Meir, I. 1973a. Data transformations in ecological ordinations. I. Some advantages of non-centering. J. Ecol. 61: 329–41.

Noy-Meir, I. 1973b. Divisive polythetic classification of vegetation data by optimized division on ordination components. J. Ecol. 61: 753–60.

Noy-Meir, I. 1974. Catenation: quantitative methods for the definition of coenoclines. Vegetatio 29: 89–99.

Noy-Meir, I. & M.P. Austin. 1970. Principal component ordination and simulated vegetational data. Ecology 51: 551–2.

Noy-Meir, I., D. Walker & W.T. Williams. 1975. Data transfor-

tions in ecological ordination. II. On the meaning of data standardization. J. Ecol. 63: 779–800.
Orlóci, L. 1966. Geometric models in ecology. I. The theory and application of some ordination methods. J. Ecol. 54: 193–215.
Orlóci, L. 1967. Data centering: a review and evaluation with reference to component analysis. Syst. Zool. 16: 208–12.
Orlóci, L. 1974. Revisions for the Bray & Curtis ordination. Can. J. Bot. 52: 1773–6.
Orlóci, L. 1975. Multivariate Analysis in Vegetation Research. W. Junk, The Hague.
Orlóci, L. 1978. Multivariate Analysis in Vegetation Research, 2nd edn. W. Junk, The Hague.
Pearson, K. 1901. On lines and planes of closest fit to systems of points in space. Phil. Mag. 6: 559–72.
Sibson, R. 1971. Some observations on a paper by Lance & Williams. Comput. J. 14: 156–7.
Sobolev, L.N. & V.D. Utekhin. 1973. Russian (Ramensky) approaches to community systematization. Ordination and Classification of Communities (Handbook of Vegetation Science Vol. 5), p 75–103. W. Junk, The Hague.
Sørensen, T. 1948. A method of establishing groups of equal amplitude in plant sociology based on similarity of species content. Biol. Skr. 5(4): 1–35.
Swan, J.M.A. 1970. An examination of some ordination problems by use of simulated vegetation data. Ecology 51: 89–102.
Swan, J.M.A., R.L.Dix. & C.F. Wehrhahn. 1969. An ordination technique based on the best possible stand-defined axes and its application to vegetational analysis. Ecology 50: 206–12.
Tansley, A.G. 1923. Practical Plant Ecology. George Allen and Unwin, London.
Whittaker, R.H. 1952. A study of summer foliage insect communities in the Great Smoky Mountains. Ecol. Monogr. 22: 1–44.
Whittaker, R.H. 1956. Vegetation of the Great Smoky Mountains. Ecol. Monogr. 26: 1–80.
Whittaker, R.H. 1960. Vegetation of the Siskiyou Mountains, Oregon and California. Ecol. Monogr. 30: 279–338.
Whittaker, R.H. 1967. Gradient analysis of vegetation. Biol. Rev. 42: 207–64.
Whittaker, R.H. (ed.) 1973. Ordination and Classification of Communities (Handbook of Vegetation Science, Vol. 5). W. Junk, The Hague.
Williams, W.T. (ed.), 1976. Pattern Analysis in Agricultural Science. CSIRO and Elsevier, Melbourne and Amsterdam.
Williams, W.T. & J.M. Lambert. 1959. Multivariate methods in plant ecology. I. Association-analysis in plant communities. J. Ecol. 47: 83–101.
Williams, W.T. & J.M. Lambert, 1960. Multivariate methods in plant ecology. II. The use of an electronic digital computer for association-analysis. J. Ecol. 48: 689–710.
Williams, W.T., G.N. Lance, M.B. Dale. & H.T. Clifford. 1971. Controversy concerning the criteria for taxonometric strategies. Comput. J. 14: 162–5.

Accepted 20 December 1979

SEARCHING FOR A MODEL FOR USE IN VEGETATION ANALYSIS

M.P. AUSTIN

Division of Land Use Research, Institute of Earth Resources, CSIRO, P.O. Box 1666, Canberra City, A.C.T. 2601, Australia

Keywords:
Direct gradient analysis, Ecological response curves, Elevation, Environmental gradient, Indirect gradient analysis, Nutrients, pH, Vegetation model

Introduction

Current methods of indirect vegetation analysis either explicitly or implicitly assume a certain ecological model of how vegetation responds to environment. Indirect or vegetational ordination methods; (Whittaker 1978) including the more recent methods; of reciprocal averaging (Hill 1973), multidimensional scaling (Fasham 1977, Prentice 1977), PARAMAP (Noy-Meir 1974), and Gaussian ordination (Gauch et al. 1974; Ihm & van Groenewoud 1975) appear sensitive to small changes in the generating model (Austin 1976a, b). These indirect methods are multivariate exploratory data analysis techniques (Noy-Meir 1971, see also Tukey 1977) whose purpose is to expose the unknown ecological dimensions associated with floristic variation; their efficacy depends on the relevance of their model of the vegetation/environment relationship (Austin 1976a, Whittaker 1978). Testing their effectiveness with artificial data sets is entirely dependent on the appropriateness of the model used to generate the artificial data.

Progress in developing an explicit model of vegetation/environment relations is needed rather than further elaboration of ordination methods. This paper is an attempt to review the assumptions of current models and the available observational and experimental evidence. In particular, attention is directed towards defining environmental gradients.

Current models

Apart from the general models of principal component analysis, reciprocal averaging and polar (Bray & Curtis 1957) ordination etc. which have no particular ecological content in their assumptions, several models of vegetation response to environment have been presented.

The majority of models are concerned with the shape of the individual species response to an environmental gradient. Most models can be characterized as having been developed from an inductive study of certain observed cases of vegetation/environment correlation and are descriptive rather than functional, i.e. do not provide a statement of the processes producing the described relationships.

Gaussian response model (Gauch & Whittaker 1972, Whittaker 1978)

This is the familiar bell-shaped curve which has figured in many tests of ordination performance since Swan's original study (1970). Gauch & Whittaker (1972) have provided the most careful statement of this model based on a subjective evaluation of species response curves from the direct gradient studies of Whittaker (1956, 1960, 1965). They assume species have Gaussian curves

$$Y = Y_0 \cdot e^{-(x-u)^2/2\sigma^2}$$

x: position on environmental gradient
Y: measure of species performance

where three parameters, position of mode (u) maximum value (Y_0) and standard deviation (σ) have certain associated distributions. They indicate that while minor species have their modes randomly scattered along the environmental gradient, major species (forest canopy species) appear to have a more even distribution of modes.

This they interpret as a form of Gause's competitive exclusion principle; an example of species niche packing along an environmental gradient (May 1974). They also suggest that the normal curves may be modified by competition. This is incorporated in their simulations by considering that the sum of species importance values are limited by an environmental carrying capacity (N.B. the data are normalized by setting the sums equal to 100). Species richness (number of species per site) is recognized as changing along environmental gradients but is not considered to be consistent between gradients (Whittaker 1972). More complex response curves than the Gaussian are recognized as occurring (Whittaker 1978) but where polymodal curves occur, the different optima are thought to represent different ecotypes and are not considered in their simulation studies. Non-Gaussian curves are ignored in the simulation studies.

No test of the data for fit to a Gaussian curve has apparently been done and Austin (1976a) has pointed out that in published accounts, non-Gaussian curves seem more frequent.

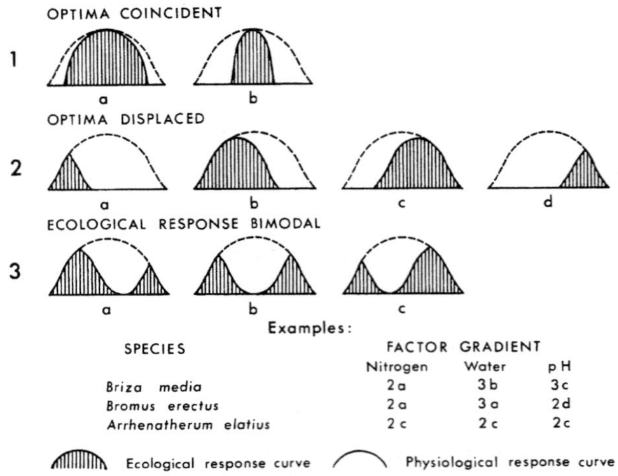

Fig. 1. Schematic representation of the types of ecological response to an environmental gradient proposed by Ellenberg (modified from Mueller-Dombois & Ellenberg 1974). Examples of species responses based on subjective interpretation of field observations.

Ecological displacement model

Ellenberg (in Mueller-Dombois & Ellenberg 1974) put forward a graphical model in which species have more or less similar physiological optima and response curves (i.e. in monoculture), though this is not essential to the model, and depending on competitive ability with respect to the environmental factor, their ecological response curves are displaced from their physiological optima. The ecological response curves may assume a variety of shapes including asymmetric bimodal curves (Fig. 1). No reference is made to vegetation properties in relation to the environmental gradient, e.g. species richness or dominance, neither is there any reference to changing resource availability (carrying capacity) along the gradient.

Austin (1976a), in an attempt to provide such an ecological response model, put forward an extended version of Ellenberg's model, which contained an explicit statement of the effect of competition and the shape of the carrying capacity response to an environmental gradient. A quadratic physiological response curve for each species is assumed, and competition is represented by a series of linear interaction coefficients between all species. Total standing crop of vegetation is determined by the position on the environmental gradient. This constraint assumes that the relationships between an environmental factor and total standing crop is parabolic. Total carrying capacity is then partitioned amongst the species in proportion to their performance after considering their physiological response and competition from other species. By suitable choices of coefficients, the full range of response categories recognized by Ellenberg can be generated (see Austin 1976a for an example of the types of response curve produced).

While the model is rational, it was not developed from any particular set of observations. No consideration is given to species richness along environmental gradients.

Environmental stress model (Grime 1973)

One exception to the generalization that models are individual species oriented and non-functional is the model of Grime first put forward in 1973. He postulates that species density (number of species per unit area) and dominance are a function of environmental stress (Fig. 2) 'under conditions of low environmental stress, productivity is high and species of high competitive index attain maximum vigour and species density is low due to competitive exclusion. Under conditions of increased environmental stress, the competitive species decline in vigour and species of lower competitive ability are able to survive. With further reductions in productivity, species density falls as conditions of extreme environmental stress are reached and species density is limited by the scarcity of

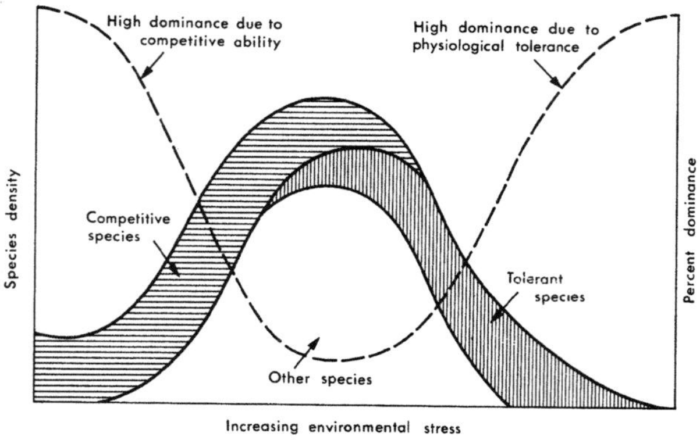

Fig. 2. Diagram of effect of increasing environmental stress on species density (number of species per stand) and dominance (proportion of biomass contributed by single species). Redrawn from Grime (1973).

species tolerant of the specific conditions limiting productivity' (Grime 1973).

Competitive ability is defined in terms of specific species characteristics, e.g. tall stature and high relative growth rate (RGR). High environmental stress is equated with low productivity. No statement is made about the shape of individual species response curves. He discusses environmental stress and disturbance as the components of the environment influencing vegetation. Species are considered as being of three types, competitors, stress-tolerators and ruderals (i.e. plants living in disturbed areas). The possibility that physiological tolerance may also involve a competitive advantage as well as a physiological advantage is not discussed (Newman 1973).

Other models

Other models of vegetation structure have been put forward by various workers. Two deductive models based on ideas of a general species competition factor (Gilbert & Wells 1966) and a two-parameter model (Dale & Anderson 1973; Dale & Webb 1975) involving a measure of site difficulty for plant growth and a parameter of species abundance have been used for classification. Fresco (pers. comm.) has developed a model for vegetation analysis based on Gaussian response curves and linear competition effects. Van der Maarel (1976) suggests a model for species response curves, where in the central regions of an environmental gradient, the responses are Gaussian while at the extremes, they are skewed. He comments, 'We only lack accurate descriptions of both species response and fluctuations [measurements] of environmental factors'!

Components of a vegetation model

Each of the models discussed emphasizes different aspects of vegetation. Can they be put together? The Gaussian response model describes only the ecological response curve of a species. When a species is absent from a gradient, other species may be expected to expand their ecological amplitude to exploit a particular zone on the gradient, thus giving rise to response curves of variable shape. There will be a limit to such expansion depending on the physiological (monoculture) response curve of the species to the particular environmental gradient. The Gaussian model simply has no way of incorporating such phenomena.

The ecological displacement model incorporates consideration of the physiological response curve, competitive displacement and types of ecological response curve. Mueller-Dombois & Ellenberg (1974) imply that physiological response curves may have varied shapes and occupy different positions on the gradient. The extension that I presented (1976a) does little, except make explicit statements about the physiological response curve shape *and* assumes that competitive effect is independent of the level of the environmental factor. However there is also a statement regarding the change in environmental carrying capacity along the gradient. This factor must be expected to influence the shape of species response curves, and raises the question how best to standardize both physiological and ecological response curves in order to compare them.

Grime's approach (1977) is distinctly different, emphasizing vegetation properties (dominance and species richness) and growth strategies rather than individual species response. He provides a competitive explanation for species behaviour at the extremes of a gradient, but what evolutionary mechanism ensures more species and greater niche packing at intermediate levels?

Observation using direct gradient analysis methods on known gradients and experimentation along gradients can provide us with information about the components of a possible conceptual model for use in vegetation analysis. There are however numerous questions that need answers. The questions can be organized under a number of headings.

1. Shape and position of ecological response curves.
2. Shape and position of physiological response curves.
3. Behaviour of carrying capacity (total biomass) along an environmental gradient.
4. Behaviour of niche capacity (species richness) along an

environmental gradient.

If we could make positive statements about these topics and their interrelationships we would have the information to assess whether a vegetation model was useful for analysis purposes or as a conceptual framework for research.

Evidence for model components

Field observations on vegetation behaviour in relation to known environmental gradients, i.e. direct gradient analysis and experimental studies of known species mixtures along controlled environmental gradients provide the only acceptable sources of evidence for a model for *indirect* vegetation analysis.

Field observations

The classic studies of Whittaker (1956, 1960, 1965) provide the clearest evidence that species and vegetation properties show complex behaviour in relation to environmental gradients. Unfortunately, the information from the topographic moisture gradient is not acceptable as the positions on the gradient are calculated from the species' weighted scores. It is a form of indirect ordination and any conclusions regarding species responses would involve a degree of circularity (Austin 1972, Whittaker 1978). Elevation (altitude) is therefore the only unequivocal gradient which can be used from these pioneering studies of Whittaker.

Fig. 3 from Whittaker (1978) shows the ecological

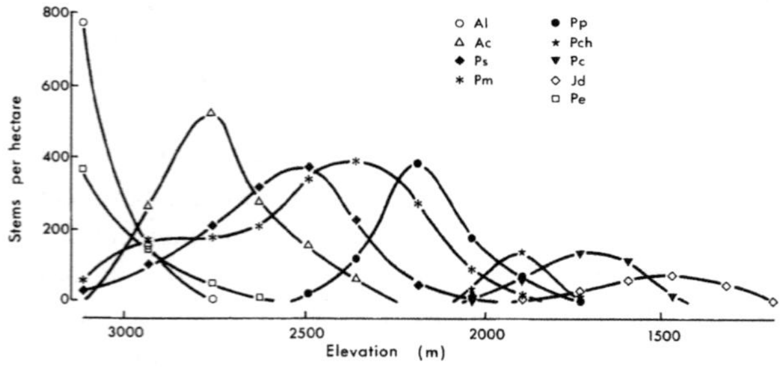

Fig. 3. Ecological response curves for coniferous tree species in relation to an elevation gradient on northfacing slopes in the Santa Catalina and Pinaleno Mountains, Arizona. *Al, Abies lasiocarpa; Ac, Abies concolor; Ps, Pinus strobiformis; Pm, Pseudotsuga menziesii; Pp, Pinus ponderosa; Pch, Pinus chihuahuana; Pc, Pinus cembroides; Jd, Juniperus deppeana; Pe, Picea engelmanni* (modified from Whittaker (1978)).

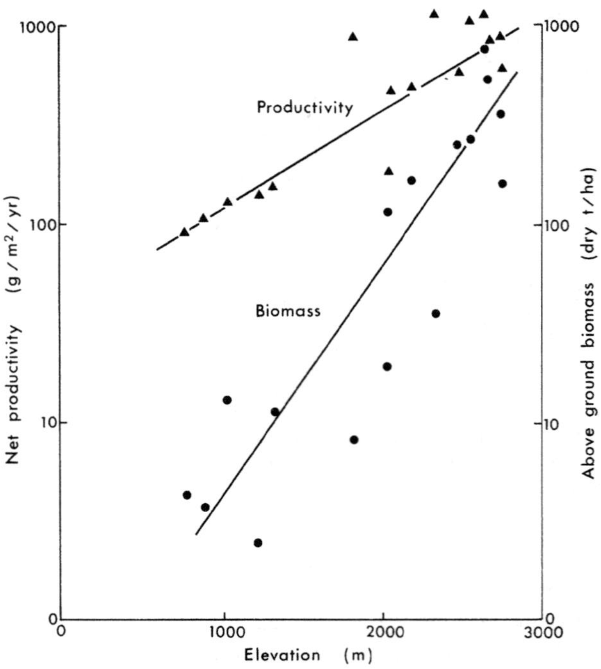

Fig. 4. Vegetation properties in relation to an environmental gradient of elevation (from Table 4 of Whittaker & Niering 1975).

response curves for a sequence of tree species (community dominants) which form a replacement series along the elevation gradient. The curves are skewed relative to the Gaussian symmetric response curve but not grossly. Whittaker (pers. comm. in Austin 1976a) has observed a number of bimodal and 'shouldered' response curves in his studies (Whittaker 1967). Total numbers of stems of all species which may be regarded as an estimate of the carrying capacity also shows a decline with elevation in these data. Fig. 4 (Whittaker & Niering 1975) shows more detailed estimates of standing biomass (carrying capacity) and productivity relationships with elevation. These results are examples of the very extensive studies of Whittaker (1956, 1960, 1965) indicating the relationships of species response curves and carrying capacity to elevation.

Whittaker (1969) has summarized his observations regarding species richness and diversity: 'There are broad trends of decreasing diversities of vascular plant communities toward environments less stable, less favourable, and more extreme but these trends are in detail strongly modified by more complex relationships involving different plant strata and life-forms and by the effects of characteristics of dominant species on composition and diversity of the communities they dominate'.

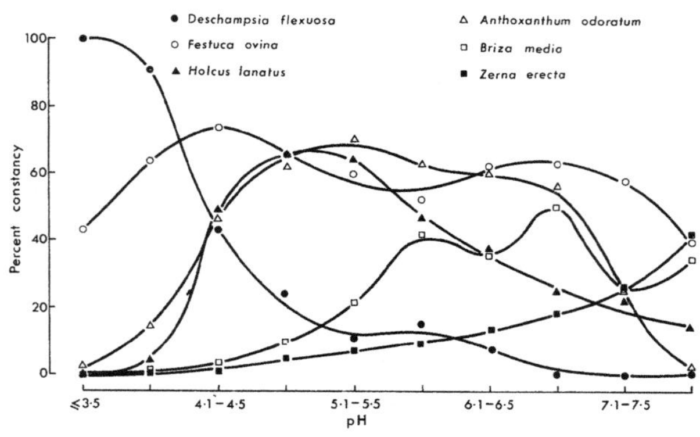

Fig. 5. Ecological response curves for selected grass species along a pH gradient. Data taken from histograms in Grime & Lloyd (1973).

The environmental gradient, elevation, is a complex-gradient in which physiologically active variables such as temperature, evaporation, and moisture availability are correlated with elevation but the correlation is a local one, changing from area to area. Local species populations may be adapted to the gradient and show consistent response curves which are unlikely to be applicable elsewhere.

Another example of direct gradient analysis based on a large number of samples and an acceptable form of survey design is the grassland survey of the Sheffield region by Lloyd et al. (1971), Lloyd (1972) and Grime & Lloyd (1973). The ecological response curves using percent constancy (percentage of sites with that species) as a measure for a selected series of grasses in relation to pH are shown in Fig. 5 (drawn from histograms in Grime & Lloyd 1973). There is skewness apparent in the curves, and the possibility of ecotypic differentiation in *Festuca ovina*. These curves can only indicate the shape of ecological response curves, they tell us nothing about the species physiological tolerance or competitive processes involved.

Grime (1973) has also examined species density in relation to pH for these samples. There is an increase in species richness with pH up to a maximum at 6.5–7.0 with a slight decline at higher pH's. pH is a physiologically active environmental variable having direct influence by means of explicit physical and chemical processes on the availability of ions such as phosphate through its effect on the solubility of aluminium etc. (Rorison 1973). This is in contrast to elevation.

The evidence from gradient analysis indicates that there are important relationships between environmental

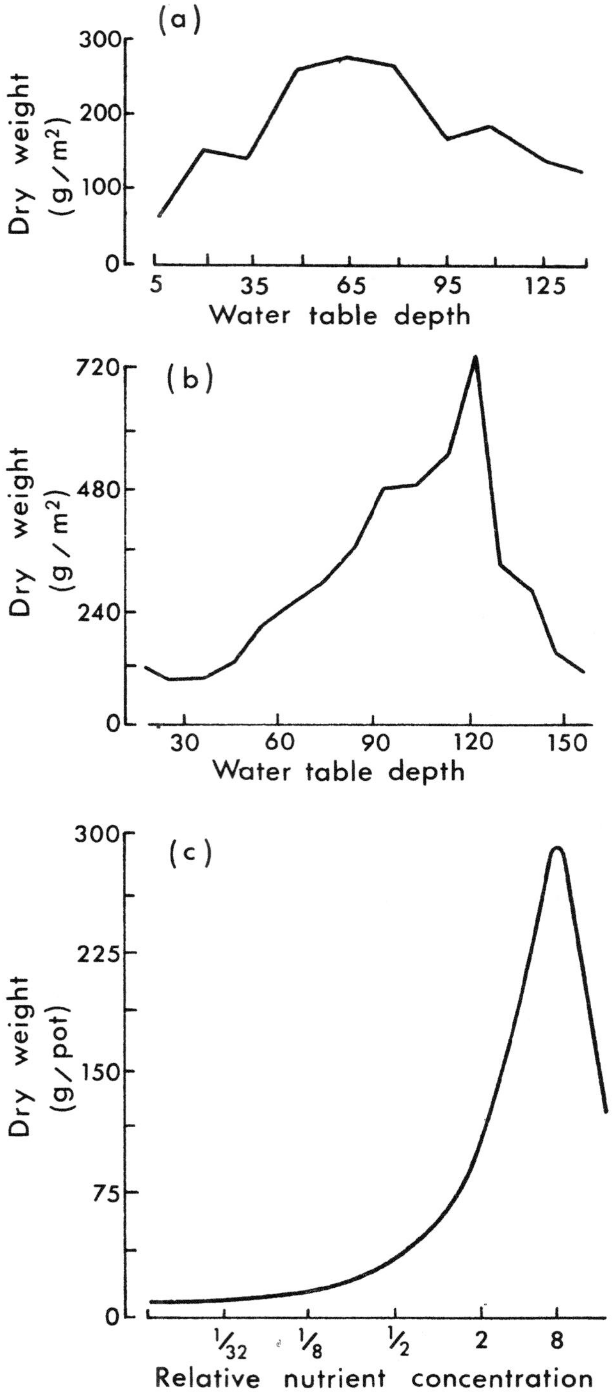

Fig. 6. Relationship of total biomass to environmental gradients under experimental conditions
(a) Ellenberg (1953), 6-species mixture,
(b) Mueller-Dombois & Sims (1966), 3-species mixture,
(c) Austin & Austin (in press), curve based on maximum values from monocultures, 5- and 10-species mixtures.

gradients and species response curves, carrying capacity (total biomass) and species (niche) density.

Experimental observations

It is only from experimental studies that clear statements can be obtained on differences between physiological and ecological responses, though their applicability to field situations may be in doubt.

Environmental carrying capacity.

Several sets of data are available for the relationship between total vegetation biomass (carrying capacity) of species mixtures and environmental gradients. Fig. 6 presents results from three experiments, two from water-table depth gradients (Ellenberg 1953, Mueller-Dombois & Sims 1966; see also Sims & Mueller-Dombois 1968) and the third from a general nutrient gradient (Austin & Austin, in press). There is little doubt that there is strong control of carrying capacity along an environmental gradient.

Species response curves.

Ellenberg (1953) (see also Austin 1979) reports results showing a bimodal ecological response in relation to a pH gradient for one species in a two-species mixture. Mueller-Dombois (1964) shows (see also Sims & Mueller-Dombois 1968) an example for height growth of tree seedlings of four species in mixture along a water-table gradient in which a variety of response curves are obtained

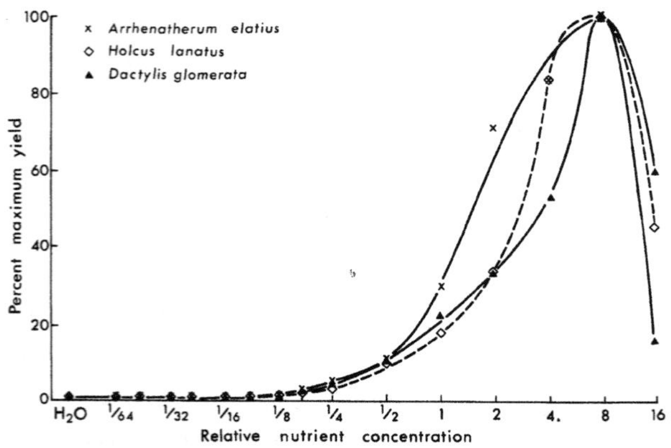

Fig. 7. Physiological response curves for three species in relation to a nutrient gradient (see Austin & Austin, in press).

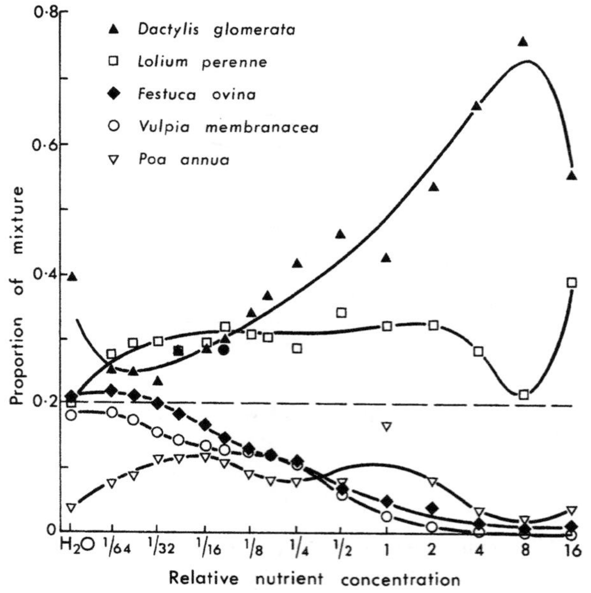

Fig. 8. Relative ecological performance curves for species grown in a 5-species mixture along a relative nutrient concentration gradient (from Austin 1979).

including a bimodal one for white spruce *(Picea glauca)*.

Austin & Austin (in press) made a study of species response curves in relation to a gradient based on a dilution series of a general nutrient solution. Physiological response curves (Fig. 7) when expressed as absolute shoot yield (dry weight per pot of ten individuals) for nine grass species show very similar shapes and optima (when nutrient concentration is expressed on a log scale) but have different maxima for biomass production. Ecological response curves based on biomass are also similar but if expressed as a proportion of the total biomass of a mixture (relative ecological performance), distinct differences between species become apparent (Fig. 8).

Ellenberg (1953, 1954) in his early experiments on a water-table depth gradient was able to show displacement of the ecological response curve relative to physiological response, though statistically there may be some queries about the reality of certain bimodal curves (Austin 1979). Ernst (1978) has questioned whether ecological optima are displaced from a species physiological optima by competition; he maintains that species are not displaced by competition. There are several problems regarding this controversy but one central issue not addressed is standardizing species performance in monoculture and multi-species stands to make the data comparable. In field studies, the ecological response curve is a probability-of-survival and growth which can be measured by a number of attributes, percent constancy (Fig. 5), stem density (Fig. 4) percent frequency or biomass. Under experimental conditions biomass may not be the most appropriate measure. It does not then measure capacity to produce viable seed or persistence in relation to perturbation by other environmental factors over a realistic period of time; experimental periods are usually too short.

Relative performance.

The relative ecological performance (proportion of total mixture biomass) represents the ability of a species to capture resources in competition with other species in a particular environment.

The carrying capacity (total biomass produced) of the environment varies along environmental gradients (Fig. 6). The relative physiological ability of a species to exploit an environment can be estimated by comparing a species performance with an estimate of carrying capacity. A relative nutrient utilization value can be defined as

$$R_{ij} = \frac{Y_{ij}}{\text{Max. } Y_{.j}}$$

where Y_{ij} is the shoot dry weight yield at nutrient level j and maximum $Y_{.j}$ is the maximum yield for a monoculture of any of the nine species studied. Fig. 9 shows some of the results obtained. *Festuca ovina* has a utilization curve similar to its performance in mixture (cf. Fig. 8). Maximum performance of *L. perenne* in mixture corresponds with maximum utilization values for the species. Whether a monoculture yield accurately measures the carrying capacity along the environmental gradient rather than the yield of a mixture requires further study (see Austin & Austin, in press). It does provide a standard for a given set of species in a given experiment.

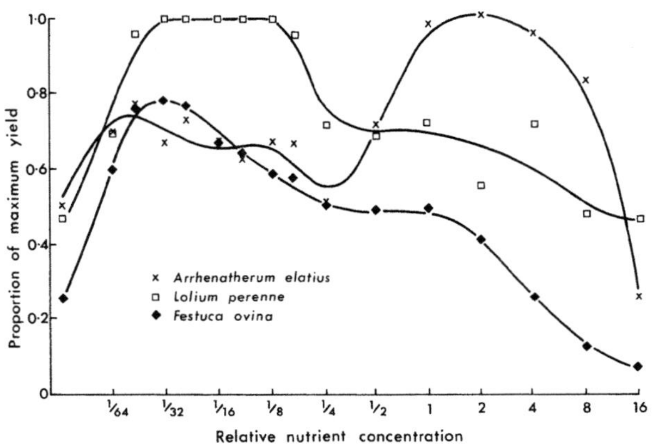

Fig. 9. Relative nutrient utilization value curves for selected species along a relative nutrient concentration gradient (from Austin and Austin unpublished data).

Austin & Austin (Fig. 9 and unpublished) have compared relative utilization values with the performance of the same species in mixtures. Some general agreement was found but problems associated with changes in the relative yields of monocultures and mixtures along the nutrient gradient (Austin & Austin, in press) complicate the picture. We (Austin & Austin, in press) suggested that the nutrient gradient could be divided into five zones:

(a) Nutrient deficient (tap water treatment)
(b) Nutrient limited (1/64 × – 1/6 × relative nutrient level)
(c) Nutrient and shading determined performance ($\frac{1}{4}$–2 × rel. nutrient levels)
(d) Shading and nutrient toxicity determined performance (4–8 ×)
(e) Nutrient toxicity (16 ×).

Correlations are only found for the nutrient limited zone (b). The nutrient gradient approximates an essential resource gradient for plants. When resource (nutrient) supply exceeds demand (approx. at a relative nutrient level of $\frac{1}{2}$ ×), some other factor becomes limiting, presumably light. Other factors such as canopy closure and associated shading effects modify any relationship between performance in mixture and relative nutrient utilization values under optimum conditions. Resource levels may increase further until they reach toxic levels. For a resource gradient which covers all these possible conditions, a species response curve may show a variety of shapes depending on the full complement of its relevant attributes, and those of its companion species.

If ecological response is measured as frequency or percent constancy, and physiological response is measured as biomass production for a resource gradient like a general nutrient concentration, then the ecological optimum will be displaced. If the physiological response is measured relative to other species by means of a relative nutrient utilization value, then ecological and physiological optima may coincide (cf. Ernst 1978).

Species richness and dominance.

Taller species *(Dactylis glomerata)* dominate in the multispecies mixtures (Fig. 8) but their dominance varies along the nutrient graduent (Fig. 10). Greatest

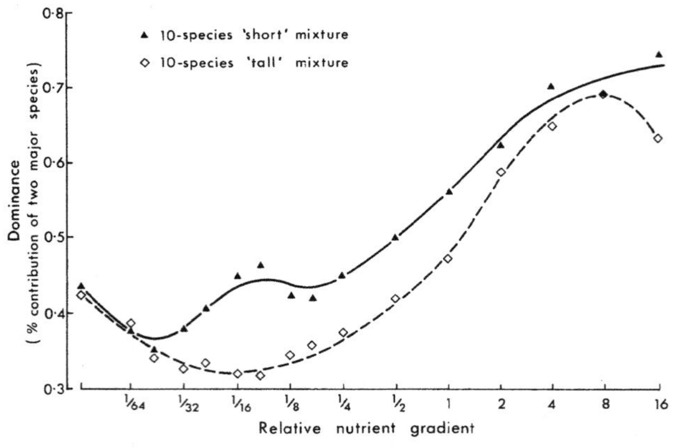

Fig. 10. Dominance behaviour along nutrient gradient for two 10-species mixtures (from Austin & Austin, in press).

dominance is observed under optimum nutrient levels. Grime's suggestion regarding the behaviour of dominance in relation to an environmental stress (gradient) is supported for low nutrient conditions. Minimum diversity occurs at low but not extreme nutrient with an increase in dominance under extreme nutrient deficiency. Supra-optimal conditions are insufficiently sampled for certainty but the data do not appear to support the hypothesis.

Discussion

Evidence for the shape of a species response curve is limited (Ellenberg 1953, 1954, Ernst 1978, Austin & Austin, in press) and mainly restricted to grasses (though see Sharitz & McCormick 1973). Current information suggests that performance needs to be expressed relative to other species for each position on the environmental gradient and particularly relative to those with which it grows. The experimental evidence does not provide support for Gaussian curves (Figs. 3 and 8). However the shape of the curve is dependent on the scaling of the environmental gradient and the type of gradient (see below). Many more field and experimental observations of performance along environmental gradients are needed.

In relation to a resource gradient such as nutrients, the competitive processes can be tentatively recognized but they are *not* the same at all levels of the resource. Under deficient resource levels, exploitative competition for the resource occurs; under optimal levels (supply exceeding demand), competition occurs for an alternative resource, light with interference (shading) competition effects contributing; under supra-optimal conditions, tolerance of toxicity is a significant component in determining performance. Response curves in relation to a resource gradient are likely to show complex shapes along such gradients unless evolution has resulted in selection of attribute combinations relevant to all processes acting along the resource gradient. Response to other environmental gradients is unlikely to be similar to that for resource gradients.

Fig. 7 and the results obtained using the relative nutrient utilization values (see also Whittaker & Niering 1975) demonstrate that the environmental carrying capacity (stand abundance, Austin in press) is a function of the environmental gradients and has an important determining role in species behaviour. The hypothesis of Grime regarding dominance and species richness may require modification as regards an explanation of the processes producing such effects (Newman 1973) but the general relationship of the properties with environmental gradients is now well-established (Whittaker 1969, Peet 1978, Austin in press). However the details of the relationship seem to vary with the environmental factor or the region studied (Whittaker 1972).

Discussion and development of better models for vegetation analysis is limited by our concept of an environmental gradient. In reviewing the evidence on species responses, three types of environmental gradient have been recognized.

1. Indirect environmental gradients, in which the environmental factor used to order the observations does not of itself have a *direct* physiological influence on plant growth.

Elevation is an example of such a gradient (Figs. 4 and 5), aspect is another. Such variables may show strong correlation with species performance in some areas and none in others depending not on a change in the limiting factor for plant growth but a change in the correlation of the limiting factor with the environmental variable.

The environmental scalars of Loucks (1962), Waring & Major (1964) and Austin (1972) have been efforts to overcome this problem. Without such transformations, models based on indirect environmental gradients will be specific to one region and one gradient.

2. Direct environmental gradients, where the environmental factor has a physiological influence on plant growth but is not a resource for plant growth for which exploitative competition might take place.

pH is an example (Fig. 6); it has a major impact on plant growth by modifying the availability of various nutrient resources, through known physio-chemical processes.

Examples of such processes are the influence of pH on the solubility of aluminium and the influence of aluminium on phosphate availability (Rorison 1973) or, the form of availability of nitrogen either as ammonium or nitrate (Gigon & Rorison 1972). Temperature is another environmental variable of the same kind influencing resources and growth by modifying reaction rates, through frost effects, or protein denaturation.

Models assuming processes and species curves for such environmental gradients will need to consider what changes in resources and behaviour will occur along the gradient. If plants are responding to a resource factor which has an abrupt change in availability at a particular pH, then abrupt changes in response curves may take place. Any models appear likely to be general with respect to location but specific for the particular environmental gradient.

3. Resource gradients where the environmental variable is actually an essential resource for plant growth, e.g. nutrients.

Nitrate nitrogen is an example. Note that this type of resource is different from the usual type of resource discussed by animal ecologists in niche studies (May 1974) where, for example, food size is regarded as a resource which can be partitioned amongst species. Nutrients can not be partitioned in the same way, they are either sufficient or deficient and response to oversupply (toxicity) is unlikely to be correlated with ability to utilize low levels of nutrient. There appears to be some possibility of developing predictive models for performance in experimental mixtures from monocultures (Fig. 9) but the shapes of response curves are likely to be sensitive to companion species.

Much of the discussion on species response curves and environmental gradients has been confused by a lack of distinction between the different types of environmental gradient. Species response curves for indirect environmental gradients may be expected to have at least one maximum but very little else is certain. Response to a direct gradient may be more predictable but the physiochemical relationship between plant resources and the gradient may have abrupt transitions with the result that response curves may be highly skewed. Resource gradients may show limiting factor type response curves until toxic levels are reached (Ellenberg, pers. comm.). The nutrient gradient results discussed here are not a single resource gradient, different nutrients may be limiting at different levels of nutrient deficiency. There may well be a continuum of environmental gradient types from pure resource to spurious statistical correlations.

Unfortunately the number of suitable direct gradient studies is so few, that a comparative study of species response curves to the different types of gradient is not possible. Van der Maarel's (1976, see p. 3) remarks regarding the lack of accurate observations is very apposite; the lack of suitable data is as big an obstacle to progress as lack of a model. Ecological carrying capacity and niche (or species) capacity are related to gradients but as yet there is insufficient information to incorporate them. Until such studies become available, we cannot expect to develop suitable models for vegetation, nor advise others on the most suitable methods of vegetation analysis, except on the basis of heuristic experience (Noy-Meir & Whittaker 1978). Indirect analysis of vegetation by means of ordination or numerical classification methods, where the environmental relationships are not understood, requires an approximate model based on direct analysis studies of vegetation.

Summary

Indirect gradient analysis methods require an explicit vegetation model which must be based on direct gradient analysis studies. Various vegetation models are reviewed. Field evidence for the models is discussed. Experimental studies of species response to environmental gradients are reviewed and discussed. Three types of gradient are recognized as important for development of models: indirect environmental gradients where the environmental factor has no direct physiological influence on plant growth e.g. elevation; direct environmental gradients where the factor has a direct physiological effect on growth but is not an essential resource, e.g. pH; resource gradients where the factor is an essential resource for plant growth. The behaviour of the ecological carrying capacity and the role of competition along such gradients are shown to be important for developing vegetation models.

References

Austin, M.P. 1972. Models and analysis of descriptive vegetation data. In 'Mathematical models in Ecology' (ed. J.N.R. Jeffers). Symp. Brit. Ecol. Soc. 12: 61–86.

Austin, M.P. 1976a. On non-linear species response models in ordination. Vegetatio 33: 33–41.

Austin, M.P. 1976b. Performance of four ordination techniques assuming three different non-linear species response models. Vegetatio 33: 43–49.

Austin, M.P. 1979. Current approaches to the non-linearity problems in vegetation analysis. In 'Satellite Program in Statistical Ecology'. S12 Contemporary Quantitative Ecology and Related Ecometrics. Ed. G.P. Patil & M. Rosenzweig, International Co-operative Publishing House, Fairland, Maryland.
Austin, M.P. in press. The role of certain diversity properties in vegetation classification. In 'Vegetation Classification in the Australian Region'. Ed. A. Gillison and D.J. Anderson, CSIRO and A.N.U. Press.
Austin, M.P. & B.O. Austin (in press). Behaviour of experimental plant communities along a nutrient gradient. J. Ecol.
Bray, J.R. & J.T. Curtis. 1957. An ordination of the upland forest communities of southern Wisconsin. Ecol. Monogr. 27: 325–349.
Dale, M.B. & D.J. Anderson. 1973. Inosculate analysis of vegetation data. Aust. J. Bot. 21: 253–276.
Dale, M.B. & L.T. Webb. 1975. Numerical methods for the establishment of associations. Vegetatio 30: 77–87.
Ellenberg, H. 1953. Physiologisches und ökologisches Verhalten derselben Pflanzarten. Ber. Deutsch. Bot. Ges. 65: 351–362.
Ellenberg, H. 1954. Über einige Fortschritte der kausalen Vegetationskunde. Vegetatio 5/6: 199–211.
Ernst, W. 1978. Discrepancy between ecological and physiological optima of plant species. A re-interpretation. Oecol. Plant. 13: 175–188.
Fasham, M.J.R. 1977. A comparison of nonmetric multidimensional scaling, principal components and reciprocal averaging for the ordination of simulated coenoclines, and coenoplanes. Ecology 58: 551–561.
Gauch, H.G. Jr. & R.H. Whittaker. 1972. Coenocline simulation. Ecology 53: 446–451.
Gauch, H.G. Jr., G.B. Chase & R.H. Whittaker. 1974. Ordination of vegetation samples by Gaussian species distributions. Ecology 55: 1382–1390.
Gigon, A. & I.H. Rorison. 1972. The response of some ecologically distinct plant species to nitrate- and ammonium-nitrogen. J. Ecol. 60: 93–102.
Gilbert, N. & T.C.E. Wells. 1966. Analysis of quadrat data. J. Ecol. 54: 675–685.
Grime, J.P. 1973. Control of species density in herbaceous vegetation. J. Environ. Manage. 1: 151–167.
Grime, J.P. 1977. Evidence for the existence of three primary strategies in plants and its relevance to ecological and evolutionary theory. Am. Nat. 111: 1169–1194.
Grime, J.P. & P.S. Lloyd. 1973. An ecological atlas of grassland plants. Edward Arnold, London, vii † 192 pp.
Hill, M.P. 1973 Reciprocal averaging: an eigenvector method of ordination. J. Ecol. 61: 237–249.
Ihm, P. & H. van Groenewoud. 1975. A multivariate ordering of vegetation data based on Gaussian type gradient response curves. J. Ecol. 63: 767–778.
Lloyd, P.S. 1972. The grassland vegetation of the Sheffield region. II. Classification of grassland types. J. Ecol. 60: 739–776.
Lloyd, P.S., J.P. Grime & I.H. Rorison. 1971. The grassland vegetation of the Sheffield region. I. General features. J. Ecol. 59: 863–886.
Loucks, D.L. 1962. Ordinating forest communities by means of environmental scalars and phytosociological indices. Ecol. Monogr. 32: 137–166.
Maarel, E. van der. 1976. On the establishment of plant community boundaries. Ber. Deutsch. Bot. Ges. 89: 415–443.
May, R.M. 1974. Stability and Complexity in Ecosystems. 2nd Ed. Princeton University Press, Princeton, New Jersey. 265 pp.
Mueller-Dombois, D. 1964. Effect of depth of water table on height growth of tree seedlings in a greenhouse. Forest Science 10: 306–316.
Mueller-Dombois, D. & H. Ellenberg. 1974. Aims and methods of vegetation ecology. Wiley, New York, 547 pp.
Mueller-Dombois, D. & H.P. Sims. 1966. Response of three grasses to two soils and a water table depth gradient. Ecology 47: 644–648.
Newman, E.I. 1973. Competition and diversity in herbaceous vegetation. Nature 244: 310.
Noy-Meir, I. 1971. Multivariate analysis of the semi-arid vegetation in south-eastern Australia: Nodal ordination by component analysis. Quantifying ecology, Proc. Ecol. Soc. Aust. 6: 159–193.
Noy-Meir, I. 1974. Catenation: quantitative methods for the definition of coenoclines. Vegetatio 29: 89–99.
Noy-Meir, I. & R.H. Whittaker. 1978. Recent developments in continuous multivariate techniques. 2nd ed. In 'Ordination of Plant Communities', Ed. R.H. Whittaker. Junk, The Hague, p. 337–378.
Peet, R.K. 1974. The measurement of species diversity. Ann. Rev. Ecol. Syst. 5: 285–307.
Prentice, I.C. 1977. Non-metric ordination methods in ecology. J. Ecol. 65: 85–94.
Rorison, I.H. 1973. The effect of extreme soil acidity on the nutrient uptake and physiology of plants. In 'Acid Sulphate Soil'. Ed. H. Dost. Proc. Int. Symp. on Acid Sulphate Soils. Publication 18, Vol. 1, Int. Inst. Land Reclamation and Improvement, Wageningen.
Sharitz, R.R. & J.F. McCormick. 1973. Population dynamics of two competing annual plant species. Ecology 54: 723–739.
Sims, H.P. & D. Mueller-Dombois. 1968. Effect of grass competition and depth to water-table on height growth of coniferous tree seedlings. Ecology 49: 597–603.
Swan, J.M.A. 1970. An examination of some ordination problems by use of simulated vegetational data. Ecology 51: 89–102.
Tukey, J.W. 1977. Exploratory data analysis. Addison-Wesley. Reading, Mass., 688 pp.
Waring, R.H. & J. Major. 1964. Some vegetation of the California coastal redwood region in relation to gradients of moisture, nutrients, light and temperature. Ecol. Monogr. 34: 167–215.
Whittaker, R.H. 1956. Vegetation of the Great Smoky Mountains. Ecol. Monogr. 26: 1–80.
Whittaker, R.H. 1960. Vegetation of the Siskiyou Mountains, Oregon and California. Ecol. Monographs 30: 279–338.
Whittaker, R.H. 1965. Dominance and diversity in land plant communities. Science, N.Y. 147: 250–260.
Whittaker, R.H. 1967. Gradient analysis of vegetation. Biol. Rev. 42: 207–264.

Whittaker, R.H. 1969. Evolution of diversity in plant communities. In 'Stability and diversity in ecological systems Eds. G.M. Woodwell & H.N. Smith. Brookhaven. Symposium in Biology 22, p. 178–196. Brookhaven National Laboratory.
Whittaker, R.H. 1972. Evolution and measurement of species diversity. Taxon 21: 213–251.
Whittaker, R.H. 1978. Direct Gradient Analysis. 2nd ed. In 'Ordination of Plant Communities'. (Ed. R.H. Whittaker). Junk, The Hague, p. 7–51.
Whittaker, R.H. & W.A. Niering. 1975. Vegetation of the Santa Catalina Mountains, Arizona. V. Biomass, production, and diversity along the elevation gradient. Ecology 46: 771–790.
Whittaker, R.H. & H.G. Gauch Jr. 1978. Evaluation of ordination techniques. 2nd ed. In 'Ordination of Plant Communities'. Ed. R.H. Whittaker. Junk, The Hague, p. 277–337.

Accepted 31 October 1979

AN ALGORITHM FOR PREDICTIVE ORDINATION

László ORLÓCI*

Department of Plant Sciences, University of Western Ontario, London, Ontario, Canada N6A 5B7

Keywords:
Linear, Non-linear, Ordination, Prediction, Vegetation

The problem

Assume that compositional variation is observed in a vegetation sample. If this variation is directed, meaning that it is not without a clearly defined trend, response to a directed environmental influence is indicated. An ordination algorithm is sought which can predict local levels of environmental influence based on observed compositional characteristics of the vegetation.

Algorithms

The predictive use of ordinations is not novel (see Orlóci 1978), but the proposed algorithms differ considerably. In one group of methods, prediction is based on simple averaging,

$$\bar{X}_j = \sum Y_{hj}X_h/Y_{.j};\ h = 1, \ldots, p \qquad (1)$$

(Whittaker 1967). $\bar{X}_j$ is a prediction of the level of environmental influence at point j of a given environmental gradient X at which the quantity of species h is Y_{hj}. The optimum of species h is at X_h. The optima $X_1, \ldots X_p$ are given a priori. $Y_{.j}$ is the sum of all Y_{hj} at X_j for all of the p species.

It is a weakness of (1) that it does not make provision for the possibility of species optima not being symmetrically dispersed about X_j on the predicted gradient, and for the unequal performance Y_{hj} of species at their optima. The predictions based on (1) can thus be very unreliable.

The algorithm, PROD, which is the subject of further discussions in the present paper, incorporates the notion of species response directly. The steps include:

1. Statement of a general hypothesis (H) which describes the exact functional form of species response.
2. Computation of ordination scores (co-ordinates for quadrats) which are completely consistent with H.
3. Interpretation of the ordination scores as predictions, considered proportional in quantity with the levels of environmental influence.

Description of PROD

In predictive ordination the computational problem is to derive t sets of co-ordinates for n points,

$$\mathbf{X} = \begin{bmatrix} X_{11} \cdot & X_{1n} \\ \cdot & \cdots \cdot \\ X_{t1} & \ldots X_{tn} \end{bmatrix}$$

The X_{ij} are assumed to be proportional to the actual levels of environmental influence, as they are also completely consistent with the measured responses on p species,

$$\mathbf{Y} = \begin{bmatrix} Y_{11} \ldots Y_{1n} \\ \cdot \quad \cdots \quad \cdot \\ Y_{p1} \ldots Y_{pn} \end{bmatrix}$$

and with a hypothesis (H) which specifies the functional form $Y_{ij} = (f\,(X_{ij}|m_i))$ of the trajectory of average response, conditional on a single parameter or a parameter set m_i.

Based on $f\,(X|m)$, the anticipated compositional distance $d(j,k|i)$, corresponding to a $\Delta(j, k|\ i) = |X_{ij} - X_{ik}|$ separation on the ith gradient is proportional to

$$d^2(j, k|i) = \int_{-\infty}^{\infty} \{f(X|m) - f(X + \Delta(j, k|i)|m)\}^2 dX \qquad (2)$$

* The project has been supported by N.S.E.R.C. of Canada funds.

The composite compositional distance is proportional to

$$d^2(j, k) = \sum d^2(j, k|i), i = 1, \ldots, t \tag{3}$$

H can of course be defined to apply to individual species responses and not just to an average. For the hth species, $Y_{hij} = f(X_{ij}|m_{ih})$ is the hypothesis and $d(j, k,| h, i)$ is the compositional distance. The composite compositional distance is proportional to

$$d^2(j, k) = \sum\sum d^2(j, k| i, h);\ h = 1, \ldots, p; i = 1, \ldots, t.$$

This assumes that the species responses are independent. Now if $d(j, k)$ is used for the ordination distances, the co-ordinate set $\boldsymbol{X}$ which minimizes stress,

$$\sigma^2(\boldsymbol{D};\ \boldsymbol{\delta}) = 1-\rho^2(\boldsymbol{D};\ \boldsymbol{\delta}), \tag{4}$$

between the ordination distances ($\boldsymbol{D}$) and the observed compositional distances ($\boldsymbol{\delta}$) is the best predictive ordination possible under the given H. $\rho(\boldsymbol{D};\boldsymbol{\delta})$ measures correlation on a zero to one scale.

The function $f(X|m)$ may be of a complex shape, in which case σ may not be stationary, but strongly varying between k consecutive segments of $f(X|m)$. This can be detrimental if excessive, and it requires remedies. The solution is conceptually simple: compute a different δ_q, $q = 1, \ldots, k$, for each segment. The computational consequences may however be severe.

Examples

A linear hypothesis (LH) is rarely appropriate, but if the gradient is short, LH may be plausible. If all p trajectories have the same sense, the function takes the very simple form of

$$f(X|m) = bX+c \tag{5}$$

In any linear case, the desired predictions are obtained as the component scores in the following algorithm:

1. Compute compositional distances $\boldsymbol{\delta}$ based on

$$\delta(j,k) = [2(1-\cos\alpha_{jk})]^{\frac{1}{2}} \tag{6}$$

where α_{jk} is the subtending angle of two quadrat vectors.

2. Extract the non-zero eigenvalues ($\boldsymbol{\lambda}$) and eigenvectors ($\boldsymbol{X}$) of Q, with elements $q_{jk} = -0.5\ (\delta^2(j, k)-\delta_j^2-\delta_k^2+\delta^2)$ (Orlóci 1978), and adjust the jth element in the ith eigenvector X_{ij} so that $X_{i1}^2+\ldots+X_{in}^2 = \lambda_i$ for any i. The adjusted X_{ij} are the predictions sought.

3. Compute correlations between the t sets of predictions and the measured environmental variables to establish the identity of the most likely influential factors.

It is generally believed that a non-linear hypothesis (NH) is more appropriate than a linear one. There is however no agreement between the proponents of NH regarding the most likely functional form of the response trajectory. That this may vary depending on species and site conditions is conceivable. It is however often felt that the trajectories conform to a universal curve which is continuous (Curtis & McIntosh 1951, McIntosh 1967), and more or less bellshaped (Whittaker 1956, van Groenewoud 1965). It has been described as Gaussian for $m = (a,1)$,

$$f(X|a, 1) = exp\ -(X-a)^2/2 \tag{7}$$

Standardization is implied. Let us assume that this is indeed the case. The average response at point j on the ith gradient is given by

$$Y_{ij} = \exp\ -(X_{ij}-a_i)^2/2 \tag{8}$$

With such a trajectory for Y, the expected compositional distance is proportional to

$$\begin{aligned} d^2(j, k|i) &= \frac{1}{\sqrt{\pi}}\ \{e-(X-a_i)^2/2- \\ &\qquad \exp-(X+|X_{ij}-X_{ik}|-a_i)^2/2\}^2 dX \\ &= 2(1-S_{ijk}) \end{aligned} \tag{9}$$

The compositional distance is a *chord* distance, since

$$S_{ijk} = e-(X_{ij}-X_{ik})^2/4 \tag{10}$$

is a similarity measure in the interval with end points zero and one.

The composite distance is given by (3). Since a_i does not appear in S_{ijk}, and since the standard deviations are assumed unity in all cases, the same H applies irrespective of species.

What has been given as (9) is what Gauch has described in 1973. He has in fact offered an algorithm to obtain a single set of predictions. Another possibility is to embed $d(j, k)$ in the Kruskal & Carmone (1972) algorithm which has the advantage of supplying not one but t independent sets of n predictions for quadrats.

Reaffirming that the objective is to obtain the predictions (which are the elements in $\boldsymbol{X}$), the steps in PROD are as follows:

1. Compute $\boldsymbol{\delta}$ with elements,

$\delta(j, k) = [2(1-\cos\alpha_{jk})]^{\frac{1}{2}}\; j < k = 2, \ldots, n$ (12)

The α_{jk} are quantities measuring subtending angles where position vectors meet in the zero origin.

2. Specify t, the number of independent environmental influences affecting compositional variation.

3. Select t sets of n arbitrary numbers to serve as first approximations for the unknown X_{ij}.

4. Compute the $d(j, k)$ values according to formulae (3), (9), (10) and examine the stress, σ, in (4).

5. If σ is stable, or less in value than a specified threshold limit, stop. If not, change the X_{ij} a little to further reduce stress and continue at step 4.

The final set of $\boldsymbol{X}$ values are the desired predictions. These have to be examined for correlations with environmental variables to establish the identity of the predicted influence.

Discussion

The Gaussian assumption underlies several formal algorithms, each capable of producing predictions ($\boldsymbol{X}$). The algorithm of Gauch (Gauch, Chase & Whittaker 1974) and that of Johnson (1973) are iterative. The latter incorporates the statistical notions of estimation. Another algorithm, devised by Ihm & van Groenewoud (1975), is based on eigenanalysis.

The advantages of the algorithm described in the present paper are several fold. First, it can be generalized to functions other than the Gaussian provided that their integral exists. Second, it relies on a distance matrix. Once the sample distances are computed, no further access is needed to the basic data. Third, it yields not one but t sets of n co-ordinates. A program is available for the linear case. One for non-linear cases is in the developmental stage.

Summary

An algorithm is described for predictive ordination. The functional form of species response is required, but it need not be Gaussian. Any integrable function is acceptable.

References

Curtis, J.T. & R.P. McIntosh. 1951. An upland forest continuum in the prairie-forest border region of Wisconsin. Ecology 32: 476–496.

Gauch, H.G. 1973. The relationship between sample similarity and ecological distance. Ecology 54: 618–622.

Gauch, H.G., G.B. Chase & R.H. Whittaker. 1974. Ordination of vegetation samples by Gaussian species distributions. Ecology 55: 1382–1390.

Groenewoud, H. van. 1965. Ordination and classification of Swiss and Canadian coniferous forests by various biometric and other methods. Ber. Geobot. Inst. ETH Stiftg. Rubel, Zurich 36: 28–102.

Ihm, P. & H. van Groenewoud. 1975. A multivariate ordering of vegetation data based on Gaussian type gradient response curves. J. Ecol. 63: 767–777.

Johnson, R. 1973. A study of some multivariate methods for the analysis of botanical data. Ph.D. thesis, Utah State Univ., Logan, Utah.

McIntosh, R.P. 1967. The continuum concept of vegetation. Bot. Rev. 33: 130–187.

Orlóci, L. 1978. Multivariate analysis in vegetation research, 2nd ed. Junk, The Hague. p. 451.

Whittaker, R.H. 1956. Vegetation of the Great Smoky Mountains. Ecol. Monogr. 26: 1–80.

Whittaker, R.H. 1967. Gradient analysis of vegetation. Biol. Rev. 42: 207–264.

Accepted 31 October 1979

VEGETATION ANALYSIS AND ORDER INVARIANT GRADIENT MODELS*

I.C. PRENTICE

Department of Plant Biology, The University, Newcastle upon Tyne, NE1 7RU, United Kingdom

Introduction

Ordination has been a major topic of recent research in vegetation analysis. A central problem has been to find a method that will recreate ecological gradients from vegetation sample data, given that species are expected to respond to gradients in a certain way: thus the current approach to ordination involves an explicit ecological response model, and differs from the purely 'descriptive' approach where efficient reduction of dimensionality is the sole objective.

Austin (1976a) and Noy-Meir & Whittaker (1978) reviewed recent developments in ordination research. One trend has been the recruitment of methods devised for other disciplines. A promising-looking method is subjected to a performance test, either with real vegetation data whose underlying gradient structure is understood, or with artificial data from coenocline/coenoplane simulation programs whose 'species' have uni- or bivariate Gaussian distributions in habitat space. But so far no one has found a method whose performance in both kinds of test is unequivocally satisfactory.

Ideally, the research procedure should be a little different. Before doing any tests we ought to know that a method has the right theoretical properties with respect to a particular gradient model. It may be necessary to devise new methods or special modifications of old ones. A good example of a specially designed method with appropriate theoretical properties is Gaussian ordination (Gauch et al. 1974, Ihm & van Groenewoud 1975). However, as Austin (1976a) pointed out, the Gaussian model is too restrictive: firstly other models can produce bell-shaped curves, and secondly species-response curves are usually *not* bell-shaped. A more generally useful method would have to be related to a less restrictive response model. The model should be defined in more than one dimension and it is particularly important that an ordination method should be able to reproduce k gradients undistorted in k-1 dimensions, e.g. one gradient as a straight line in a two-dimensional ordination (Kendall 1971).

The search for an optimal ordination is also relevant to classification. 'Ordination versus classification' is no longer seen as a controversy: we now realise that ordination can reveal clusters and that cluster analysis can usefully partition a multidimensional continuum. Two- or three-dimensional ordination is commonly used as an interactive classification method, the group boundaries being drawn by the user. If more axes are needed the co-ordinates might usefully be clustered by a Euclidean method, e.g. Ward's (agglomerative sum-of-squares), or Lefkovitch's (1976) which would start by rotating the ordination configuration to its principal axes. Hill et al.'s (1976) 'indicator species analysis' also derives a classification from ordination axes.

Order invariant gradient models

I use this term for any species-response model that is unaffected by non-linear but monotone (order-preserving) transformations of the species abundance measure. Such models are more in keeping with our current understanding of vegetation than anything more sophisticated.

* The ordination experiments reported here were carried out while I held a NERC research studentship under Dr. H.J.B. Birks at the Botany School, University of Cambridge and subsequently a Royal Society European Exchange fellowship at the Botanisk Museum, Universitetet i Bergen, Norway. The article was written during my tenure of a Sir James Knott fellowship at Newcastle. I am grateful to Dr. H.J.B. Birks for stimulating my interest in the subject, for supplying the pteridophyte data, for providing computer resources at Cambridge for the later experiments and for his comments on the manuscript. Computer programs were provided by Dr. H.G. Gauch, Jr. (gradient simulation) and Prof. R. Sibson (multidimensional scaling).

There are fewer plausible order invariant models than there are strictly quantitative models. The archetype of an order invariant response model is Curtis & McIntosh's (1951) model of a single gradient, in which the species are assumed to be (weakly) unimodal: i.e. each species' values rise to a peak or plateau, then fall (there may or may not be flat 'shoulders'). The equivalent model for presence-absence data specifies that each species' occurrence would be concentrated into one region of the gradient. These models are conveniently general. Most ecological response curves *are* unimodal although they may differ greatly in other respects. The models have the additional property of order invariance with respect to the gradient axis. (Austin (1976a) illustrated how the form of a response curve can be affected by a monotone transformation of the environmental axis.)

M.O. Hill (pers. comm.) has pointed out that the Curtis-McIntosh model can be generalized to more than one gradient. In the generalized model each species' abundance contours are convex in habitat space. An alternative one-gradient model would also insist on symmetry in the response curves; this model would more readily generalize to two or more dimensions – the abundance contours would then have to be circular, spherical or hyperspherical in habitat space – but the model would also be a more drastic simplification of the real situation. In the rest of the paper I shall mainly be concerned with the Curtis-McIntosh one-gradient model ('model 1') and its presence-absence equivalent ('model 2'). The Gaussian coenocline model is, of course, subsumed by model 1.

Bimodal ecological response curves can occur as a result of competition and are observed in direct gradient analyses (Austin, 1976a, 1976b). The Curtis-McIntosh model does not allow for bimodality. It is probably necessary to overlook this defect; an ordination method that 'allowed' species to be bimodal would almost certainly give nonunique and ecologically misleading solutions (Austin 1976b). Real data will not in any case conform exactly to the model. Some species will appear to be bi- or multi-modal because of random effects on sampling and plant micro-distribution and because of minor habitat differences orthogonal to the main underlying gradients. The robustness of an ordination method with respect to such deviations is a matter for experiment.

Reciprocal averaging

Reciprocal averaging (RA), alias correspondence analysis, is equivalent to non-centred principal components analysis (PCA) of (0, 1) data after simultaneous normalization by the square roots of the sample and species totals (Hill 1973, 1974). Abundance data can be handled in RA by treating each species as a number of 'pseudo-species' (Hill et al. 1976) corresponding to abundance classes. Thus the occurrence of species A in abundance class 4 in a given sample is coded as 'occurrences' there for pseudo-species A_1 through A_4. Hill (1974) proved that the order of samples from a model 2 gradient is always correctly reproduced by the first (non-trivial) RA axis: this property is not shared by ordinary PCA. The same holds for a model 1 gradient if the pseudo-species method is used. RA treats species and samples symmetrically. A parallel property therefore holds for species order, but this is less interesting ecologically – model 2 does not necessarily imply a *two-way* concentration of occurrences in the manner of Hill's (1974) Fig. 1. RA in two dimensions yields arch shapes from simulated coenoclies, without the involution caused by ordinary PCA; it reproduces well the grid sampling pattern of simulated coenoplanes with equal 'x' and 'y' beta-diversities, but in the case of unequal beta-diversities the 'long' gradient assumes an arch shape while the 'short' gradient is obscured (Gauch et al. 1976, Fasham 1977). Austin (1976b) found RA susceptible to involution with more complex data. RA has all the computational advantages of PCA without the worst gradient curvature problems, but it is not an ideal method.

Non-metric multidimensional scaling

Non-metric (or non-parametric) multidimensional scaling (NMDS) is an order invariant method (Sibson 1972) since it maps the rank order of sample similarities into the (inverse) order of distances between points in the ordination. In other words only the rank order of the similarity values between samples is used to generate the ordination. In this respect NMDS contrasts with PCA and principal co-ordinates analysis (= principal axes analysis), which may be considered as methods that approximate a linear relationship between sample dissimilarities and ordination distances – ordinary PCA implying the use of Euclidean distance as a dissimilarity measure, principal co-ordinates analysis allowing a wider choice of coefficients – and with other metric ordination methods, such as Kruskal and Carmone's method as used by Orlóci (this volume), Anderson's (1971) quadratic loss function method

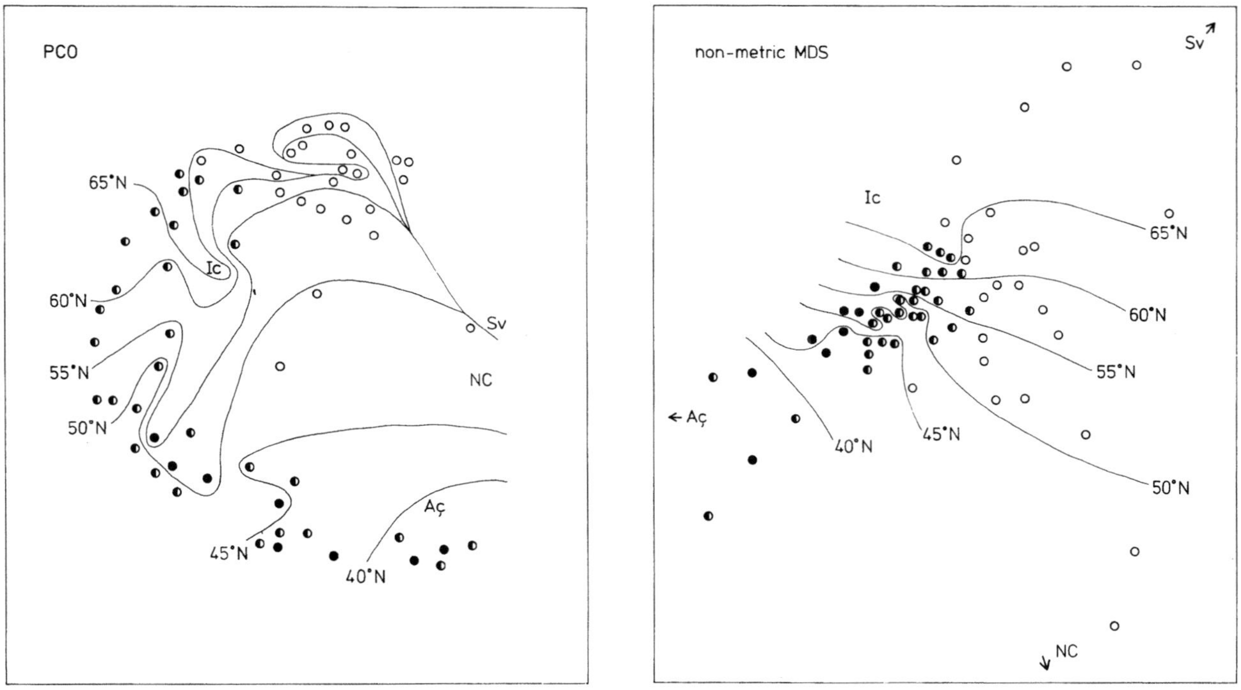

Fig. 1. Principal co-ordinates analysis (left) and non-metric MDS (right) of European pteridophyte distribution data from Birks (1976). Each symbol represents a unit area, most of which are 5° (latitude) × 6° (longitude) blocks. Latitude divisions are indicated; three longitude classes are symbolized by black dots (west of 0°), half-black dots (0°–30°E) and white dots (east of 30°E), except for the four outlying areas of Açores (Aç), the N. Caspian sea area (NC), Iceland (Ic) and Svalbard (Sv). Ic is shown as being north of 65°N although it lies over the 65th parallel. Aç, NC and Sv are distant outliers in the non-metric ordination. PCO residual variance, 80.0 %; non-metric MDS stress, 1.6 %.

and Sammon's (1969) 'non-linear mapping', all of which minimize the departure from a *linear* relationship between input dissimilarities and output distances.

This property of order invariance does not in itself imply a relationship to any particular gradient model. However, in a previous paper (Prentice 1977) I recommended the use of a particular form of NMDS: Sibson's (1972) 'local' variant of the method, with 'primary' tie treatment (i.e. disregarding ties between similarity values) and a particular similarity coefficient. The coefficient is:

$$s(i,j) = \sum_k \min(a_{ik}, a_{jk})$$

(where a_{ik} denotes the abundance of species k in sample i) and Sibson's ordination criterion is

$s(i,j) > s(i,k)$ implies that $d(i,j) \leqq d(i,k)$ for all i, j, k

where the d's denote interpoint distances corresponding to the s's. The more usual 'global' variant specifies a stronger ordination criterion,

$s(i,j) > s(k,l)$ implies that $d(i,j) \leqq d(k,l)$ for all $i, j, k, l,$

and the alternative 'secondary' tie treatment specifies '$\geqq$' rather than '$\geqq$' in the lefthand inequalities, thus constraining tied similarities to be represented by equal distances. The best-known version of NMDS proceeds by steepest-descent minimization of a stress function,

$$S = 100\sqrt{\frac{\sum[d(i,j)-\hat{d}(i,j)]^2}{\sum[d(i,j)]^2}}$$

where the $\hat{d}$'s are numbers chosen so as to minimize S, subject to the ordination criterion, by Kruskal's 'monotone regression' technique; the summations are over all i, j (local) or for $i < j$ (global). In this paper, as in the earlier one, I present results obtained with R. Sibson's unpublished program MDSCAL 015, which is loosely based on Kruskal's (1964a, 1964b) algorithm and implemented on the Cambridge University IBM 370/165

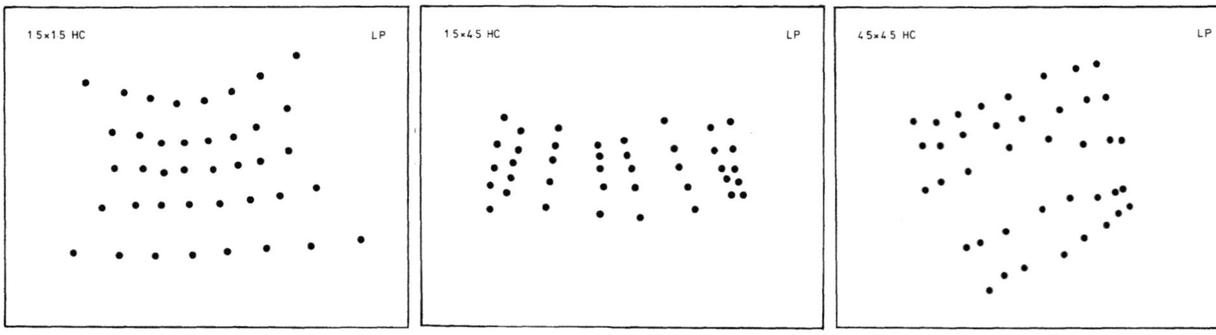

Fig. 2. The non-metric method (local primary) applied to (left to right) 1.5 × 1.5 HC, 1.5 × 4.5 HC and 4.5 × 4.5 HC coenoplanes. Stresses 4.7 %, 2.1 %, 5.7 %.

computer. The theoretical strength of the method that I recommended lies (a) in Kendall's (1971) proof that the information required to reconstruct the order of samples from a model 1 gradient is contained in the inequalities between $s(i, j)$ and $s(i, k)$ for all i, j, k, and (b) in the robustness of NMDS towards imperfect data (Kendall 1971, Sibson 1972).

There is, however, a weakness in the theory. Local NMDS is preferred to global NMDS because the latter imposes some irrelevant constraints: the 'tetrad' inequalities between $s(i, j)$ and $s(k, l)$ are not all specified by model 1. Local NMDS uses only the 'triad' inequalities between $s(i, j)$ and $s(i, k)$. But even some of these are irrelevant to the problem in hand. It follows from model 1 that

$$s(i, j) > s(i, k) \text{ implies that } \Delta(i, j) < \Delta(i, k)$$

(where Δ denotes ecological separation) for j and $k > i$ or j and $k < i$, and the same applies for model 2 if the data are in (0, 1) form. The above expression does *not* necessarily hold for $j > i$ and $k < i$ or $j < i$ and $k > i$ (Kendall 1971). In other words, similarities remain constant or decrease with increasing gradient distance to left or right of i, but the fall-off is generally at different rates to left or right. This asymmetry applies whether or not the response curves are symmetric. Thus even local NMDS attempts to preserve the sign of some inequalities – about a quarter of the total used – that are actually not specified by model 1 or 2. One may foresee that distortion could result, although it is hard to predict what kind of distortion it would be, or how serious. A similar weakness affects the relationship of the method to Hill's convexity model. Hill's model implies that similarities (s) remain constant or decrease with gradient distance in any one direction from any sample. The method, however, makes similarities fall away equally in all directions. One-dimensional NMDS could be modified so as to distinguish 'left' and 'right' in the current ordering and ignore the offending inequalities (D.G. Kendall, pers. comm.) but it is not obvious how an analogous modification could be built into two-dimensional NMDS.

My recommended method proved successful in 'unbending' curved principal co-ordinates analysis (PCO) configurations obtained from two data sets, one dominated by a single, long succession, the other by two orthogonal habitat gradients of unequal beta-diversity (Prentice 1977). As a further illustration of the method's capacity to reproduce gradients 'straight' I have included here a biogeographic example (Fig. 1): a comparison of PCO (the dissimilarity coefficient used was the one-complement of Jaccard's coefficient) and NMDS (local, primary; similarity coefficient as before) on Birks' (1976) data on the pteridophyte floras of 65 unit areas of Europe. The PCO is redrawn from Birks (1976). With 80.0 % of variance *not* accounted for by the first two principal axes, the PCO is inefficient; it shows extreme curvature of geographic boundaries (Fig. 1) and of the minimum spanning tree (Birks 1976). NMDS is much more successful, with far less curvature and only 1.6 % stress. It is evident from the non-metric ordination that longitude and, especially, latitude are major determinants of the pattern of floristic variation. Some specific deviations can be seen: the C. European pteridophyte flora between 45 ° and 55 °N is not strongly dependent on latitude; the S. Mediterranean pteridophyte flora (south of 40 °N) is not strongly dependent on longitude. The object of this biogeographic example is to show how well the non-metric method can cope with high beta-diversity: in this respect the pteridophyte data set is probably at least as extreme as anything normally encountered in vegetation analysis.

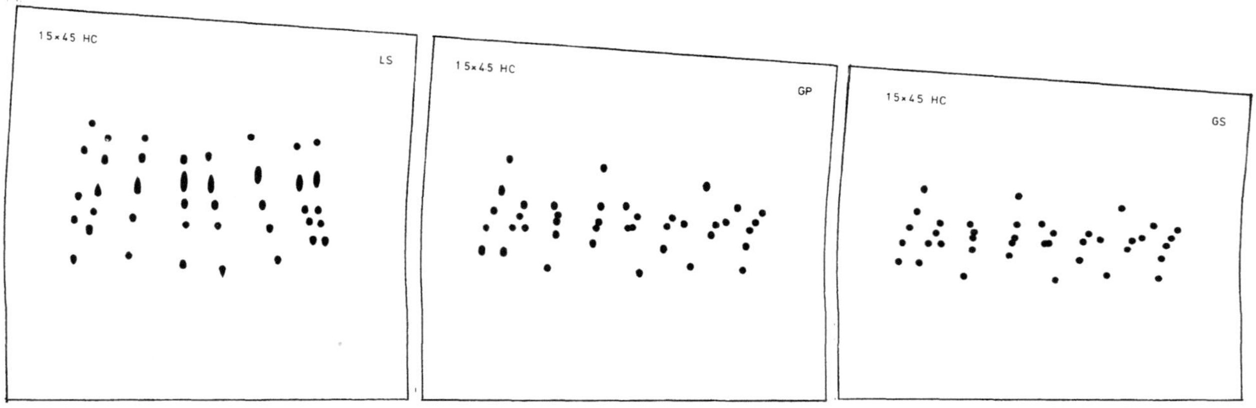

Fig. 3. Left to right: local secondary, global primary and global secondary non-metric MDS applied to the 1.5 × 4.5 HC coenoplane. Stresses 2.1 %, 9.8 %, 9.8 %.

The method is also capable of convincing (though not totally undistorted) representation of simulated coenoplane data (Fig. 2). These data were produced using the program CEP-21 (Gauch & Whittaker 1976) and consist of 30 species and 40 samples placed on a regular 8 × 5 grid. The parameter settings were identical with those used by Gauch et al. (1976) and Fasham (1977); the three coenoplanes have beta-diversities of 1.5 × 1.5 half-changes (HC), 1.5 × 4.5 HC and 4.5 × 4.5 HC. In contrast with RA (Fasham 1977) the result for the 1.5 × 4.5 HC coenoplane is as good as the other two results. Fasham (1977) also obtained good results from a NMDS analysis (with a different coefficient) of these three coenoplanes. The 1.5 × 4.5 HC coenoplane was also subjected to local secondary and global primary and secondary scaling, all with the similarity coefficient *s* (Fig. 3): the results confirm the superiority of the local ordination criterion, at least with this coefficient.

The method was also applied to simulated coenoclines. Four coenoclines were generated, using the program CEP-1 of Gauch & Whittaker (1972), under the same parameter settings as used by Fasham (1977). The coenoclines have beta-diversities 2.25, 4.5, 9 and 22 HC. Each began with 65 'initial species' and each is represented by 50 evenly spaced samples. All four coenoclines were subjected to local primary scaling and the 2.25 and 22 HC coenoclines were subjected to local secondary and global primary and secondary scaling, all with the similarity coefficient *s* as before. The main, local primary results (Fig. 4) are similar to Fasham's (1977) results from NMDS with a different coefficient. They indicate that a single gradient (in the absence of outliers?) is distorted by the non-metric method into a curve, different in form from the characteristic PCA or RA curves. At low beta-diversities the degree of curvature is comparable with that produced by RA. Curvature increases and becomes more complex with increasing beta-diversity, but at high beta-diversities the configuration is nevertheless 'flatter' than that produced by RA. Incorporation of a limited amount of noise into the coenoclines (results not shown) did not alter the general form of the curves. At 2.25 HC (Fig. 5) primary and secondary tie treatments gave identical results but global scaling caused an undesirable looping-back at one end of the gradient; at 22 HC (Fig. 6), on the other hand, local and global methods gave equally acceptable results with primary tie treatment, but both secondary methods resulted in 'horseshoes', presumably because of there being many zero, hence tied, similarity values (Kendall 1971, Prentice 1977).

Conclusions

NMDS with Kendall's similarity coefficient *s* has some theoretical justification in terms of the Curtis-McIntosh one-gradient model, and has shown itself capable of convincing reconstruction of real and simulated two-gradient data, but its performance is less good with simulated one-gradient (Gaussian coenocline) data; presumably this weakness is attributable to using some spurious information, although the precise mechanism that creates the curvature is not understood. The experiments of Fasham (1977, see also Austin 1976b) and those presented here establish that NMDS is one of the best ordination methods currently in use in vegetation analysis, but at the same time give cause for doubt whether the computationally demanding non-metric method should be recommended in preference to RA. RA also has some justification in terms of order invariant

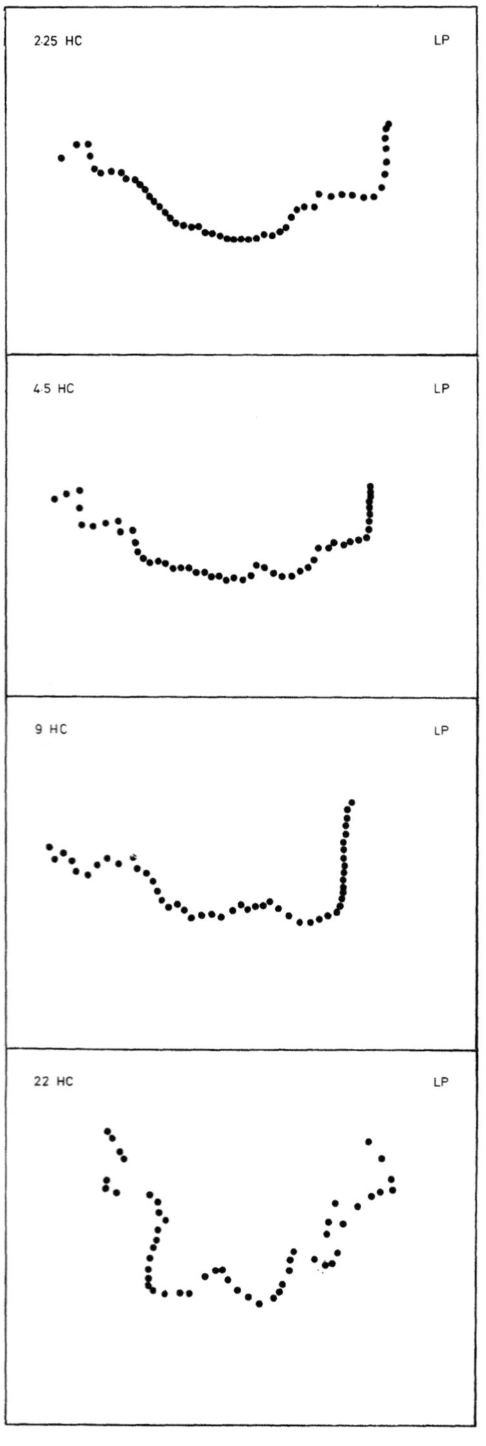

Fig. 4. The non-metric method (local primary) applied to (top to bottom) 2.25 HC, 4.5 HC, 9 HC and 22 HC coenoclines. Stresses 0.4 %, 0.6 %, 0.9 %, 1.7 %.

models, it performs as well as NMDS in many situations, and it is fast and convenient. The search continues for a method with exactly the right theoretical properties, even with respect to a one-gradient model.

Fasham (1977) suggested that intrinsically one-dimensional data could be recognized in a distorted two-dimensional representation by superimposing the minimum spanning tree, then re-analysed by NMDS in one dimension. This seems to me unnecessary effort: one could equally well use RA, or Sibson's (1971) sequencing method with the coefficient *s*, either of which methods should be able to recover the precise order of samples from a model 1 or 2 gradient. But (more importantly) part of the reason for ordinating data dominated by a single habitat gradient may be to expose deviations from the model or interesting outliers – compare the Yukon example in Prentice (1977), where NMDS in two dimensions isolated a small group of samples apparently belonging to a succession parallel to the main one.

A possible alternative approach to the problem discussed in this paper would involve dispensing with similarity coefficients and using a non-metric method directly operating on species abundances. For example, Guttman and Lingoes' program SSAR-II (= MDA-RC) (Lingoes 1972) could be used to provide joint ordinations of species and samples, the ordination criterion being

$$a_{ik} > a_{jk} \text{ implies that } d(i, k) \leq d(j, k) \text{ for all } i, j, k$$

where $d(i, k)$ denotes the distance between sample-point i and species-point k. This kind of ordination is known as 'multidimensional unfolding': see e.g. Dale (1975). The implicit species-response model assumes that species abundances fall away symmetrically on both sides of the species' optimum (in the one-gradient case) or that each species' abundance contours are concentric circles or spheres (in the two-or three-gradient case). The method is directly related to the model and no spurious information is used. Some more general unfolding models are considered by Carroll (1972). However, the more general the model, the greater the chance of degeneracy – the attainment of low-or zero-stress solutions that convey no useful information – and Carroll was not optimistic about algorithms for fitting such models without external constraints. Ultimately a choice may have to be made between a realistic model that leads to degeneracy and a simpler, more artificial one that does not, but it is not clear at the moment where the line would have to be drawn.

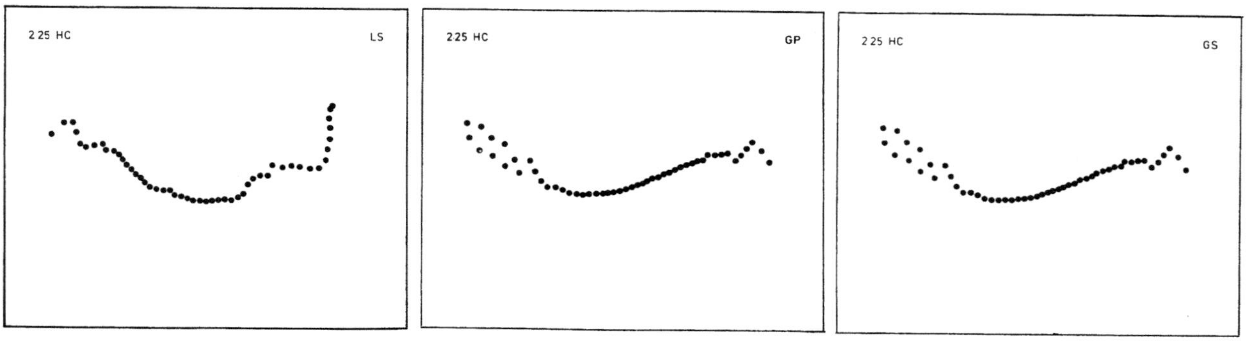

Fig. 5. Left to right: local secondary, global primary and global secondary non-metric MDS applied to the 2.25 HC coenocline. Stresses 0.4 %, 2.4 %, 2.4 %.

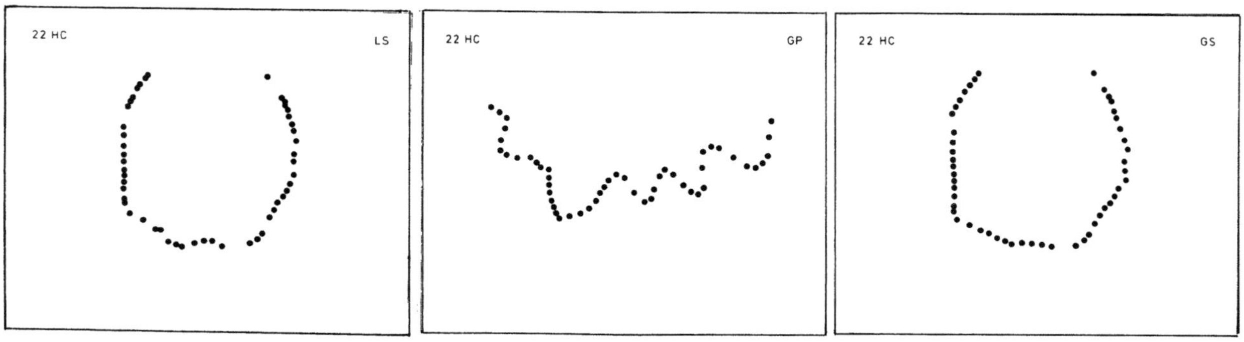

Fig. 6. Left to right: local secondary, global primary and global secondary non-metric MDS applied to the 22 HC coenocline. Stresses 9.7 %, 3.2 %, 11.1 %.

Summary

An ideal ordination method would have known properties with respect to an explicit, order invariant ecological response model defined in more than one dimension. The Curtis-McIntosh model (each species weakly unimodal) is a good general-purpose model of a single gradient, but no current method is guaranteed to dispose samples from such a gradient along a straight line. There is some theoretical justification for using reciprocal averaging (RA), or local non-metric multidimensional scaling (NMDS) with Kendall's simple similarity coefficient, but the former tends to produce arches rather than straight lines and the latter can produce erratic curves (as illustrated here by applying the method to simulated coenocline data). The non-metric method is nevertheless shown to perform well (a) with high beta-diversity plant distribution data and (b) with simulated coenoplane data, where its performance is better than that of RA. Results with coenoclines and coenoplanes concur with those of Fasham (1977) who tested NMDS with a different coefficient. Local scaling is shown to be preferable to global, and primary tie treatment to secondary, in tests on coenocline and coenoplane data. A possible alternative non-metric approach is mentioned.

References

Anderson, A.J.B. 1971. Ordination methods in ecology. J. Ecol. 59: 713–726.

Austin, M.P. 1976a. On non-linear species response models in ordination. Vegetatio 33: 33–41.

Austin, M.P. 1976b. Performance of four ordination techniques assuming three different non-linear species response models. Vegetatio 33: 43–49.

Birks, H. J. B. 1976. The distribution of European pteridophytes: a numerical analysis. New Phytol. 77: 257–287.

Carroll, J.D. 1972. Individual differences and multidimensional scaling. In R.N. Shepard, A.K. Romney & S.B. Nerlove (eds.): Multidimensional Scaling. Volume 1. Theory, p. 105–155. Seminar Press, New York.
Curtis, J.T. & R.P. McIntosh. 1951. An upland forest continuum in the prairie-forest border region of Wisconsin. Ecology 32: 476–496.
Dale, M.B. 1975. On objectives of methods of ordination. Vegetatio 30: 15–32.
Fasham, M.J.R. 1977. A comparison of nonmetric multidimensional scaling, principal components and reciprocal averaging for the ordination of simulated coenoclines, and coenoplanes. Ecology 58: 551–561.
Gauch, H.G., Jr. & R.H. Whittaker. 1972. Coenocline simulation. Ecology 53: 446–451.
Gauch, H.G., Jr. & R.H. Whittaker. 1976. Simulation of community patterns. Vegetatio 33: 13–16.
Gauch, H.G., Jr., G.B. Chase & R.H. Whittaker. 1974. Ordination of vegetation samples by Gaussian species distribution. Ecology 55: 1382–1390.
Gauch, H.G., Jr., R.H. Whittaker & T.R. Wentworth. 1976. A comparative study of reciprocal averaging and other ordination techniques. J. Ecol. 65: 157–174.
Hill, M.O. 1973. Reciprocal averaging; an eigenvector method of ordination. J. Ecol. 61: 237–249.
Hill, M.O. 1974. Correspondence analysis: a neglected multivariate method. J. Roy. Statist. Soc. Ser. C (Applied Statistics) 23: 340–354.
Hill, M.O., R.G.H. Bunce & M.W. Shaw. 1975. Indicator species analysis, a divisive polythetic method of classification, and its application to a survey of native pinewoods of Scotland. J. Ecol. 63: 597–614.
Ihm, P. & H. van Groenewoud. 1975. A multivariate ordering of vegetation data based on Gaussian type gradient response curves. J. Ecol. 63: 767–777.
Kendall, D.G. 1971. Seriation from abundance matrices. In F.R. Hodson, D.G. Kendall and P. Tăutu (eds.): Mathematics in the Archaeological and Historical Sciences, p. 215–251. Edinburgh University Press, Edinburgh.
Kruskal, J.B. 1964a. Multidimensional scaling by optimizing goodness of fit to a nonmetric hypothesis. Psychometrika 29: 1–27.
Kruskal, J.B. 1964b. Nonmetric multidimensional scaling: a numerical method. Psychometrika 29: 115–129.
Lefkovitch, L.P. 1976. Hierarchical clustering from principal co-ordinates: an efficient method for small to very large numbers of objects. Mathematical Biosciences 31: 157–174.
Lingoes, J.C. 1972. A general survey of the Guttman-Lingoes program series. In R.N. Shepard, A.K. Romney & S.B. Nerlove (eds.): Multidimensional Scaling. Volume 1. Theory, p. 49–68. Seminar Press, New York.
Noy-Meir, I. & R.H. Whittaker. 1978. Recent developments in continuous multivariate techniques. In R.H. Whittaker (ed.): Ordination of Plant Communities, p. 337–378. Junk, The Hague.
Prentice, I.C. 1977. Non-metric ordination methods in ecology. J. Ecol. 65: 85–94.
Sammon, J.W., Jr. 1969. A nonlinear mapping for data structure analysis. IEEE Transactions on Computers C-18: 401–409.
Sibson, R. 1971. Some thoughts on sequencing methods. In F.R. Hodson, D.G. Kendall and P. Tăutu (eds.): Mathematics in the Archaeological and Historical Sciences, p. 263–266. Edinburgh University Press, Edinburgh.
Sibson, R. 1972. Order invariant methods for data analysis. J. Roy. Statist. Soc. Ser. B 34: 311–349.

Accepted 31 October 1979

EVALUATION OF ORDINATION METHODS THROUGH SIMULATED COENOCLINES: SOME COMMENTS*

Enrico FEOLI & Laura FEOLI CHIAPELLA**

Istituto ed Orto Botanico, Università degli Studi, Via A. Valerio 30, Trieste, Italy 34100

Keywords:
Coenocline, Linearity, Ordination, Seriation

Introduction

The ordinations based on resemblance matrices (Orlóci 1978b) received special attention in plant ecology during the past decade. Many papers dealing with this topic could be mentioned. It should however be satisfactory to refer to the general survey by Dale (1975). He distinguished two main approaches in ordination (cf. also Beals 1973): One is associated with pattern recognition and pattern analysis (ordination sensu lato, Noy-Meir & Whittaker 1977), which started with Goodall (1954) and is still dominant in phytosociology (see Westhoff & van der Maarel 1978 and Orlóci 1978a for references). The other one is more strictly related to Whittaker's work (ordination sensu stricto, Noy-Meir & Whittaker 1977) and concerned with gradient analysis (Whittaker 1967).

As stated by Austin (1976a) these two approaches are complementary since the former is the basis for the latter. The fact that both are productive and not competitive has been emphasized by van der Maarel (1971), and Gauch, Chase & Whittaker (1974). However, a strong tendency exists to evaluate the ordination methods mainly on the basis of their capacity to give axes easily interpretable in terms of gradients. It has in fact been suggested to evaluate the performance of different ordination methods through simulated coenoclines (Swan 1970, Noy-Meir & Austin 1970, Austin & Noy-Meir 1971, Gauch & Whittaker 1972a, b, Kessel & Whittaker 1976, Austin 1976a, b, Gauch, Whittaker & Wentworth 1977, Whittaker & Gauch 1978).

* The research was supported by a grant from the Italian CNR (Programma Finalizzato 'Promozione Qualità dell'Ambiente – sottoprogetto Metodologie Matematiche e Basi di Dati').
** We would like to thank dr. M. Dale, dr. E. van der Maarel and prof. L. Orlóci for useful comments on the manuscript and Mr. A. Zampar for technical assistance.

In the present paper we wish to comment on the investigations of the performances of the ordination methods based on the use of simulated coenoclines.

Coenocline simulation

A coenocline may be defined as a directed sequence of given species responses along a gradient (Fig. 1). The simulation of coenoclines, sometimes extended to more than one gradient, known as a coenoplane (Austin & Noy-Meir 1971, Gauch & Whittaker 1976), is often based on the assumption that the species responses (score) along a gradient may be accurately described as a bell-shaped curve, frequently of the Gaussian type. The simulation of a coenocline was recently used by Kessel & Whittaker (1976) to estimate the interaction of noise and beta diversity (cf. Whittaker 1972b) in ordination methods, including discriminant function analysis, Bray & Curtis ordination, and principal component analysis (PCA). Austin (1976b) has

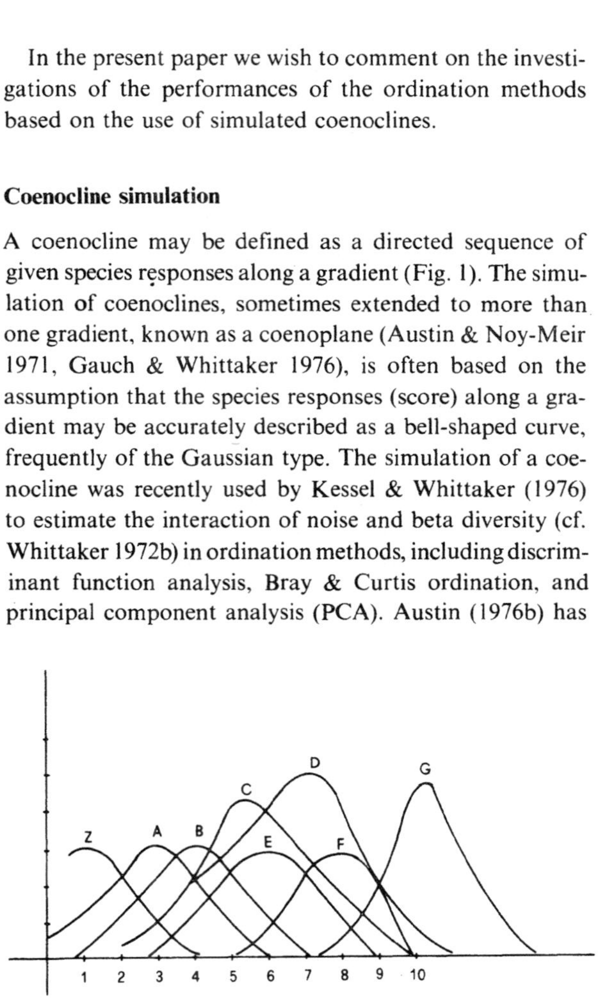

Fig. 1. A simulated coenocline with eight species (Z, A, B, C, D, E, F, G). The numbers on the gradient axis represent the points of ten hypothetical relevés. The scores of the species in these relevés are given in Table 1.

compared reciprocal averaging (Hill 1973, cf. also Hill & Smith 1976), Gaussian ordination (Gauch, Chase & Whittaker 1974, Ihm & van Groenewoud 1976, Van Groenewoud 1976), parametric mapping (cf. Noy-Meir 1974), and non-metric multidimensional scaling (Shepard & Carrol 1966, Kruskal 1964a,b, Anderson 1971), in terms of their sensitivity to different curvilinear responses of the species within simulated coenoclines.

The problem of linearity seems to play a fundamental role in these attempts for the evaluation of the ordination results. The basic objection against the application of linear methods (PCA and related methods) lies in the fact that the non-linear correlations produce distortion and loss of information in the ordinations (type A distortion, Orlóci 1974). However, before proceeding any further, it may be useful to clearly distinguish between linear and non-linear models. We prefer to define as linear all the models based on linear matrix algebra, or those approximating this (cf. van der Maarel 1969). It is noted that the data transformations (non-centering, centering, double centering, standardization, normalization and so on; Orlóci 1967, 1978a, Noy-Meir 1973 Noy-Meir & al. 1975, Feoli 1977) has been shown to lead to different performances of the linear models, and a very rich terminology, e.g. centered PCA, non-centered PCA, principal coordinate analysis (Gower 1966), reciprocal averaging (Hill 1973, Hill & Smith 1976).

With the term linear ordination we designate all methods involving the extraction of the eigenvalues (λ) and eigenvectors (α) of resemblance matrices. In this context it is preferable to define the ordination method on the basis of the resemblance function used to obtain the symmetric matrix. So, the ordination produced by the eigenvectors of a resemblance matrix of, for example, Sørensen's index will be referred to as a linear ordination based on Sørensen's index, and so on. PCA is a special case of linear ordination based on covariance (correlation coefficient, euclidean distance). It has the property of reproducing exactly the relative positions of the points in a metric space. The two examples of Fig. 2a,b and the examples given by Orlóci (1978) illustrate graphically such a property. Nichols (1977) presented an analytical discussion. When PCA is applied to the relevé of a coenocline, PCA re-establishes the relevés, as points, in a metric space, and if the relevé points in an ordination scatter diagram have a horseshoe type dispersion, this will mean that in the metric space the relevé points do in fact have such a dispersion. The requirement that a coenocline should be always represented by a straight line in a PCA ordination is unrealistic. There is of course a good reason to prefer another method: in PCA two or more axes may be necessary to reproduce an essentially unidirection trend, associated with a dispersion of relevés along a coenocline, instead of only one (Orlóci 1979). Feoli & Feoli Chiapella (1976), Feoli Chiapella & Feoli (1977) and Werger (1978) have identified gradients by interpolation based on species or relevé points in ordinations produced by PCA. Phillips (1978) suggested polynomial regression of principal components to reveal gradients (ordination axes). These are examples of the profitable use of PCA in indirect gradient analysis. The fact that the interpretability of the axes is difficult, and the statistical use of PCA in most cases is unrealistic is well known in phytosociology (Orlóci 1978a).

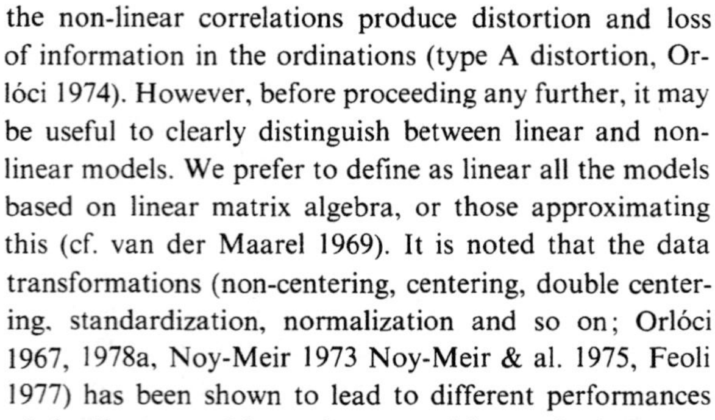

Fig. 2. Ordination by Orlóci's method (1966) of a set of points arranged on a circle (a) and on a perimeter of a square (b). The co-ordinates of the points are given in Table 2a, b. The dotted lines are the first two principal components.

Coenoclines and resemblance of relevés

A set of relevés may be related to one or several coenoclines. Each relevé may be described by a set of abiotic variables and by a set of species. In such a setup, the ecological problem amounts to find the correlation between individuals or sets of abiotic variables and the vegetation. This can be done by using different graphic methods, i.e., by superimposing on ordination diagrams, based on the species, the scores of abiotic variables in the corresponding relevés (Gittins 1965) or using numerical methods, both parametric and non-parametric ones (cf. Gounot 1969, Orlóci 1972a, Stanek, Jeglum & Orlóci 1977, Feoli 1976).

The resemblances of relevés may be calculated by different coefficients (Orlóci 1972b, 1978a, Goodall 1978) in two basically different ways: (a) considering the sets of abiotic and biotic variables jointly and (b) considering them separated. Recently, in his discussion on the covariation of plant species along ecological gradients, van Groenewoud (1976) stated that since the responses of species to a gradient are conceived as bell-shaped curves, the covariance, product moment correlation coefficient (*r*) and euclidean distance have little ecological value. This supposition could be considered true only if we consider an analysis of the relevés based on the joint use of biotic and abiotic variables.

However, if only the species are used as relevé descriptors, the argument of van Groenewoud becomes much weaker, and his interpretation of the correlation coefficient seems to be too restricted. It is true that the use of the product moment correlation coefficient in statistical tests of covariation is conditional on the assumption that the set of points:

$$G_R = \{(x, y): x \in X \wedge y \in Y, xRy\} \qquad (1)$$

is such that the Cartesian product of two sets of measures on the variables *X* and *Y* will fit to a straight line, it is also true that the points may be so dispersed around this line that the correlation coefficient has no statistical meaning. Furthermore, the fact that the points (x, y) in G_R are lying on opposite sides of a straight line (see van Goenewoud 1976) may have a definite ecological meaning it is not a problem to be blamed on the mere application of the product moment correlation coefficient.

Since in an ecosystem different species do not tend to occupy the same niche (cf. Volterra 1931, Lotka 1932, Whittaker 1972b among others), it is obvious that for two species the same bell-shaped curve cannot be expected. The areas delimited by the curves should be more or less overlapping but not coincident. The problem to establish if *r* is a really inefficient measure of covariation, may be faced in the context of a simulated coenocline. The coenocline of Fig. 1 and Table 1 and the one presented by Austin (1976a, Fig. 4) are quite suitable for this purpose. The relationship of *r* and the relative overlapping areas of the bell-shaped curves of the species is important. The relative overlap (*oR*) of two species may be calculated by the following formula:

$$oR_{ij} = \frac{S_{ij}}{S_i + S_j - S_{ij}} \qquad (2)$$

where S_i is the integral of the bell-shaped curve of species *i*, S_j is the integral of the bell-shaped curve of species *j*, and S_{ij} is the overlap area. The integrals in this case have been calculated by a planimeter. Formula (2) is analogous to the similarity ratio coefficient (cf. Westhoff & van der Maarel 1978). Fig. 3 gives the graphics of regression for *r* on *oR* based on the data of coenoclines presented in Fig. 1 and by Austin (1976a, Fig. 4). The regression is clearly linear, meaning that the determinristic use of *r* should not be too problematic in gradient analysis.

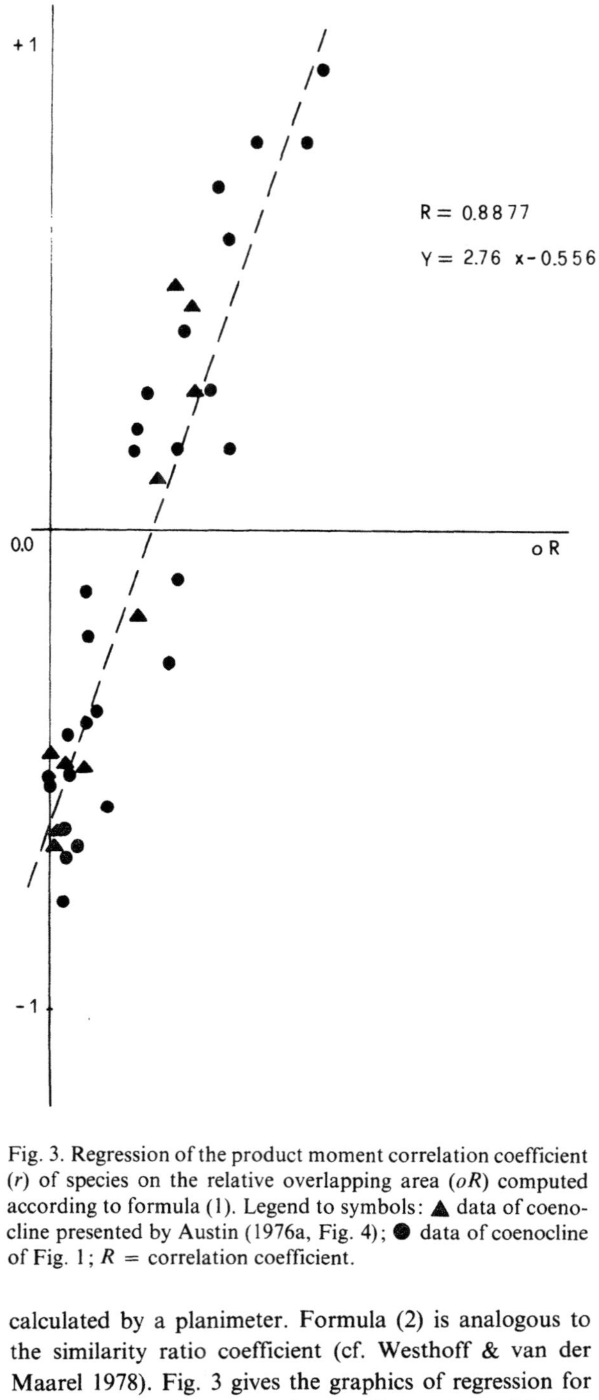

Fig. 3. Regression of the product moment correlation coefficient (*r*) of species on the relative overlapping area (*oR*) computed according to formula (1). Legend to symbols: ▲ data of coenocline presented by Austin (1976a, Fig. 4); ● data of coenocline of Fig. 1; *R* = correlation coefficient.

Table 1. Scores of the species in ten hypothetical relevés of simulated coenocline presented in Fig. 1. The numbers identify relevés, the letters identify species.

	1	2	3	4	5	6	7	8	9	10
Z	3	2	1	0	0	0	0	0	0	0
A	1	2	3	2	1	0	0	0	0	0
B	0	1	2	3	2	1	0	0	0	0
C	0	0	1	2	4	4	3	2	1	0
D	0	0	0	2	3	4	5	4	2	0
E	0	0	0	1	2	3	2	1	0	0
F	0	0	0	0	0	1	2	3	2	1
G	0	0	0	0	0	0	0	1	2	3

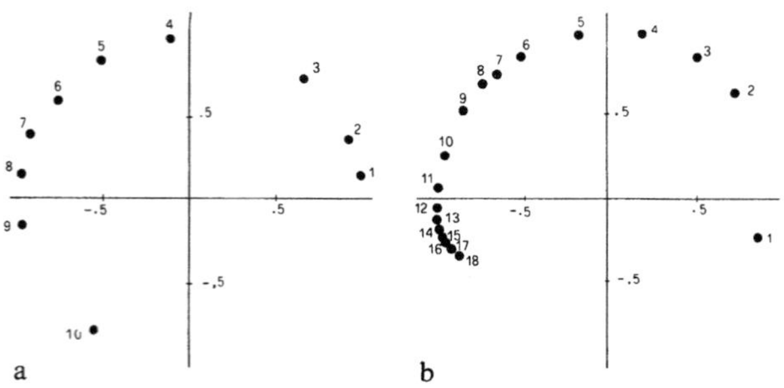

Fig. 4. a) Ordination of the coenocline of Fig. 1. b) ordination of the coenocline presented by Austin (1976a, Fig. 4). In both cases the axes are given by the method proposed by Feoli Chiapella & Feoli (1977).

Gradient analysis by a linear method

Feoli Chiapella & Feoli (1977) proposed a method of ordering relevés based on PCA applied to the log-similarity ratio matrix between them. The log-similarity between the relevés constitutes a special data transformation (see van der Maarel 1979 for other transformations of phytosociological data) which in PCA tends to give very regular disposition of relevés along arcs. As shown by van der Aart & Smeenk-Enserink (1975) the simple log-transformation of the normalized data has the same effect, however by using the log-similarity ratio each relevé is redescribed based on new data i.e. the similarities with other relevés. Although these data are not independent, they allow to apply more correctly the product moment correlation coefficient to measure the similarity between the relevés, since each relevé is now described by the same variable: similarity. The application of the product moment correlation coefficient produces a hyperspherical space where the relevés are settled on the surface of the hypersphere. If in a data set there is a dominant trend which can be explained by a gradient or by a compositional gradient the relevés will be arranged on an arc or a more or less twisted line. If no dominant trend is present, then the relevés will be arranged on the surface of the hypersphere in more or less dense clouds.

The application of Feoli Chiapella & Feoli's (1977) method to the simulated coenocline of Fig. 1 and of Fig. 4 of Austin (1976a) has led to ordination patterns in which

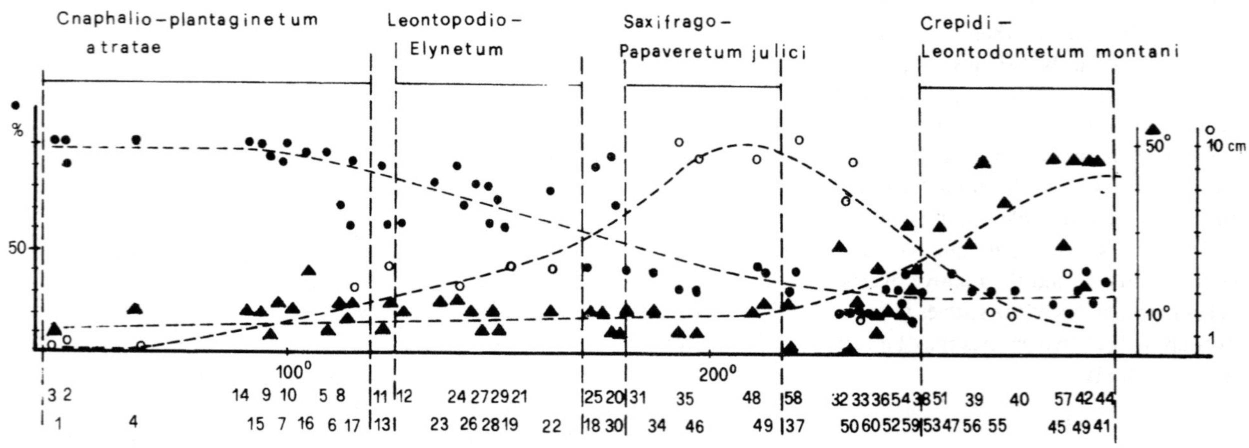

Fig. 5. Seriation of relevés from the summit of the Majella massive (Feoli Chiapella & Feoli 1977, Fig. 5) according to the angle of their position vectors given by the first two principal components (see the text for explanation). Legend to symbols: ○ = mean dimension of stones in each relevé; ▲ = slope angle; ● = percentage of ground surface covered by vegetation. The numbers along the abscissa identify the relevés as in Tables 1, 2, 3, 4 of Feoli Chiapella & Feoli (1977).

the fit of the first two axes to the arc is almost perfect (Fig. 4a, b). Along the arcs the rank order of the relevés is quite correct. In this case, the linear ordination has given the same results as a non-linear ordination (Austin 1976b, Fig. 4), and the preference for one or another is only a matter of taste. When the arrangement of the relevés fits to an arc, two axes can account for the trend of variation and the angles that the relevé vectors enclose can be used to arrange the relevés on a straight line. The angles have been already used to seriate sites by van der Aart & Smeenk-Enserink (1975).

The problem arises how to choose the starting point. This can be chosen indifferently among one of the two extreme points of the arc fitting the relevé points in the ordination scatter diagram. An example of such a seriation is shown in Fig. 5 for the relevés of the Majella's summit (Feoli Chiapella & Feoli 1977). In this way curved trends can be simply unfolded on straight lines. Such a seriation allows us to give a clear graphical description of the relationships between the four associations and the surface area of ground covered by the vegetation, the slope angle, and the mean dimension of stones. Furthermore, the seriation of the relevés can be useful to quantify the degree of overlap between the associations along the gradient. From Fig. 5, by considering the segments where the relevés belonging to two successive associations are mixed, it appears that the associations with the highest degree of overlap are *Saxifrago-Papaveretum julici* and *Crepidi-Leontodontetum montani*. This observation corresponds with the results of cluster analysis which show that the similarity between these two associations is higher than the similarity between *Gnaphalio-Plantaginetum atratae* and *Leontopodio-Elynetum* (cf. Feoli Chiapella & Feoli 1977, Fig. 4).

Table 2. Co-ordinates of ten points arranged on a circle (a) and on a perimeter of a square (b).

A

1	2	3	4	5	6	7	8	9	10	11	12	13	14	15	16
17	19	20	19	17	15	10	5	3	2	0	2	3	5	10	15
3	5	10	15	17	19	20	19	17	15	10	5	4	2	0	0

B

1	2	3	4	5	6	7	8	9	10	11	12	13	14	15	16
0	5	10	15	20	20	20	20	20	15	10	5	0	0	0	0
20	20	20	20	20	15	10	5	0	0	0	0	0	5	10	15

Conclusions

What Thurstone (1947) says in his introduction to Factor analysis should be considered by many data analysts:

'If the scientist takes his numerical coefficients very seriously at the exploratory stage he may be lacking in a desirable sense of humor about the crudeness of all his tools in spite of their polished appearance'.

The application of multidimensional scaling to plant ecology may be considered to be still at an exploratory stage since it is very difficult to postulate a general species response pattern to gradients (Austin 1972, 1976a). However the application of linear methods has always given positive results for those who have a proper understanding of them and consequently interpreted the results. The knowledge that linear methods arrange coenoclines in an involute horseshoe-shaped curves can indeed help the ecologist to use the results in a profitable way. The basic misunderstanding of those who on the basis of simulated coenoclines suggest the inadequacy of linear ordinations in gradient analysis relies on the fact that they have not considered some of the intrinsic aspects.

Summary

The evaluation of ordination methods through simulated coenoclines is discussed. The inconsistency of some arguments against the use of linear ordinations is stressed. The regression between the product moment correlation coefficient and the relative overlapping area of the bell-shaped curves of the species of coenoclines may be fitted to a linear function. An example of a seriation of relevés along a straight line based on a linear method is given using field data.

References

Aart, P.J.M. van der & N. Smeenk-Enserink. 1975. Correlation between distributions of hunting spiders (Lycosidae, Ctenidae) and environmental characteristics in a dune area. Neth. J. Zool. 25: 1–45.

Anderson, A.J.B. 1971. Ordination methods in ecology. J.Ecol. 59: 713–726.

Austin, M.P. 1972. Models and analysis of descriptive vegetation data. In: J.N.R. Jeffers (ed.): Mathematical models in ecology, p. 61–86. Blackwell, Oxford.

Austin, M.P. 1976a. On non-linear species response models in ordination. Vegetatio 33: 33–41.

Austin, M.P. 1976b. Performance of four ordination techniques

assuming three different non-linear species response models. Vegetatio 33: 43–49.
Austin, M.P. & I. Noy-Meir. 1971. The problem of non-linearity in ordination: experiments with two gradient models. J.Ecol. 59: 762–773.
Beals, E.W. 1973. Ordination: mathematical elegance and ecological naïveté. J.Ecol. 61: 23–35.
Dale, M.B. 1975. On objectives of methods of ordination. Vegetatio 30: 15–32.
Feoli, E. 1976. Correlation between single ecological variables and vegetation by means of cluster analysis. Not. Fitosoc. 12: 77–82.
Feoli, E. 1977. On the resolving power of principal component analysis in plant community ordination. Vegetatio 33: 119–125.
Feoli, E. & L.Feoli Chiapella. 1976. Due associazioni rupicole della Majella. Not. Fitosoc. 12: 67–75.
Feoli Chiapella L. & E. Feoli. 1977. A numerical phytosociological study of the summits of the Majella massive (Italy). Vegetatio 34: 21–39.
Gauch, H.G.Jr. & R.H. Whittaker. 1972a. Coenocline simulation. Ecology 53: 446–451.
Gauch, H.G.Jr. & R.H. Whittaker. 1972b. Comparison of ordination techniques. Ecology 53: 868–875.
Gauch, H.G.Jr. & R.H. Whittaker. 1976. Simulation of community patterns. Vegetatio 33: 13–16.
Gauch, H.G.Jr., G.B. Chase & R.H. Whittaker. 1974. Ordination of vegetation samples by Gaussian species distribution. Ecology 55: 1382–1390.
Gauch, H.G.Jr., R.H. Whittaker & R.T. Wentworth. 1977. A comparative study of reciprocal averaging and other ordination techniques. J.Ecol. 65: 157–174.
Gittins, R. 1965. Multivariate approaches to a limestone grassland community. I. A stand ordination. J.Ecol. 53: 385–401.
Goodall, D.W. 1954. Objective methods for the classification of vegetation. III. An essay on the use of factor analysis. Austral.J.Bot. 2: 304–324.
Goodall, D.W. 1978. Sample similarity and species correlation. 2nd ed. In: R.H. Whittaker (ed.): Ordination of plant communities, p. 99–149. Junk, The Hague.
Gounot, M. 1969. Méthodes d'étude quantitative de la végétation. Masson, Paris, 314 pp.
Gower, J.C. 1966. Some distance properties of latent root and vector methods used in multivariate analysis. Biometrika 53: 325–338.
Groenewoud, H. van. 1976. Theoretical considerations on the covariation of plant species along ecological gradients with regard to multivariate analysis. J.Ecol. 64: 837–847.
Hill, M.O. 1973. Reciprocal averaging: an eigenvector method of ordination. J.Ecol. 61: 237–249.
Hill, M.O. & A.J.E. Smith. 1976. Principal component analysis of taxonomic data with multi-state discrete characters. Taxon 25: 249–255.
Ihm, P. & H. van Groenewoud. 1975. A multivariate ordering of vegetation data based on Gaussian type gradient response curves. J.Ecol. 63: 767–777.
Kessel, S.R. & R.H. Whittaker. 1976. Comparison of three ordination techniques. Vegetatio 32: 21–29.
Kruskal, J.B. 1964a. Multidimensional scaling by optimizing goodness of fit to a non-metric hypothesis. Psychometrika 29: 1–27.
Kruskal, J.B. 1964b. Non-metric multidimensional scaling: a numerical method. Psychometrika 29: 115–129.
Lotka, A.J. 1932. The growth of mixed populations: two species competing for a common food supply. J. Washington Acad. Sci. 22: 461–469.
Maarel, E. van der. 1969. On the use of ordination models in phytosociology. Vegetatio 19: 21–46.
Maarel, E. van der. 1971. Basic problems and methods in phytosociology. Rep. 14th Int. Symp. Rinteln 1970. Vegetatio 22: 269–273.
Maarel, E. van der. 1979. Transformation of cover-abundance values in phytosociology and its effects on community similarity. Vegetatio 39: 97–114.
Nichols, S. 1977. On the interpretation of principal component analysis in ecological contexts. Vegetatio 34: 191–197.
Noy-Meir, I. 1973. Data transformation in ecological ordination. I. Some advantage of non-centering. J.Ecol. 61: 329–341.
Noy-Meir, I. 1974. Catenation: quantitative methods for the definition of coenoclines. Vegetatio 29: 89–99.
Noy-Meir, I. & M.P. Austin. 1970. Principal component ordination and simulated vegetation data. Ecology 51: 551–552.
Noy-Meir, I., D. Walker & W.T. Williams. 1975. Data transformation in ecological ordination. II. On the meaning of data standardization. J.Ecol. 63: 779–800.
Noy-Meir, I. & R.H. Whittaker. 1977. Continuous multivariate methods in community analysis: some problems and developments. Vegetatio 33: 79–98.
Orlóci, L. 1966. Geometric models in ecology. I. The theory and application of some ordination methods. J.Ecol. 54: 193–215.
Orlóci, L. 1967. Data centering: a review and evaluation with reference to component analysis. Syst. Zool. 16: 208–212.
Orlóci, L. 1972a. On information analysis in Phytosociology. In: E. van der Maarel & R. Tüxen (eds.), Grundfragen und Methoden der Pflanzensoziologie, Ber. Int. Symp. Rinteln 1970, p. 75–88. Junk, Den Haag.
Orlóci, L. 1972b. On objective functions of phytosociological resemblance. The American Mid. Natur. 88: 28–55.
Orlóci, L. 1974. On information flow in ordination. Vegetatio 29: 11–16.
Orlóci, L. 1978a. Multivariate analysis in vegetation research. Junk, The Hague, 2nd ed. 451 pp.
Orlóci, L. 1978b. Ordination by resemblance matrices. 2nd ed. In: R.H. Whittaker (ed.) Ordination of plant communities, p. 239–275. Junk, The Hague.
Orlóci, L. 1979. Non-linear data structures and their description. International Statistical Ecology Program. Parma 1978.
Phillips, D.L. 1978. Polynomial ordination: field and computer simulation testing of a new method. Vegetatio 37: 129–140.
Shepard, R.N. & J.D. Carrol. 1966. Parametric representation of non-linear data structure. In: P.R. Krishnaiah (ed.), Multivariate analysis, p. 516–592. Academic Press.
Swan, J.M.A. 1970. An examination of some ordination problems by use of simulated vegetation data. Ecology 51: 89–102.
Stanek, W., J.K.Jeglum & L.Orlóci. 1977. Comparisons of peatland types using macronutrient contents of peat. Vegetatio 33: 163–173.
Volterra, V. 1931. Variations and fluctuations of the number

of individuals in animal species living together. In: Chapman (ed.), Animal ecology with special reference to insects, pp. 409–448. Mc Grow-Hill, New York.

Werger, M.J.A. 1978. Vegetation structure in the southern Kalahari. J. Ecol. 66: 933–941.

Westhoff, V. & E. van der Maarel. 1978. The Braun-Blanquet approach. 2nd ed. In: R.H. Whittaker (ed.), Classification of plant communities, Junk, The Hague. p. 287–399.

Whittaker, R.H. 1967. Gradient analysis of vegetation. Biol. Rev. 49: 201–264.

Whittaker, R.H. 1972a. Convergences of ordination and classification. In: E. van der Maarel & R. Tüxen (eds.), Grundfragen und Methoden der Pflanzensoziologie, Ber. Int. Symp. Rinteln 1970. p. 39–57.

Whittaker, R.H. 1972b. Evolution and measurement of species diversity. Taxon 21: 213–251.

Whittaker, R.H. & H.G. Gauch. 1978. Evaluation of ordination techniques. 2nd ed. In: R.H. Whittaker (ed.), Ordination of plant communities, p. 277–336. Junk, The Hague.

Accepted 4 December 1979

ON THE INTERPRETABILITY OF ORDINATION DIAGRAMS

Eddy VAN DER MAAREL

Division of Geobotany, Toernooiveld, 65 25 ED Nijmegen, The Netherlands

Keywords:
Distortion, Gradient, Horseshoe effect, Ordination

In my review of multivariate methods in phytosociology (van der Maarel 1979) I briefly commented on the interpretability of Type A distortions (see Orlóci 1974b, 1978), i.e. 'horseshoe', ('arch') effects in ordination diagrams of axes 1 and 2. Such effects are now generally recognized as the result of curvilinearities intrinsic to community data i.e. (1) the bellshaped species response curves along environmental gradients and (2) the non-linear decrease of sample similarity with increasing sample separation (Gauch, Whittaker & Wentworth 1977, see also Swan 1970, Noy-Meir & Austin 1970, Gauch & Whittaker 1972, Austin 1976, Noy-Meir & Whittaker 1978 and Whittaker & Gauch 1978 to mention a few papers I consulted again for this purpose). Hill & Gauch (1980) even say more straightforwardly: 'the arch effect is simply a mathematical artifact, corresponding to no real structure in the data', when discussing faults of reciprocal averaging.

The demonstration of this drawback with simulated coenoclines of varying β-diversity, based on series of overlapping Gaussian species response curves is convincing. My concern has been for some time that in many ordination diagrams produced in phytosociological studies of our group or with the same methods as we use (mostly ORDINA, a PCA on a Euclidean distance matrix, Roskam 1971 and also RA, Hill 1973) we found either no horseshoe at all, or we could easily interpret the arch appearing in the diagram of axes 1 and 2. This makes me somewhat doubtful about the realism of simulated coenoclines as reflections of real environmental gradients. Such coenoclines could be unrealistic in two ways: (1) the Gaussian species response model is not as universal as it is supposed to be by users of this type of simulated data. (2) the length of the coenocline in terms of the simulated β-diversity, or maybe rather the relation between β-and α-diversity (local concentration of species in a narrow part of the gradient) does not reflect the amplitude of environmental gradients or variation in community composition phytosociologists usually describe.

Before discussing this further I confirm that part of the relatively positive results we obtained may be due to the choice of resemblance function and that of the data transformation involved. As to the resemblance function: we normally use Euclidean distance. Although this measure was found to produce more deviation from expected ordination patterns of simulated data (e.g. Whittaker & Gauch 1978) it has some advantages, certainly in combination with some sort of transformation.

First it is a fully metric resemblance function which is moreover directly related to the geometric sample space ('phytosociological space', van der Maarel 1979). Second it describes sample separation along a coenocline better than similarity measures (van der Maarel & Leertouwer 1967): with similarity measures ranging from zero to one, the resemblance profile along the coenocline decreases from one to zero rather rapidly and remains zero. With Euclidean distance the profile shows an increase which can be followed by a decrease in case of species poor extremes of the cline. In any case there is no problem with high numbers of zero values in the resemblance matrix.

As to data transformation I may add that we usually apply an ordinal transformation to Braun-Blanquet cover-abundance values (van der Maarel 1979a) which we found of intermediate impact, i.e. between dominance weighting and presence weighting (cf. Gauch et al. 1977, van der Maarel 1979) and thus comparable to standardization by species (cf. Noy-Meir, Walker & Williams 1975).

This would mean that our ordinations are relatively 'favoured' as are PCA's with standardization by species (cf. Whittaker & Gauch 1978). I agree with the latter authors that relativation by relevé totals would be a useful transformation before calculating a resemblance function, especially if this would be Euclidean distance (van der Maarel 1979), but we did not try this out yet.

Let me now discuss the reality of simulated coenoclines. First I may confirm the findings of Austin (1976) that Gaussian response curves are both theoretically questionable and less frequently found in reality. Austin (1980) discusses various other models, in which the intervals along the species response is measured as a relative performance, and the intervals along the environmental gradient are taken logarithmically. He shows how the ecological displacement model of Ellenberg (see Mueller-Dombois & Ellenberg 1974) and the environmental stress model of Grime (1973, 1977) can predict bimodal response curves and wedge-type curves. I also suggested the latter type in relation to environmental dynamics (van der Maarel 1976).

If we look at phytosociological tables arranged with the help of ordination techniques (e.g. van der Maarel, Janssen & Louppen 1978, and also in Gauch et al. 1977) we observe that, indeed, the wedge-type is frequently occurring in various data-sets, whereas the Gaussian-type is almost an exception.

So in conclusion to these considerations we may state that in phytosociological data-sets as used in European ordination studies the supposed general basis of type A distortion in ordination diagrams in the form of non-linear Gaussian-type species behaviour along environmental – or phytosociological gradients does not exist.

Let us now look at the arch effect in some detail. First I confirm that in some of the ordinations I referred to above (van der Maarel et al. 1978) the diagram of axes 1 and 2 definitely showed a horseshoe shape despite the non-Gaussian behaviour of most species involved. Here Hill & Gauch's (1980) remark on the independence of axes may be relevant. However, in some clear cases of arch shape configurations representing long coenoclines the result was still interpretable, but in a special way.

The first example I examined was that of Werger, Wild & Drummond (1978). This study deals with plant community types of the Great Dyke, Rhodesia, forming a long coenocline, but with discontinuities between types according to sharp environmental boundaries due to contrasts between different bed rock types. Both PCA and RA produced an arch shape configuration in the diagram of axes 1 and 2. The underlying environmental gradient appeared to be very complex and in fact consists of three parts: (a) a part with relatively favourable soils as to toxicity (high Mg content) but with decreasing moisture content, (b) a part with a rapid increase in unfavourable heavy metal conditions, combined with a decrease in moisture conditions, and (c) a part on serpentine soils with increasing soil moisture conditions. The ordination diagram clearly shows the relative position of these three parts.

The second example is taken from Jensén's (1978) study of a Swedish lake shore vegetation. A PCA was applied to cover data of *helophyte* species in 38 subsequent 1 sq. m. plots lying in a belt transect. A clear horseshoe pattern arises with some involution, this depending on the transformation used. The sequence of relevés according to the actual position along the belt transect is acceptably reflected in cases of transformation which reduce the influence of dominant species. The transect itself is described by Jensén and appears to consist of three parts: (a) a part above low water level in which the shrub *Myrica gale* appears and which is probably characterized by an increase in moisture content of the soil; (b) a part in which the period of inundation increases from short to almost the year round; and (c) a portion which is permanently inundated with an increase in average water depth.

I think we have a rather clear case of what may be much more generally occurring in nature than so far suggested: the singular topographical gradient consists of various environmental gradients which differ as to the master factor. PCA ordination does reveal this differentiation and any ordination which might be better from other, theoretical, points-of-view but does not produce this result in two dimensions, may be called less effective!

The practical conclusion I would draw here, is to split the data set into subsets and apply an ordination to each of them. This was done in a similar study of a topographical limestone gradient in grassland (Louppen in prep., see Louppen, Werger & Eppink 1978) and – as Austin (pers. comm.) presumed – the ordination diagrams of parts of the entire sequence again produced horseshoe configurations, be it less pronounced.

The latter remark brings me to a differentiated conclusion: ordination diagrams should not be disregarded simply because they seem to produce ineffective results. As long as an ordination diagram is interpretable it should be used, and of course the ordination technique being used, be welcomed. In this respect the user should be careful about the real dimensionality of the coenocline under study. What looks like a one dimensional variation in the phytosociological space (belt transect) may well appear as a more – dimensional partly dependent variation in the ecological space. On the other hand: mathematical artifacts due to linear constraints in a method applied to non-linearly varying variables should be overcome. Therefore we are anxious to see results of the new detrended correspondence analysis of Hill (1979, Hill & Gauch 1980) when applied to phytosociological data sets of the kind discussed here.

References

Austin, M.P. 1976. On non-linear species response models in ordination. Vegetatio 33: 33–41.

Austin, M.P. 1980. Searching for a model for use in vegetation analysis. In: E. van der Maarel (ed.) Advances in vegetation science: Classification and ordination. Vegetatio 42: 11–21.

Gauch, H.G. & R.H. Whittaker. 1972. Coenocline simulation. Ecology 53: 446–451.

Gauch Jr. H.J., R.H. Whittaker & T.R. Wentworth, 1977. A comparative study of reciprocal averaging and other ordination techniques. J. Ecol. 65: 157–174.

Grime, J.P. 1973. Control of species density in herbaceous vegetation. J. Envir. Manage. 1: 151–167.

Grime, J.P. 1977. Evidence for the existence of three primary strategies in plants and its relevance to ecological and evolutionary theory. Amer. Natur. 111: 1169–1194.

Hill, M.O. 1973. Reciprocal averaging: An eigenvector method of ordination. J. Ecol. 61: 237–249.

Hill, M.O. 1979. Decorana. A Fortran program for detrended correspondence analysis and reciprocal averaging. Ecology and Systematics, Cornell University, Ithaca. 52 pp.

Hill, M.O. & H.G. Gauch Jr. 1980. Detrended correspondence analysis, an improved ordination technique. In: E. van der Maarel (ed.). Advances in vegetation science: Classification and ordination. Vegetatio 42: 47–58.

Jensén, S. 1978. Influences of transformation of cover values on classification and ordination of lake vegetation. Vegetatio 37: 19–31.

Louppen, J.W.M., M.J.A. Werger & J.H.M. Eppink. 1978. Vegetation boundaries and species performances along an environmental gradient. Abstracts 2nd. Int. Congr. Ecology, Jerusalem, p. 217.

Maarel, E. van der. 1972. Ordination of plant communities on the basis of their plant genus, family and order relationships. In: E. van der Maarel & R. Tüxen (ed.). Grundfragen und Methoden der Pflanzensoziologie. Ber. Int. Symp. Rinteln 1970, Junk, Den Haag. p. 183–192.

Maarel, E. van der. 1976. On the establishment of plant community boundaries. Ber. Deutsch. Bot. Ges. 89: 415–443.

Maarel, E. van der. 1979. Multivariate methods in phytosociology, with reference to the Netherlands. In M.J.A. Werger (ed.) The study of vegetation, p. 161–225.

Maarel, E. van der. 1979a. Transformation of cover-abundance values in phytosociology and its effects on community similarity. Vegetatio 39: 97–114.

Maarel, E. van der, J.G.M. Janssen & J.M.W. Louppen. 1978. TABORD, a program for structuring phytosociological tables. Vegetatio 38: 143–156.

Maarel, E. van der & J. Leertouwer. 1967. Variation in vegetation and species diversity along a local environmental gradient. Acta Bot. Neerl. 16: 211–221.

Mueller-Dombois, D. & H. Ellenberg. 1974. Aims and methods of vegetation ecology. Wiley, New York, XI + 547 pp.

Noy-Meir, I. & P. Austin. 1970. Principal component ordination and simulated vegetation data. Ecology 51: 551–552.

Noy-Meir, I.D. Walker & W.T. Williams. 1975. Data transformations in ecological ordination. II. On the meaning of data standardization. J. Ecol. 63: 779–800.

Noy-Meir, I. & R.H. Whittaker. 1978. Recent developments in continuous multivariate techniques. In R.H. Whittaker (ed.) Ordination of plant communities, p. 337–378.

Orlóci, L. 1974. On information flow in ordination. Vegetatio 29: 11–16.

Orlóci, L. 1978. Multivariate analysis in vegetation research. 2nd. ed. Junk, The Hague, Boston, IX + 451 pp.

Roskam, E. 1971. Programme ORDINA: Multidimensional ordination of observation vectors. Programme Bull. 16, Psychology Lab. Nijmegen, 8 pp.

Swan, J.M.A. 1970. An examination of some ordination problems by use of simulated vegetational data. Ecology 51: 89–102.

Werger, M.J.A., H. Wild & B.R. Drummond. 1978. Vegetation structure and substrate of the Northern part of the Great Dyke, Rhodesia: Gradient analysis and dominance-diversity relationships. Vegetatio 37: 151–161.

Whittaker, R.H. & H.G. Gauch Jr. 1978. Evaluation of ordination techniques, 2nd ed. In R.H. Whittaker (ed.) Ordination of plant communities, p. 277–336. Junk, The Hague, 388 pp.

DETRENDED CORRESPONDENCE ANALYSIS: AN IMPROVED ORDINATION TECHNIQUE

M.O. HILL* & H.G. GAUCH, Jr.**

Section of Ecology and Systematics, Cornell University, Ithaca, New York 14850, USA

Keywords:
Correspondence analysis, Multivariate technique, Nonmetric multidimensional scaling, Ordination, Reciprocal averaging

Introduction

Studies by ourselves and others (Swan 1970, Austin & Noy-Meir 1972, Beals 1973, Hill 1973, 1974, Austin 1976a, b, Fasham 1977, Gauch Whittaker & Wentworth 1977, Noy-Meir & Whittaker 1977, Orlóci 1978, Gauch, Whittaker & Singer 1979) have found faults with all ordination techniques currently in use, at least when applied to ecological data specifying the occurrences of species in community samples. These faults certainly do not make existing techniques useless; but they mean that results must be interpreted with caution. Even with the best techniques, the underlying structure of the data is often poorly expressed.

Of existing ordination techniques, perhaps the most satisfactory are reciprocal averaging (Hill 1973) and nonmetric multidimensional scaling (Fasham 1977, Prentice 1977, Gauch et al. 1979). [An alternative name for reciprocal averaging, preferred in the mathematical literature, is 'l'analyse factorielle des correspondances' (Benzécri 1973), translated into English as 'correspondence analysis' (Hill 1974).] Of these two techniques, nonmetric multidimensional scaling has the disadvantage that it is computationally demanding when applied to large problems. Moreover, in spite of its apparently innocuous assumptions, it cannot be relied on to extract the correct configuration, even when applied to artificial data (Fasham 1977, Gauch et al. 1979). Part of the difficulty lies in the first stage of the analysis, for which the data are summarized as a matrix of between-sample distances. When there is large habitat-diversity, many of the larger distances will approach an upper limit for two samples with no species in common. Distances near this upper limit are all about equal with the result that, while the local configuration of the ordination may be correct, the global configuration can be seriously distorted.

Reciprocal averaging, on the other hand, makes no use of the concept of compositional distance and is closely related to the weighted-averages ordination of Whittaker (1956) and Curtis (1959). It is computationally much less demanding than nonmetric multidimensional scaling, and has a similar reliability in practice (Gauch et al. 1979). Detrended correspondence analysis – the method described here – is a development of reciprocal averaging that avoids its two major drawbacks. Tests indicate that this new technique is better than others currently available.

Description of the technique

In order to grasp the rationale of detrended correspondence analysis (DCA) it is necessary to appreciate the faults of reciprocal averaging (RA). Correction of these faults will produce an improved technique.

Faults of reciprocal averaging

Reciprocal averaging has two main faults, both of which hinder interpretation of results. Of these, perhaps the worst is the 'arch effect' (Gauch et al. 1977), sometimes also known as the 'horseshoe effect' (Kendall 1971). The arch effect (Fig. 1) is simply a mathematical artifact, corresponding to no real structure in the data. It arises because the

* Permanent address: Institute of Terrestrial Ecology, Penrhos Road, Bangor, Gwynedd LL57 2LQ, Wales, UK.

** This research was supported by the Institute of Terrestrial Ecology, Bangor, Wales, and by a grant from the National Science Foundation to R.H. Whittaker. We thank R.H. Whittaker for encouragement and comments, S.B. Singer for assistance with the Cornell computer, and H.J.B. Birks, S.R. Sabo, T.C.E. Wells, and R.H. Whittaker for data sets used for ordination tests.

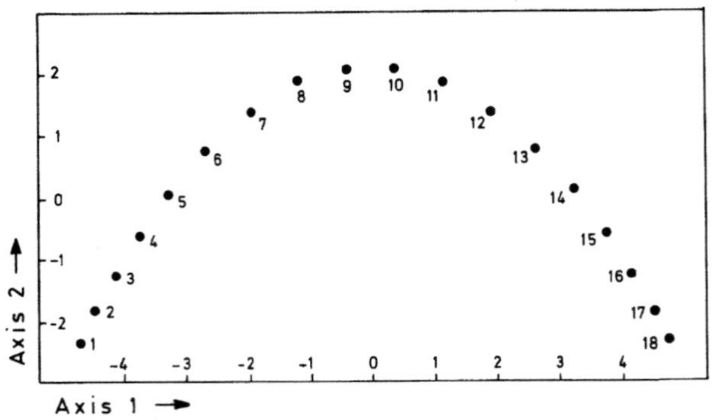

Fig. 1. The 'arch effect' exhibited by reciprocal averaging. The diagram shows the first two axes of a sample ordination of the data set shown in Fig. 2. There is only one natural gradient, so the two-dimensional configuration of points is a mathematical artifact having no basis in the structure of the data. Scaling of the axes is such that $\sum\sum a_{ij}(x_i - y_j)^2 = \sum\sum a_{ij}$, where a_{ij} is the abundance of species j in sample i, x_i is the score of sample i in the ordination, y_j is the score of sample j, and $x_i = \sum_j a_{ij}y_j / \sum_j a_{ij}$.

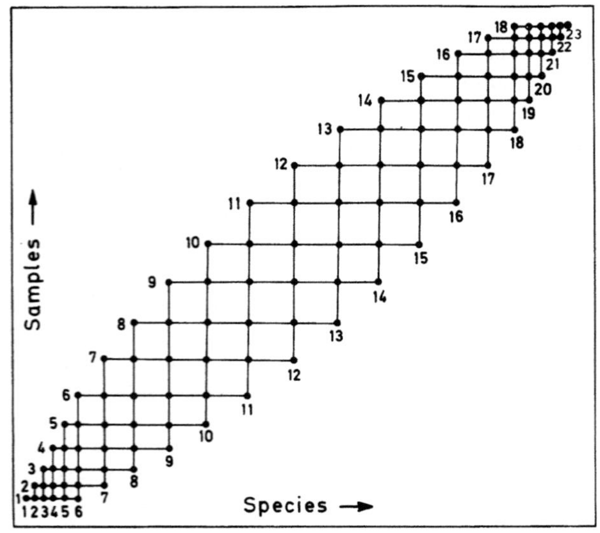

Fig. 2. Presence-and-absence data with a regular structure, ordinated by reciprocal averaging, first axis (same data as Fig. 1). Points represent the occurrence of species in samples. For example, sample 6 contains species 6, 7, 8, 9, 10, and 11. Samples and species are arranged according to their position on the first axis of an ordination by reciprocal averaging. Note the tendency for the within-sample standard deviation to be smaller at the ends of the gradient than in the middle.

second axis (canonical variate) of RA is constrained to be uncorrelated with the first axis, but is in no way constrained to be independent of it. For the axes to be separately interpretable, they need to be independent, not merely uncorrelated.

The other main fault of RA is that it does not preserve ecological distances. Sample pairs with equivalent compositional distances appear farther apart in the middle of, than toward the ends of, the first reciprocal averaging axis. For example, in Fig. 2, depicting the same data as Fig. 1, samples 15 and 18 differ by exactly the same amount as samples 9 and 12; yet on the first axis of the ordination, this equality is misrepresented.

Elimination of the arch effect

The basis of the arch effect, noted above, is that the second and sometimes the higher axes of RA can have a strong relation to the first axis while at the same time still being uncorrelated with it. To eliminate the arch effect, it is necessary to ensure that subsequent axes do not show an arch-shaped, or any other systematic relation to the first axis.

One possibility for the elimination of such systematic relations is to demand that subsequent axes are orthogonal not only to the first axis, but also to a quadratic and a cubic on that axis. Simpler and better than this, however, is to insist that the subsequent axes are drawn so that at any point along the first axis, the mean value of the subsequent axes is approximately zero (Fig. 3).

Given a vector of scores, the process of subtracting a local mean value for an appropriate segment of the first axis is here referred to as 'detrending' with respect to an independent variable. For example, in Fig. 3, if the x-axis were time and the y-axis were soil temperature, then the process of detrending would replace the y-variable by a

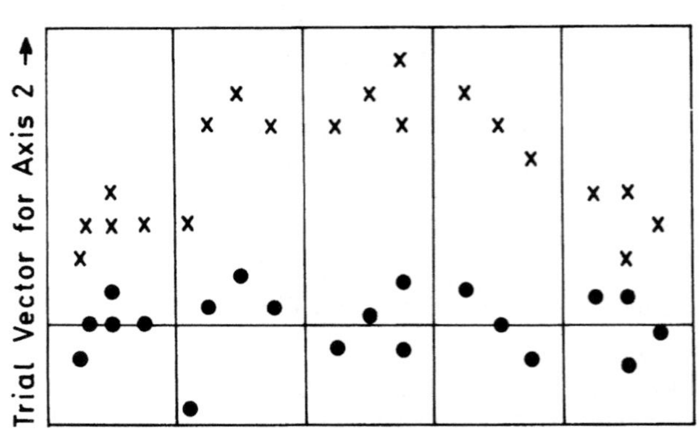

Fig. 3. Method of detrending used in detrended correspondence analysis. The gradient along axis 1 is divided into a number of segments. Within each segment the values on axis 2 are adjusted by centering them to zero mean. Running segments are used in the program (Hill 1979).

with disappointing results. Its chief disadvantage is that it is unreliable in the case where one end of the gradient incorporates substantial variability in a different dimension whereas the other end does not.

A method of rescaling that has proved to be more successful is based on the within-sample dispersion of species scores. With RA there is generally a smaller dispersion of the species scores within samples towards the ends of the gradient than at the middle. In Fig. 2, this means that samples at the middle of the gradient are represented by longer horizontal lines than those at the ends. If the scaling were regular, then the dispersion of species within samples should be approximately constant along the gradient.

To see why this should be so, the problem needs to be turned on its side. One very plausible definition of the correct scaling from an ecological point of view is that the rate of appearance and disappearance of species along the gradient should be a constant. If species are appearing and disappearing at a constant rate, then at least for those species whose range is not truncated, the within-species dispersion of the samples should be the same at all points of the gradient. In fact, it can be seen in Fig. 2 that at any one position along the gradient, the length of the vertical lines (species) is roughly equal to the length of the horizontal lines (samples). Hence, to demand that the dispersion of the species scores within samples shall not vary along the gradient is almost equivalent to demanding that the dispersion of the sample scores within species shall not vary along it.

There is very good reason to concentrate on the within-sample rather than on the within-species dispersion, although the latter might appear at first sight a more natural criterion to adopt. The reason is simply that while nature may ensure that there is roughly the 'right' species composition for each sample, she can do nothing of the kind for the sample composition of species. Indeed, in Fig. 2 it can be seen that some species have very low within-species dispersions because they have their ranges truncated. This has nothing to do with the way species arrange themselves in the world, but is merely a result of the investigator's choice of samples.

Rescaling is achieved, then, by trying to equalize the mean within-sample dispersion of the species scores at all points along the gradient. The relation between dispersion and position along the gradient is considered (Fig. 4), and where the dispersion is high the scaling is contracted, and where it is low the scaling is expanded. Note that it is the *species* ordination that is expanded, not the sample ordination. The sample ordination is defined at all times so that the sample scores are weighted mean values of the scores of the species that occur in them.

Given that the within-sample variance is to be set to a constant, it is natural to set this constant to 1 so as to achieve a standard scaling. With this standard scaling, the

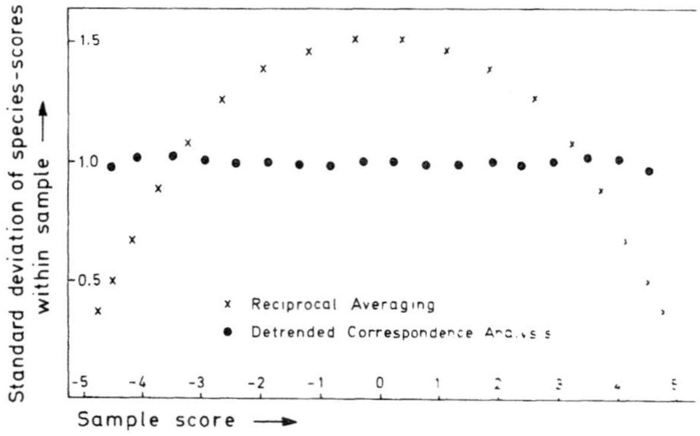

Fig. 4. Within-sample dispersion of species scores in relation to position along the gradient (same data as Fig. 2, first axis). It can be seen that with detrended correspondence analysis the dispersion of species scores within samples is effectively constant, while with reciprocal averaging it is much higher in the middle than at the ends.

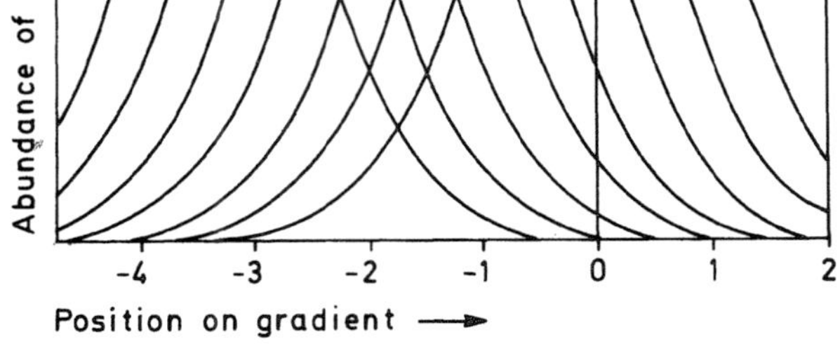

Fig. 5. Hypothetical gradient, scaled so as to have unit within-sample standard deviation. Species packing is constant along the gradient; each species is given a score equal to the sample score of the position of its mode. Note that in this case, the species-abundance profiles (i.e. the curves shown in the diagram) also have unit standard deviation. Species whose modes are more than 2 standard deviations from a given sample are absent or very poorly represented. (See, for example, species with mode at F in sample at A.)

new variable that could be interpreted as relative warmth or cold for the season.

It might be supposed that elimination of the arch effect could be achieved by performing the calculations of RA as usual, and then at the end detrending the second and subsequent axes with respect to the first axis. In this way, systematic relations could certainly be removed; but there would be the drawback that, if detrending were performed for the sample scores, then the new sample scores would no longer be the average scores of the species that occur in them. The ordination would become an abstract sample ordination, with no good relationship to any species ordination. A second and more serious drawback is that the axes would appear in an undesirable order. Thus, if the undetrended axis 2 has a strong relation to axis 1, then it will be of little significance even after detrending; whereas axis 3 could present valuable independent information. It would be much better to generate axes in descending order of significance, and this cannot be achieved if the detrending is postponed to the end, after the axes have been fixed.

To avoid these drawbacks, detrending must be built into the calculation of the axes. That is to say, it must be incorporated into the two-way averaging algorithm of RA, which is as follows (Hill 1973).

1. Select an arbitrary set of scores $y_j (j = 1, \ldots, J)$ for the species.

2. Let the abundance of species j in sample i be a_{ij}, and let $\sum\sum$ denote summation over all species and all samples in the data matrix. Standardize the species scores to zero mean and unit length, using weights defined by the data matrix; i.e. set

$$\sum\sum a_{ij}y_j = 0; \sum\sum a_{ij}y_j^2 = 1.$$

3. Ordinate the samples by weighted averages so that the score x_i of sample i is the mean score of the species that occur in it; i.e. set

$$x_i = \sum_i a_{ij}y_j / \sum_i a_{ij} \ (j = 1, \ldots, J).$$

4. Re-ordinate the species so that the score of each species is the (weighted) mean score of the samples that it occurs in. Let the new species scores be $y'_j \ (j = 1, \ldots, J)$. Then

$$y'_j = \sum_j a_{ij}x_i / \sum_j a_{ij} \ (i = 1, \ldots, I).$$

5. Using the new scores y'_j as a basis, return to stage 2, and continue going round the loop until the scores stabilize. The scores y_j are referred to as a 'trial vector.' A trial vector is deemed to be an eigenvector if $y_j = y'_j (j = 1, \ldots, J)$. In Fig. 3, the corresponding sample scores x_i are also referred to as a 'trial vector,' but this is a slightly loose way of speaking, as they are not used to test whether an eigenvector has been found. The algorithm outlined above will converge to the first axis of a RA ordination. To calculate the second axis, stage 3 needs to be augmented by an additional stage.

3a. Orthogonalize with respect to the first axis by subtracting a (weighted) linear regression on it. If $x'_i (i = 1, \ldots, I)$ are the scores of the sample on the first axis, then the desired condition of orthogonality is

$$\sum\sum a_{ij} x'_i x_i = 0.$$

It should be noted that this is a weighted condition of orthogonality, with the same weights as those used at stage 2. The logic of RA demands that the weights are applied, but they are irrelevant to the main issue, which is that orthogonalization ensures that there is no correlation, but does not guarantee independence.

To calculate axis 3, it is necessary to orthogonalize also with respect to axis 2 as well as with respect to axis 1, and so on for the remaining axes.

Clearly, the place to do detrending of the sort outlined in Fig. 3 is at stage 3a. The orthogonalization is replaced by a detrending, but otherwise the two-way averaging proceeds as usual. (For technical reasons that need not concern us here, it is computationally convenient to do a weighted detrending, with the same weights as were used at stage 3a.) By this means the arch effect is eliminated.

In order to ensure that each sample score is the mean score of the species that occur in it, stage 3a is omitted at the very end, after the eigenvector has been found. Hence the algorithm leads to an axis which consists of a set of species scores y_j, and a corresponding set of sample scores x_i which are weighted averages of the species scores.

Rescaling the axes

In addition to the arch effect, the other major fault of RA is its distortion of relative distances between samples (and species) on its axes. Indeed, as seen in Fig. 2, samples differing ecologically by an identical amount may nevertheless be separated by different distances in the ordination. The question arises of how to prevent this.

One plausible method would be to summarize the species composition of samples in various segments of the gradient (i.e. to form composite samples) and to try to arrange that equal differences in species composition correspond to equal differences along the gradient. This method was tried out during the development of DCA, 'average' species-abundance profile has unit standard deviation (Fig. 5). The resulting unit of ordination-length may therefore be called a 'standard deviation,' abbreviated 'sd.' [Gauch & Whittaker (1972) have used a capital letter Z for this unit; but this usage has not been widely followed, and 'sd' is more explicit.] One can then speak of a gradient of length 1.3 sd, 5.0 sd, etc. A species may be expected to appear, to rise to its mode, and to disappear again in about 4 sd (Fig. 5); and a full turnover in species composition of samples should also occur in about 4 sd. A 50 % change in sample composition, which is a half-change (Gauch 1973) will, however, occur in about 1 sd or somewhat more.

Detrended correspondence analysis

The ideas outlined above – of RA with detrending in place of orthogonalization, followed by standardization to unit within-sample variance – combine to characterize the method of DCA. A computer program to perform the calculations has been written in FORTRAN, and is called DECORANA (DEtrended CORrespondence ANAlysis). The program description (Hill 1979) discusses a number of technical details that cannot be dealt with here and is available at cost as a Cornell Ecology Program from the second author. The calculations are not especially complicated, and their magnitude rises only in proportion to the number of non-zero items in the data matrix. They do not depend on the square or cube of the number of species or samples, and there is consequently no difficulty in analyzing large data sets.

Tests of the method on simulated data

We have tested the method extensively both on field data and on simulated data. As in previous tests (e.g. Gauch et al. 1977, 1979), data sets have been simulated by setting up an underlying, 'true' space, relative to which species have normally distributed curves of abundance. [The underlying space has been called the 'ecological space' by Austin (1976a).] In the one-dimensional case (Fig. 5) this means that species have normally distributed species-abundance profiles. In the two-dimensional case, species have bivariate normal distributions, and so on. Where there is only one underlying gradient of variation, the data are said to form a 'coenocline'; where there are two underlying gradients, they form a 'coenoplane.' Analogous terms are used for data sets of higher dimensionality.

Simulated coenoclines

A simple coenocline of length 6.75 (or 5 HC) was simulated. (Note that in terms of the 'half-change' unit used in previous tests of methods at Cornell, 1 sd = 0.741 HC, and 1 HC = 1.349 sd for an ideal coenocline like that of Fig. 5. For real data the ratio of sd to HC lengths is smaller than 1.349 and may be near unity.) Twenty-one samples were placed at uniform intervals along the length of the gradient, and 21 species were placed with their modes at uniform intervals from −3.37 sd to 10.12 sd. The data were then analyzed by DCA and RA using the Cornell Ecology Program DECORANA (Hill 1979). [RA can be carried out either by the Cornell Ecology Program ORDIFLEX (Gauch 1977) or DECORANA, which includes an option for plain RA.]

Relative distortion of the scaling was measured by mean percentage displacement of sample positions on the first axis of the ordination (Kessell & Whittaker 1976). Relative distortion was 4 % for RA and 0.2 % for DCA. The absolute length of the gradient was estimated rather less accurately by DCA, as 6.03 sd, as compared with the true value of 6.75 sd, a discrepancy of 11 %. The discrepancy in the length of the species ordination was much greater, the length being estimated as 8.65 sd as opposed to 13.5 sd. Most of this discrepancy was due to species whose modes were much outside the range of the samples; for these species the estimated modes were too close in. Analogous discrepancies should not be too damaging in practice, since length of an axis (beta diversity) is determined for the samples, not the species. However, for DCA (in contrast to RA) the apparent axis length for species is greater than the actual axis length for samples. Indicated modal positions of extreme (truncated) species are more realistic in DCA than in RA.

The simulated coenocline had only one dimension of variation in ecological space, so that any variation on axes other than the first is a mathematical artifact. The first four axes recovered by RA were approximately linear, quadratic, cubic, and quartic derivatives of the true scaling in ecological space. For DCA they were approximately linear, cubic, quartic, and quintic, whereas the quadratic axis of the 'arch effect' was completely missing. Although DCA has replaced a spurious quadratic axis by a spurious cubic axis, the latter is a far weaker distortion. In the example, the eigenvalue of the second axis is so low (0.027) for DCA, as opposed to RA (0.493), that its effect in the ordination of field data would be trivial.

Several other coenoclines were simulated with lengths

ranging from 4 to 14 sd and more complex structures, incorporating variation in the heights and dispersions of the curves (cf. Gauch & Whittaker 1972). Results were similar to those for the simple coenocline of length 6.75 sd. Estimates of gradient-length were generally accurate to within 10–20 %. Distortion of the scaling was markedly less with DCA than with RA, and the eigenvalues of the spurious axes for DCA were about 1/10 to 1/20 of those with RA.

Partial disjunctions

Partial disjunctions arise when one subset of the samples contains very few species in common with the remaining samples. For complete disjunctions, when one subset of the samples has no species in common with the remainder, RA detects the structure well, always representing the disjunction by an eigenvector whose eigenvalue is 1 (Hill 1974). However, with partial disjunctions, RA gives a poor indication of how well separated the two subsets are. All that appears is a strongly polarized eigenvector corresponding to a high eigenvalue.

DCA was tested on data with partial disjunctions by taking a simple coenocline of length 13.5 sd and removing a segment of the gradient. The length of the removed segment was estimated accurately if it was less than about 3 sd. For longer gaps (4 sd), the length of the missing segment was overestimated by a factor of up to 3. The lengths of the two segments remaining after removal of the missing one were always estimated accurately, even when the length of the missing segment was overestimated.

It is easy to see why there is difficulty in estimating the length of gaps greater than about 3 sd in length. The reason is simply that gaps greater than this constitute near-complete disjunctions, so that the problem begins to become indeterminate. (A complete disjunction could correspond to a gap of any length from 4 sd upwards.) It is a fault of the technique, or perhaps of the program DECORANA, that estimates of gaps much greater than 4 sd can be obtained at all.

The lengths of gaps greater than about 3 sd, like the modes of species that are scarcely represented in the data, can be estimated only by heavy extrapolation; such estimates will seldom be of much value. Indeed, the ability of the method to estimate lengths of missing data up to about 3 sd correctly is of much greater significance than its inability to estimate greater lengths. Similar tests were performed on several coenoclines derived from field data, with the same result; the configuration of points in the ordination was almost unchanged, even when moderately long segments of the intermediate samples were removed.

This is an encouraging result, for it implies that the scaling derived by DCA depends largely on the underlying relations of the samples, and relatively little on particular accidents of the sampling scheme. What can be changed by particular accidents, however, is the order of the eigenvectors. In particular, if there are two independent gradients of variation in the data, and if the longer gradient is shortened by removal of samples that are more extreme on it, then there will come a point when the dominant eigenvector does a quantum jump, and flips to the other independent direction of variation.

Care is therefore needed in comparing analyses of differing subsets of a large data set. Depending on which subset is considered, the same gradient can be represented as axis 1 or a higher axis, or can have its direction reversed.

Simulated data with 2, 3, or 4 independent directions of variation

The two-dimensional analogue of a coenocline is a coenoplane. When RA is applied to a simulated coenoplane with a square or nearly square configuration of sampling points in ecological space, the first two axes of the ordination correspond to the two independent directions of variation. However, when the underlying structure is elongated, the eigenvalue for the quadratic distortion of the longer axis takes precedence over the eigenvalue for the shorter axis. Consequently a rectangular coenoplane will often be represented correctly by axes 1 and 3 rather than axes 1 and 2 (Gauch et al. 1977).

DCA eliminates the arch effect, and therefore represents the independent direction of variation on axis 2, even when the underlying structure is elongated. A coenoplane was simulated with sampling points distributed regularly over a rectangle whose sides were in the ratio 3:1. DCA represented the true structure on axes 1 and 2 with very little distortion. The estimated sides of the rectangle were 2.71 sd and 0.86 sd, which are in the ratio 3.15:1. The first four eigenvalues were 0.400, 0.037, 0.009, and 0.003, so that the 'energy' in the spurious third and fourth axes was very small. For RA the corresponding values were 0.400, 0.089, 0.039, and 0.019, and the spurious axes were the second and the fourth.

DCA was also applied to three- and four-dimensional simulated data sets (described in Gauch et al. 1979). In all cases the structure of the data was recovered reasonably well. With data sets of relatively simple structure there was little to choose between the quality of ordinations by DCA, RA, and nonmetric multidimensional scaling. However, with some of the complex data sets, DCA achieved markedly better results than the other two methods.

With one of the simulated data sets, the basic structure of the data was three-dimensional, and was well represented on axes 1, 2, and 3. Axis 4 was spurious, corresponding to the bilinear interaction of axes 1 and 2, i.e. to a curve of the form

$$y = ax_1x_2 + bx_1 + cx_2 + d.$$

For suitably chosen constants a, b, c, d, such a curve can be uncorrelated with x_1 or x_2 while at the same time being linear in x_1 for constant x_2 and vice versa. Such 'interaction axes' are possible with DCA because the later axes are detrended with respect to each axis taken singly, but not with respect to their interaction. However, though interaction axes can have eigenvalues somewhat greater than those of the spurious axes found for coenoclines, they are still much less than those for significant non-spurious axes, and are unlikely to confuse the picture in practice.

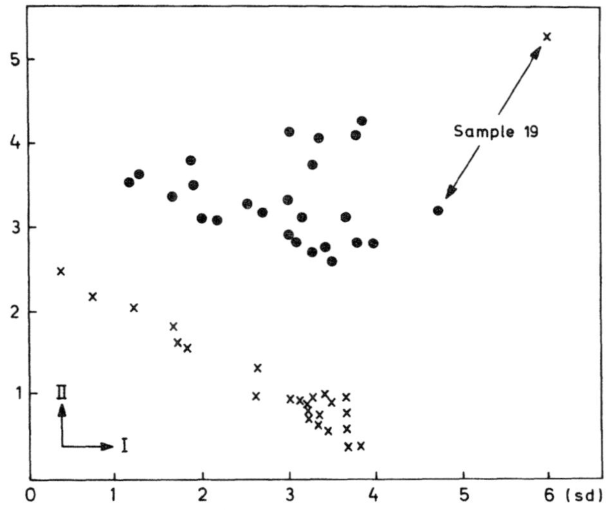

Fig. 6. Meadow data ordinated by DCA (●) and RA (x) (*Arrhenatheretum* data from Ellenberg 1956). The first axis distinguishes dry meadows (left) from wet meadows (right). The second DCA axis distinguishes grassy meadows (bottom) from meadows with more herbs (top); this distinction is absent from the RA second axis (but appears in the RA third axis, not shown). Sample 19 is unusually wet and somewhat aberrant as indicated in the DCA ordination; however RA greatly exaggerates its distinctiveness, and a quadratic distortion of this exaggerated axis then appears in the RA second axis.

Tests of the method on field data

DCA has been tested on numerous sets of field data. Results have been consistently easier to interpret than with other techniques, both because the axes are more effectively related to environmental gradients, and because the scaling is interpretable in terms of species-turnover along gradients. Four examples are considered below.

Arrhenatheretum

The *Arrhenatheretum* data of Ellenberg (1956) were ordinated by DCA and RA (Fig. 6). This data set was selected because it has been used as an example in several other publications (e.g. Mueller-Dombois & Ellenberg 1974, van der Maarel, Janssen & Louppen 1978).

The first axis of the ordination distinguishes dry meadows from wet meadows. *Bromus erectus* was dominant in samples at the dry end of the gradient (i.e. at the end that has a low score on axis 1). At the other end of the gradient *Deschampsia caespitosa*, *Cirsium oleraceum*, and *Holcus lanatus* occur with high values. At the extreme wet end of the gradient there was an aberrant sample (No. 19), which did admittedly contain *Arrhenatherum elatius* and *Festuca pratensis*, but in which the other dominant plants were the subaquatic grasses *Glyceria fluitans* and *Phalaris arundinacea*, neither of which occurred in any other sample.

With any data set the first axis of DCA is only a rescaled version of that derived by RA. With the *Arrhenatheretum* data the most striking difference between the two scalings was the much greater isolation accorded to the aberrant stand (No. 19) by RA. To get an independent estimate of how far removed from the other samples it should be, an allometric relation (not shown here) was set up between the percentage similarity of the samples and their separation on axis 1 of the DCA ordination (where percentage similarity PS for a sample pair equals the sum over species of the minimum of the abundances of each species in the sample pair, and here the sample totals were first relativized to 100). Using the 12 samples (other than 19) with highest scores on axis 1, the allometric relation estimated from the PS values gives a mean separation of sample 19 from these points of 1.4 sd, whereas the mean distance in the DCA ordination was 1.2 sd. The separation of sample 19 on the DCA ordination is therefore probably a slight underestimate; but it is substantially better than the overestimate by RA, which would make sample 19 very aberrant indeed. Sample 19 has a similarity coefficient of 44 % with sample

21, contributed mainly by *Arrhenatherum elatius* 22 %, *Festuca pratensis* 10 %, and *Trisetum flavescens* 5 %, indicating at least a moderately close relationship.

The second axis of the DCA ordination distinguishes grassy and less grassy meadows. At the lower end of the species ordination were *Arrhenatherum elatius*, *Deschampsia caespitosa*, *Festuca pratensis*, *Helictotrichon pubescens*, *Lychnis flos-cuculi*, *Medicago lupulina*, *Senecio jacobaea*, *Trifolium repens*, *Trisetum flavescens*, and *Vicia sepium*. At the upper end of the species ordination were *Centaurea jacea*, *Cirsium oleraceum*, *Daucus carota*, *Galium mollugo*, *Heracleum sphondylium*, *Lathyrus pratensis*, *Lysimachia nummularia*, *Plantago media*, *Ranunculus acer*, and *Rumex acetosa*. Evidently these herbs tend to be less abundant when there is a higher abundance of grass.

In the RA ordination, the grassy/non-grassy direction of variation appears in much the same form, but as the third, rather than as the second axis. The advantages of using DCA here are therefore exactly those that would be expected from the simulated data sets, namely that DCA gives a better indication of scale on the first axis and avoids the spurious second axis.

Siskiyou Mountains, Oregon

A data set of R.H. Whittaker from the Siskiyou Mountains, Oregon, incorporated a substantially greater range of vegetation, and was ordinated by DCA (Fig. 7). There were three main sources of variation, which were, however, partially confounded. A primary gradient ran from 600 m to 1800 m elevation on soils derived from diorite rock. At low elevations (500–700 m) there was also variation in soil parent material, some of the soils being derived from gabbro or serpentine instead of diorite. Within each

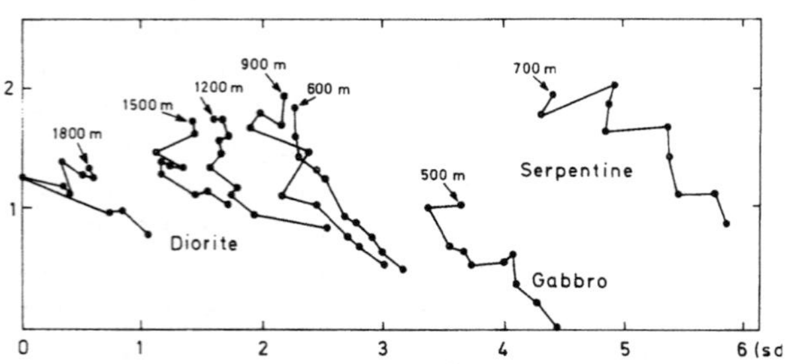

Fig. 7. DCA ordination of vegetation of the Siskiyou Mountains, Oregon (data of R.H. Whittaker). The first axis goes from high- to low-elevation diorite soils to gabbro to serpentine. The second axis expresses topographic moisture conditions from mesic to xeric (which gradient extends from upper left to lower right). See text for further discussion of the axes.

category of elevation and soil parent material the samples were arranged along a topographic moisture gradient from mesic (high on axis 2 in Fig. 7) to xeric (low on axis 2).

Because samples on gabbro and serpentine occurred only at the lower elevations, it has been possible for the ordination to display the variation in soil parent material as well as altitude on a single gradient. This is not a fault of the technique. Without external information there is no way of inferring from the data that differences found between the flora on gabbro and that on diorite are due to a different factor from differences between the flora at various altitudes. The flora on serpentine is, of course, more distinct than the others, but the gap in the ordination could in principle be due to inadequate sampling of intermediate habitats, rather than to a genuine discontinuity in nature.

In addition to reflecting variation in soil type and elevation, the first axis of the ordination also expresses some of the variation in moisture status. Just as the variation in soil type could be interpreted as a continuation of the altitudinal trend, so the variation in moisture status at the lower elevations could also be interpreted in this way; mesic sites at lower elevations supported vegetation more similar to that at higher elevations than did the xeric sites. Furthermore, the vegetation on gabbro more closely resembled that at the xeric end of the range on diorite than that at the mesic end, because the effects of mafic (gabbro) and ultramafic (serpentine, peridotite) soils on community structure and composition simulate those of drought (Whittaker 1954).

All these types of variation are included on the first axis. It is a good axis in that it arranges the floristic data well on the diagonal of a two-way table, but a bad axis in that it has no consistent interpretation. The existence of a powerful trend in the data need not imply a single controlling factor. Here there are three controlling factors that vary along the one axis. It is a complex gradient indeed!

The second axis corresponds largely to variation in moisture status, with mesic sites having high values and xeric sites low values on axis 2. Axis 2 also expresses the serpentine/gabbro distinction, which appears partly on axis 1 and partly on axis 2. These axes need to be taken together for the gap to be appreciated fully.

Despite the difficulties with the axes, the ordination embodies much information: (a) The triangular arrangement of diorite samples on the left correctly represents major features of this pattern, including decreasing beta diversity toward higher elevations (Whittaker 1960); and interrelations of elevation and topography such that low-elevation xeric and high-elevation mesic samples are more unlike one another than low-elevation mesic and high-elevation xeric. (b) The topographic moisture gradient is oriented from upper left (mesic) to lower right (xeric) throughout the ordination field, and is consistent for the sample sets from the three parent materials. (c) A gradient of parent material chemistry (from ultramafic serpentine through mafic gabbro to intermediate-felsic diorite) is oriented from upper right to lower left. (d) Along this axis the ordination shows that the serpentine flora is more distinct from that of gabbro than the gabbro flora is from that of diorite, and that the beta diversity of the serpentine sample set is high. This can be considered a successful ordination of a complex data set. As in many other ordinations, environmental data are necessary for effective interpretation of the axes, and ecological gradients as usually recognized may be oblique in relation to those axes.

Chalk grassland, England

A large data set from chalk grassland in England (data of T.C.E. Wells) was ordinated by DCA. There were 2336 relevés and 197 species. In view of the large number of relevés, there was remarkably little overall variation. The first axis was of length only 3.7 sd, and contrasted weedy communities at one end with stable short-turf communities at the other. The second axis was of length 2.8 sd, and contrasted open communities at one end with communities of woodland margins and thick, sheltered turf at the other.

In this case, RA gave a very similar result. The range of variation in the data was so low that the only difficulty in their analysis was their sheer bulk. With large data sets such as this, a computer program such as DECORANA that has linear requirements in relation to the size of the data is essential. On an IBM 370/168 the analysis of the chalk grassland data required only one minute, much of which was occupied by input of the data and output of the solution.

Vegetation survey of southeast England

Data comprising 876 species in 3270 relevés from southeast England were made available to us by Dr. H.J.B. Birks. These data were extremely heterogeneous and contained numerous outliers. Neither RA nor DCA was effective in analyzing the raw data, because each of the first few axes separated off a few aberrant samples and left the bulk of the data unanalyzed. To make the problem manageable, the

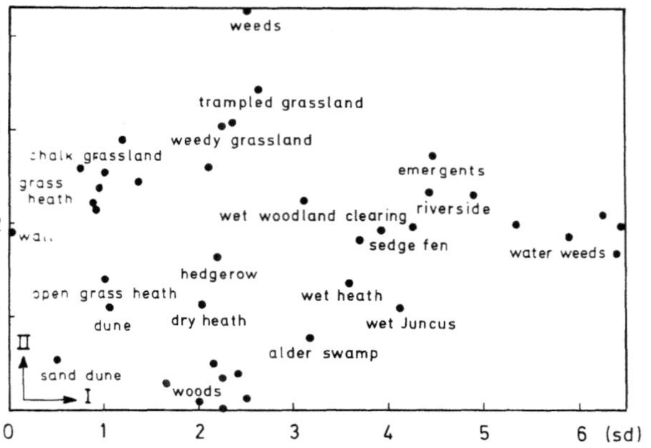

Fig. 8. DCA ordination of a vegetation survey of southeast England (data of H.J.B. Birks, 876 species in 3270 relevés, clustered into 40 composite samples). The first axis goes from dry to wet conditions, and the second axis from woodland to weed communities.

data were clustered by a rapid non-hierarchical algorithm (Gauch 1980) into about 40 composite samples with 50 outliers. The composite samples were then ordinated by DCA, omitting the outliers (Fig. 8). The same composite samples were ordinated by RA and nonmetric multidimensional scaling with broadly similar but somewhat less satisfactory results (due to partial obscuring of the second axis because of excessive emphasis on the wall communities). The species were too numerous for MDS analysis.

It is, of course, not possible to compress such wide variation completely into two dimensions. In the diagram (Fig. 8) the first axis is a dry-to-wet gradient and the second axis is a woodland-to-weed gradient. Vegetational relationships are then expressed in these terms. Soil pH has been almost completely ignored. Thus (acidic) wet heath appears between alder swamp and sedge fen because it is intermediate in water status and degree of woodedness; likewise (acidic) heathland and (mainly calcareous) hedgerows appear between woodland and grassland, because that is indeed their seral status.

Whether an ordination as wide-ranging as this is of much use is open to question. It does, however, offer an effective, synoptic view of relationships among a wide range of vegetation types. The fact that DCA can summarize major directions of variation for such diverse plant communities is a favorable indication for the strength of the technique.

Conclusions and discussion

Detrended correspondence analysis has proved itself to be

better than other techniques of ordination known to us. In no case did it produce results that were less easy to interpret than reciprocal averaging or nonmetric multidimensional scaling. In most cases the results were better. Favorable results of the technique are not only with vegetation data. For a niche ordination of birds by foraging position and behavior (Sabo 1979), DCA had the advantage (greater freedom from involutions on axes, better scaling) over reciprocal averaging and multidimensional scaling.

Good features of the technique

DCA has these good features: It provides an interpretable species ordination as well as a sample ordination. The axes are scaled in units (sd) that have a definite meaning. The arch effect is avoided. The computing time rises only linearly with the amount of data to be analyzed; very large data sets present no special difficulty.

Bad features of the technique

The good features of the technique should not obscure its drawbacks, most of which have been illustrated in the foregoing examples. The most persistent difficulties are in coping with outliers and discontinuities. The only way to cope with extreme outliers is to remove them. (The program DECORANA has an option for removal of unwanted samples from the data.) With large discontinuities, the width of the gap may be badly estimated, though estimates are reasonably reliable for small and medium-sized gaps.

A problem that can never be eliminated is how to interpret the axes (van der Maarel 1979). With the data from the Siskiyou Mountains, there was a very long principal axis of the ordination, but it had no simple physical interpretation. There may be no way of clarifying such compound gradients, because they reflect genuine structure in the data matrix. In practice, however, environmental data may permit either interpretation of the compound axis or division of the data set such that the axes have simpler meaning.

The species ordinations obtained by DCA are generally less satisfactory than the stand ordinations. The placing of species whose optima lie outside the range of habitats sampled is unreliable. Species ordinations can also be unreliable when there is a strong crossed gradient at one end of the axis but not at the other – e.g. when a moisture gradient expresses itself only at lower elevations. In this case the species ordination at the variable end of the gradient can be excessively polarized. It should be noted, however, that DCA is one of only two techniques, both devised by ecologists, that estimate modal positions for truncated species outside the range of sample positions on an axis. The other technique, Gaussian ordination (Gauch et al. 1974), can be applied only to single-axis data.

Data transformations

With multivariate analyses – and detrended correspondence analysis is no exception – it generally pays to allow a larger number of attributes (in this case, species), rather than a smaller number to play a part. This is achieved by transforming the data so that samples are mostly not dominated by the values for single or a very few species. But it also generally pays not to transform from naturally arising values without good reason, so there can be uncertainty about when and how to transform data.

Fortunately detrended correspondence analysis is more robust, in the consistency of its results with differing transformations of the same data, than most methods. Except with the Siskiyou data, the analyses described above were all made with the data in the form in which they were supplied, without transformation. With the Siskiyou data, the importance values for trees were so much larger than those for herbs that the latter had little effect on the analysis. The ordination of untransformed data was broadly similar to that illustrated above (Fig. 7), but failed to emphasize the distinctness of the gabbro, whose dominant trees were mostly the same as those on the diorite. A logarithmic (octave) transformation was therefore applied, to reduce the effect of the dominant trees.

From experience with field data so far we suggest: (a) a first ordination with untransformed data, (b) a second ordination with an octave scale (Gauch 1977), Domin scale, etc., reducing the effect of the dominant species, and, if necessary, (c) a third ordination with each species rescaled by its maximum value (see van der Maarel 1979). If the first ordination is deemed successful, no other may be needed. There is often advantage, however, in comparing it with ordination (b), which may express a wider range of species relationships, and ordination (c), which makes maximum use of quantitative information on all species. DCA, like RA, is also effective with presence/absence data. Although the DECORANA program includes an option for downweighting rare species, some ordinations are improved by removing species occurring in fewer than a threshhold number of samples (e.g. 5 % of the total sample number).

Outlook

An ideal ordination technique could be trusted with almost any data set. Detrended correspondence analysis is not ideal but represents a step in the right direction. To improve it, attention should be given to the species ordination, and to the question of how to handle large gaps in the sample ordination. The aim should be to converge by stages towards a representation of the data that conforms as well as possible with a valid model of its underlying structure.

The basis of ordination as a method for ecological research has been discussed by Beals (1973), Dale (1975), Austin (1976a), Noy-Meir & Whittaker (1977, 1978), Orlóci (1978), and Whittaker & Gauch (1978). Austin (1976a) has formulated a model of the ordination process, according to which there is an underlying 'ecological space' with environmental gradients as axes, in which species populations are dispersed with overlapping distributions approaching multivariate Gaussian forms. Indirect ordinations take the species composition of community samples as data, and derive axes that relate samples and species to the environmental gradients and to one another. The main technical difficulty lies in the complex, curvilinear and non-monotonic distributions of species in the ecological space. Methods that assume linear responses (e.g. principal components analysis) are unreliable in the presence of high beta-diversity. Noy-Meir (1974, Noy-Meir & Whittaker 1977, 1978) refers to ordinations that are free of linear assumptions and effective in recognition of directions of compositional variation in that space as catenations; an axis of a catenation should be a linear track, or coenocline, in ecological space.

A number of techniques are serviceable, none fully satisfactory, as catenations in this sense. Polar ordination is effective because it defines a coenocline by end-point samples and approximates sample positions along it fairly well (Gauch 1973, Whittaker & Gauch 1978, Gauch & Scruggs 1980); but polar ordination requires choice of end-point samples and uses only similarities to those end-points for its calculations. Gaussian ordination (Gauch et al. 1974, Ihm & Groenewoud 1975) is successful in basing ordination on a Gaussian model but is limited by its dependence on that model and inability to relate samples and species to more than one axis. Reciprocal averaging is more tolerant of curvilinearity than other eigenvector techniques, but its results can be much affected by distortions on higher axes and by outlier samples. Multidimensional scaling is dependent on a secondary matrix of sample distances, does not ordinate species well, and has only marginal advantage over reciprocal averaging for sample ordinations (Gauch et al. 1979). Detrended correspondence analysis is a dual ordination of samples and species, realistically scaled, relatively tolerant of species truncation and outlier samples, and with corrections for the effects of curvilinearity on sample positions and higher axes. We think it most appropriate to the model and most successful in application of ecological ordinations so far.

Summary

Detrended correspondence analysis (DCA) is an improvement upon the reciprocal averaging (RA) ordination technique. RA has two main faults: the second axis is often an 'arch' or 'horseshoe' distortion of the first axis, and distances in the ordination space do not have a consistent meaning in terms of compositional change (in particular, distances at the ends of the first RA axis are compressed relative to the middle). DCA corrects these two faults. Tests with simulated and field data show DCA superior to RA and to nonmetric multidimensional scaling in giving clear, interpretable results. DCA has several advantages. (a) Its performance is the best of the ordination techniques tested, and both species and sample ordinations are produced simultaneously. (b) The axes are scaled in standard deviation units with a definite meaning. (c) As implemented in a FORTRAN program called DECORANA, computing time rises only linearly with the amount of data analyzed, and only positive entries in the data matrix are stored in memory, so very large data sets present no difficulty. However, DCA has limitations, making it best to remove extreme outliers and discontinuities prior to analysis. DCA consistently gives the most interpretable ordination results, but as always the interpretation of results remains a matter of ecological insight and is improved by field experience and by integration of supplementary environmental data for the vegetation sample sites.

References

Austin, M.P. 1976a. On non-linear species response models in ordination. Vegetatio 33: 33–41.

Austin, M.P. 1976b. Performance of four ordination techniques assuming three different non-linear species response models. Vegetatio 33: 43–49.

Austin, M.P. & I. Noy-Meir. 1972. The problem of non-linearity in ordination: experiments with two-gradient models. J. Ecol. 59: 763–773.

Beals, E.W. 1973. Ordination: mathematical elegance and ecological naïveté. J. Ecol. 61: 23–35.

Benzécri, J.P. 1973. L'Analyse des données (vol. 2: L'analyse des Correspondances). Dunod, Paris, 619 pp.

Curtis, J.T. 1959. The Vegetation of Wisconsin: An Ordination of Plant Communities. University of Wisconsin, Madison, 657 pp.

Dale, M.B. 1975. On objectives of ordination. Vegetatio 30: 15–32.

Ellenberg, H. 1956. Aufgaben und Methoden der Vegetationskunde. Ulmer, Stuttgart, 136 pp.

Fasham, M.J.R. 1977. A comparison of nonmetric multidimensional scaling, principal components and reciprocal averaging for the ordination of simulated coenoclines, and coenoplanes. Ecology 58: 551–561.

Gauch, H.G. 1973. The relationship between sample similarity and ecological distance. Ecology 54: 618–622.

Gauch, H.G. 1977. ORDIFLEX – A flexible computer program for four ordination techniques: weighted averages, polar ordination, principal components analysis, and reciprocal averaging, Release B. Ecology and Systematics, Cornell University, Ithaca, New York 14850, 185 pp.

Gauch, H.G. 1980. Rapid initial clustering of large data sets. In: E. van der Maarel (ed.) Advances in vegetation science: Classification and ordination. Vegetatio 42: 103–111.

Gauch, H.G. & W.M. Scruggs. 1980. Variants of Bray-Curtis polar ordination. Vegetatio 40: 147–153.

Gauch, H.G. & R.H. Whittaker. 1972. Comparison of ordination techniques. Ecology 53: 868–875.

Gauch, H.G., G.B. Chase & R.H. Whittaker. 1974. Ordination of vegetation samples by Gaussian species distributions. Ecology 55: 1382–1390.

Gauch, H.G., R.H. Whittaker & S.B. Singer. 1979. A comparative study of nonmetric ordinations. J. Ecol. (in press).

Gauch, H.G., R.H. Whittaker & T.R. Wentworth. 1977. A comparative study of reciprocal averaging and other ordination techniques. J. Ecol. 65: 157–174.

Hill, M.O. 1973. Reciprocal averaging: an eigenvector method of ordination. J. Ecol. 61: 237–249.

Hill, M.O. 1974. Correspondence analysis: a neglected multivariate method. J. Roy. Stat. Soc., Ser. C 23: 340–354.

Hill, M.O. 1979. DECORANA – A FORTRAN program for detrended correspondence analysis and reciprocal averaging. Ecology and Systematics, Cornell University, Ithaca, New York 14850, 52 pp.

Ihm, P. & H. van Groenewoud. 1975. A multivariate ordering of vegetation data based on Gaussian type gradient response curves. J. Ecol. 63: 767–777.

Kendall, D.G. 1971. Seriation from abundance matrices. In: F.R. Hodson, D.G. Kendall & P. Tautu (eds.). Mathematics in the archeological and historical sciences, p. 215–252. Edinburgh University Press.

Kessell, S.R. & R.H. Whittaker. 1976. Comparisons of three ordination techniques. Vegetatio 32: 21–29.

Maarel, E. van der. 1979. Transformation of cover-abundance values in phytosociology and its effects on community similarity. Vegetatio 39: 97–114.

Maarel, E. van der, J.G.M. Janssen & J.M.W. Louppen. 1978. TABORD, A program for structuring phytosociological tables. Vegetatio 38: 143–156.

Mueller-Dombois, D. & H. Ellenberg. 1974. Aims and methods of vegetation ecology. John Wiley & Sons, New York, 547 pp.

Noy-Meir, I. 1974. Catenation: quantitative methods for the definition of coenoclines. Vegetatio 29: 89–99.

Noy-Meir, I. & R.H. Whittaker. 1977. Continuous multivariate methods in community analysis: some problems and developments. Vegetatio 33: 79–98.

Noy-Meir, I. & R.H. Whittaker. 1978. Recent developments in continuous multivariate techniques. In: R.H. Whittaker (ed.). Ordination of plant communities, p. 337–378. Junk, The Hague.

Orlóci, L. 1978. Multivariate analysis in vegetation research. Junk, The Hague, 451 pp.

Prentice, I.C. 1977. Non-metric ordination methods in ecology. J. Ecol. 65: 85–94.

Sabo, S.R. 1979. Niche and habitat relations of birds in subalpine forests, New Hampshire. Ecology (in press).

Swan, J.M.A. 1970. An examination of some ordination problems by use of simulated vegetational data. Ecology 51: 89–102.

Whittaker, R.H. 1954. The ecology of serpentine soils. IV. The vegetational response to serpentine soils. Ecology 35: 275–288.

Whittaker, R.H. 1956. Vegetation of the Great Smoky Mountains. Ecol. Monogr. 26: 1–80.

Whittaker, R.H. 1960. Vegetation of the Siskiyou Mountains, Oregon and California. Ecol. Monogr. 30: 279–338.

Whittaker, R.H. & H.G. Gauch. 1978. Evaluation of ordination techniques. In: R.H. Whittaker (ed.). Ordination of plant communities, p. 277–336. Junk, The Hague.

Accepted 20 December 1979

NON-CENTRED COMPONENT ANALYSIS OF VEGETATION DATA: A COMPARISON OF ORTHOGONAL AND OBLIQUE ROTATION

T.J. CARLETON*

Department of Botany and Department of Forestry, University of Toronto, Toronto, Ontario, M5S 1A1, Canada

Keywords:
Clustering, Components, Rotation, Oblique, Unipolar, Vegetation

Introduction

Factor analysis embraces a suite of related continuous multivariate statistical methods whose prime function is to reduce a set of complex data to some simpler form. This objective may be achieved through the use of a direct analysis or a derived solution in which direct factor methods play an initial role (Harman 1976). Of the direct methods, principal component analysis (PCA) has received most attention in vegetation study (Goodall 1954, Orlóci 1966, Gittins 1969). PCA is based upon a 'closed' model in which no a priori assumptions need be made concerning the decomposition of total variance into unique and covariant portions. Of the derived factor approaches, rotation to simple structure is the most common objective (Thurstone 1947). Simple structure criteria ensure that variables form saliently high or low weights (loadings) on components and this may be achieved through the use of orthogonal or oblique methods. To date only orthogonal rotation by the varimax criterion (Kaiser 1958) has been applied to vegetation data (Ivimey-Cook & Proctor 1967, Noy-Meir 1971, Hinneri 1972, Shmida 1972, Seligman 1973, Carleton 1978, 1979, Carleton & Maycock 1980).

Traditionally PCA has been applied in vegetation study for the production of ordination scatter plots and trend seeking. In this latter respect its performance is poor compared with reciprocal averaging (Gauch et al 1976) and non-metric multidimensional scaling (Austin 1976, Fasham 1977, Prentice 1977). However, Nichols (1977) points out other uses of PCA, notably for the summary of large similarity matrices and the replacement of species/site resemblances with component loadings/scores for a great reduction in the number of variables. Of particular interest has been the use of PCA upon non-centred similarity matrices (Noy-Meir 1973a, Feoli 1977). Here the underlying geometric model emphasises cluster seeking rather than trend seeking as in the centred mode. Cluster recognition with non-centred PCA depends upon the emergence of unipolar components. In this case each component is oriented from the common origin in data space to the centroid of a local cluster (i.e. a submatrix group within the data table). Such perfect correspondence between unipolar components and clusters occurs only when the submatrix groups are completely disjoint (Feoli 1977). In reality, vegetation data shows continuity and submatrix groups will overlap to varying degrees. On variance extraction criteria, principal components become more bipolar when operating among and within these overlapping clusters (Orlóci 1967, Noy-Meir 1971, 1973a). Subsequent rotation to simple structure emphasises cluster recognition at the expense of the variance contribution by individual components and results in more of these becoming unipolar.

In theory the orthogonal solution by rotation should apply only to data consisting of completely independent clusters, in which case a varimax solution is identical to the corresponding principal components. The detection of overlapping clusters requires an oblique solution in which the unipolar components that each identify a single cluster are more or less correlated. Noy-Meir & Whittaker (1977) defend the orthogonally rotated solution for vegetation analysis by arguing that it prevents the assignment of separate unipolar components to groups (vegetational noda) which are highly overlapping. However, no evidence is presented to support their argument.

* This work was conducted with the assistance of N.S.E.R.C. of Canada Grant No. A6234 and computer funding by the Department of Botany, University of Toronto.

The purpose of this communication is to report on the derived non-centred component analysis of some artificial data tables of known structure and to compare the efficiency of orthogonal and oblique rotation methods at recapturing this structure through the emergence of unipolar components. In addition, derived orthogonal and oblique component analyses are carried out on a set of boreal forest understorey data and the results are discussed in the light of findings from analysis of the artificial data tables.

The data

Four artificial, sorted data tables, representing 30 species in 30 sites are shown in Fig. 1. Table entries can be separated into three regions of magnitude; i) species absent, ii) species entries of 0–49 and iii) entries of 50–99. No distinction is made between absence and an entry of zero. Within the bounds 0–49 and 50–99 numbers were chosen at random (Fisher & Yates 1963) so that no further assumptions need be made concerning structure within the data tables. A shift from data table I via II and III to table IV (Fig. 1) represents a progression in decreasing data structure through the expansion and coalescence of six disjoint submatrix groups. These species x site submatrix dimensions are 5 × 5, 7 × 7, 3 × 5, 7 × 3, 2 × 8 and 6 × 2 succesively, following the principal diagonal of table I.

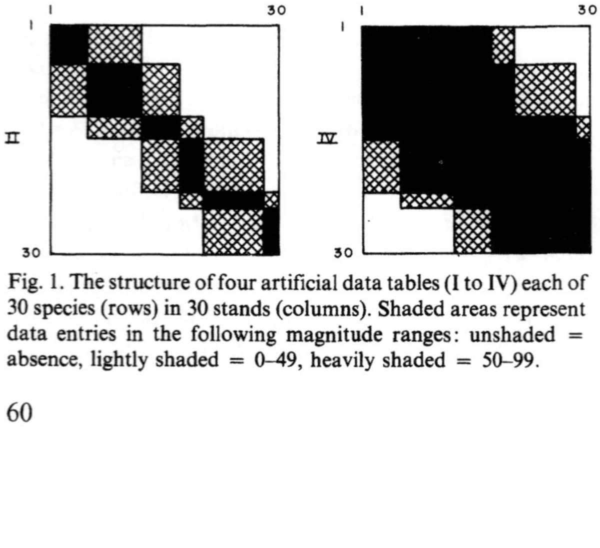

Fig. 1. The structure of four artificial data tables (I to IV) each of 30 species (rows) in 30 stands (columns). Shaded areas represent data entries in the following magnitude ranges: unshaded = absence, lightly shaded = 0–49, heavily shaded = 50–99.

Both symmetric and asymmetric groups were chosen to test the resolution of the methods below.

A single dataset representing the frequency (15×1 m^2 quadrats) of 93 vascular plant species in 197 boreal forest stands in eastern Ontario and western Quebec, Canada was subjected to similar methods of analysis.

Methods

PCA

Unstandardized cross-product matrices were calculated between species and subjected to eigenanalysis. Following postnormalization of component loadings (eigenvectors) stand scores were projected onto each component of interest.

Orthogonal rotation

The varimax method of Kaiser (1958) was used to rotate principal components. Unlike principal components, rotated components are highly dependent upon those principal components used for rotation. No simple criterion has been derived by which to exclude principal components beyond a certain number. The problem is circumvented here by rotating different numbers of initial principal components. Indeed, by comparing the similarity of unipolar components at different levels of rotation useful insight is gained into the structure of a dataset and the stability of groups (Noy-Meir, 1971; Carleton 1979). Similarity or conjunction is assessed according to a normalized cross-product between the component loadings:

$$C_{fg} = \frac{\sum_{k=1}^{m} a_{fk} \cdot a_{gk}}{\sqrt{\sum_{k=1}^{m} a_{fk}^2 \cdot \sum_{k=1}^{m} a_{gk}^2}}$$

In this conjunction coefficient a represents the m component loadings on axes f and g.

Oblique rotation

An advantage of orthogonal rotation methods is their ease of interpretation. Both the factor pattern (species regression weights) and factor structure (species correlations with components) matrices are identical but, within an oblique reference co-ordinate frame, these diverge. In addition, a third matrix of correlations between component pairs must be derived to complete the solution of which all three matrices are a part (Harman 1976). Because species regression weights are of prime interest in characterizing a component as a vegetation unit, attention is here restricted to the factor pattern matrix.

The ortho-oblique rotation method of Harris & Kaiser (1964) is used here as it subsumes the varimax criterion. The method takes its name from the fact that a sequence of intermediate orthogonal transformations are employed in arriving at an objective oblique solution. By setting a parameter, γ, in advance it is possible to specify extremely oblique ($\gamma = 0.0$) to orthogonal, varimax solutions ($\gamma = 1.0$).

Computation of principal components, varimax and ortho-oblique rotation were carried out using the NT-SYS programme package (Rohlf et al. 1974).

Unipolarity criterion

The polarity of all unrotated and rotated components is assessed by a simple coefficient of asymmetry (Noy-Meir 1971), $\alpha = 1 - \Sigma^a/\Sigma^b$, where Σ^a and Σ^b are the sums of squares of negative and positive loadings on a single component. The larger sum is taken as the denominator. For unipolar components, α approaches 1.0 while α approaches 0.0 for bipolar components. Unipolar coefficients are here recognized where $\alpha > 0.9$ (Noy-Meir 1971).

Problems arise for this criterion with extremely oblique factor pattern matrices. Figure 2 compares the oblique coordinates of four data points within the moderately oblique (a) and the extremely oblique (b) plane of two components. The points exterior to the angle in (b) must have coordinates on the negative side of components with which they are not closely associated. The more oblique the solution, the more bipolar are components forced to become. Efficient ortho-oblique unipolar component analysis therefore depends upon finding the optimum level of γ at which the sensitivity of oblique rotation is not offset by imposed bipolarity. Table 1 shows the number of unipolar components emerging from ortho-oblique rotation of the first 23 principal components extracted from the boreal forest data. Clearly the extremely oblique ($\gamma = 0.0$) and orthogonal ($\gamma = 1.0$) solutions produce fewer unipolar components than the moderately oblique solution ($\gamma = 0.5$). The parameter $\gamma = 0.5$ has therefore been assigned to all oblique rotations reported here.

Table 1. The number of unipolar components to emerge from PCA of the boreal forest vegetation data followed by ortho-oblique rotation of the first 23 principal components. Rotation is compared for five different values of the obliqueness parameter, γ.

	0.0	0.25	0.5	0.75	1.0
Number of unipolar components	3	6	18	16	14

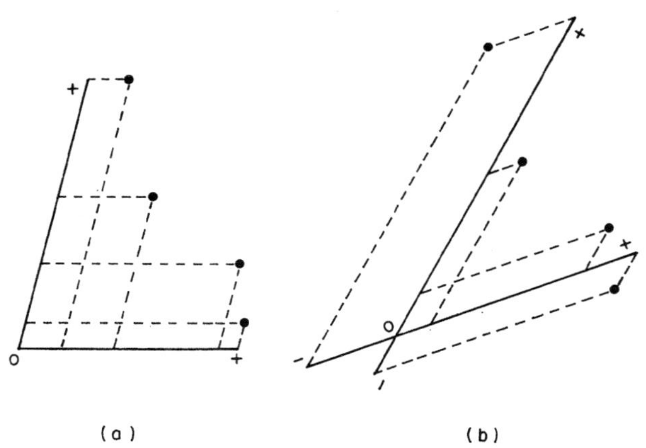

Fig. 2. Four data points in a plane showing coordinate projections onto (a) mildly oblique reference axes, and (b) strongly oblique reference axes. Note the imposed negative coordinates in (b).

Group membership criterion

Noy-Meir (1971, 1973b) considered species and stands on the terminal portion of a bisected unipolar varimax component as constituting the cluster of interest. This practice can lead to the inclusion of species and stands that are not core members of a cluster, especially where a large number of components are being rotated. The terminal third of a unipolar component has been used as the cut-off length here on the basis of a more efficient resolution of the artificial data tables.

Results

Artificial data tables: basic PCA

'Feoli (1977) has pointed to the value of the percentage variance contributions of principal components at indicating inherent structure in a dataset. Those he presents summarize resemblance matrices of similar dimension as is the case with the artifical data here. However, comparisons among datasets of different dimension are poorly understood on this basis.

Table 2 summarizes the percent variance and the polarity

Table 2. Summary of the first 6 principal components from analysis of four artificial data tables (I to IV) and a set of boreal forest vegetation data. Numbers are the percentage contribution of each component toward the total variance in each dataset. The numbers in brackets are the percentage contribution for each component in the boreal dataset, scaled to a subtotal (97 %) that is comparable to the artificial data tables. Unipolar components are indicated by asterisks.

Principal component	Artificial tables I	II	III	IV	Boreal data	
1	*35.3	*43.6	*60.2	*78.3	*38.3	(52.3)
2	*16.8	*23.9	30.2	15.9	12.4	(17.0)
3	*15.4	12.9	5.0	1.4	9.0	(12.3)
4						
5	*10.4	3.9	0.6	0.7	3.8	(5.2)
6	* 8.2	2.4	0.5	0.4	3.0	(4.0)
Subtotal	98.0	95.0	97.4	97.4	70.9	(97.0)

* = Unipolar components

of the first 6 principal components for each of the artificial data tables. These components account for at last 95 % of the total variance in each case but of most interest is the way in which variance is apportioned. In Table 2-I each of the six intrinsically unipolar components is associated with a disjoint submatrix. These are intercepted in decreasing sequence of submatrix order. The only non-zero component species loadings and site scores are those of the corresponding submatrix elements.

For Table 2-II, two unipolar components emerge. The second corresponds to a solid 12 × 12 block of entries at the upper left while the first component identifies a centroid for the remainder of the data table. For tables 2-III and 2-IV only a single unipolar component appears and this seeks out the centroid of the single homogeneous block of data in each case. Thus with increasing submatrix overlap, as seen in Tables 2-I to 2-IV, there is a corresponding reduction in intrinsic unipolarity and an increase in the proportion of variance concentrated in initial components. These results conform to the findings and predictions of Feoli (1977).

Simple structure rotation: Cluster recognition

The most informative levels of varimax and ortho-oblique rotation are expressed in terms of data table recapture in Figure 3. Since this rotation for Table 2-I is trivial, the varimax solution being identical to principal components, it is omitted. It is clear that in only Tables 2-II and 2-III can varimax methods approximate the original table. In the case of Table 2-IV, poor resolution and much overlap characterize the three unipolar varimax components that emerge. At all levels of structure represented by tables 2-II to 2-IV, ortho-oblique rotation performs more effectively than the varimax criterion and identifies largely the non-overlapping core regions of overlapping submatrices. Indeed, analysis of Table II by this method reconstructs data Table I perfectly.

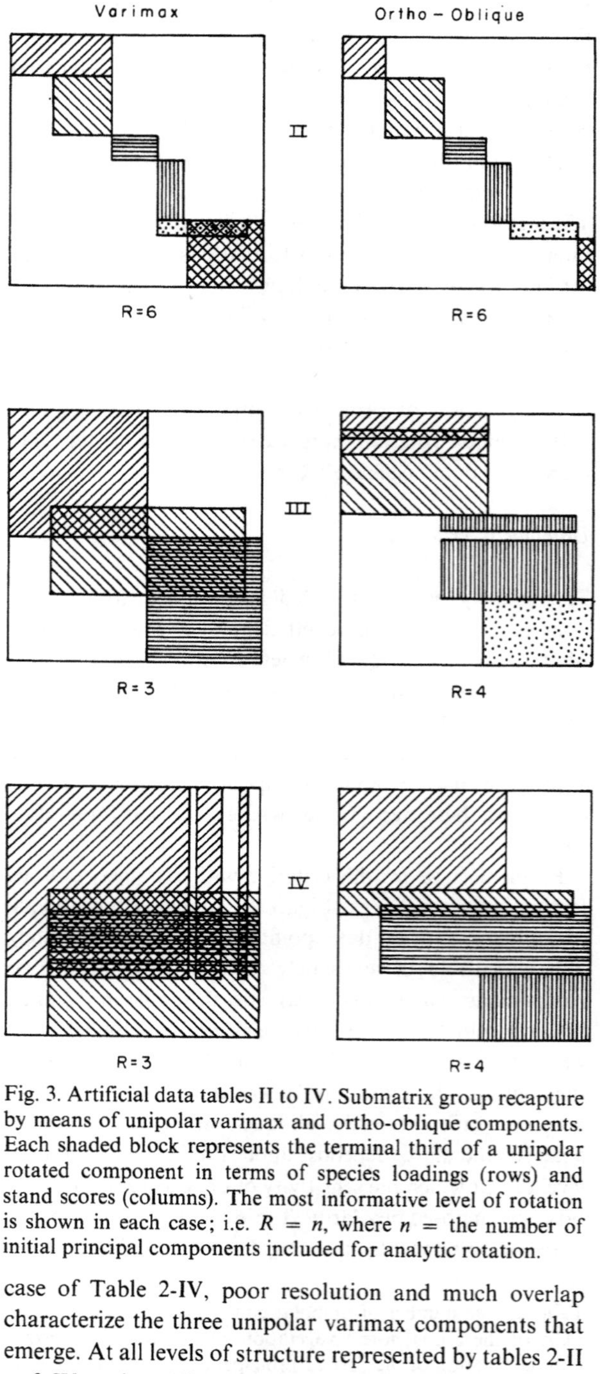

Fig. 3. Artificial data tables II to IV. Submatrix group recapture by means of unipolar varimax and ortho-oblique components. Each shaded block represents the terminal third of a unipolar rotated component in terms of species loadings (rows) and stand scores (columns). The most informative level of rotation is shown in each case; i.e. $R = n$, where n = the number of initial principal components included for analytic rotation.

At levels of rotation greater than those shown, results were more confusing, especially with the oblique method. For this reason group stability was investigated by the construction of dendrograms based upon the relationship between unipolar components at a series of rotational levels.

Group stability

Fig. 4 summarizes orthogonal and oblique analyses in the form of dendrograms for each data table. Strength of similarity between unipolar components at successive rotation levels is assessed according to the conjunction coefficient described above. The varimax dendrograms are all vertical in appearance. That is, once a maximum number of clusters appear, little further recruitment or division occurs. There can be no doubt, for example, that the fundamental structure of Table 2-II consists of six submatrix groups. Cluster stability is less marked with oblique rotation. The fundamental number of groups is well shown for Table 2-II and at a rotation of 20 principal components ($R = 20$), nearly all unipolar components are hierarchically related to the six basic clusters. However, in Tables 2-III and 2-IV this relationship breaks down and numerous fortuitous components appear at $R = 20$. These derive from the random magnitude of entries, within and large overlapping submatrix blocks, to which oblique rotation has been unduly sensitive.

In summarizing derived unipolar component analysis of the artificial data tables, varimax and ortho-oblique rotation possess complementary strengths and weakness. Whereas ortho-oblique rotation provides a more accurate detection of overlapping clusters, those identified by varimax rotation are less susceptible to single and few species/stand outliers which here arise through random effects. These results suggest that the use of both rotation techniques would aid in the elucidation of structure in an unknown dataset.

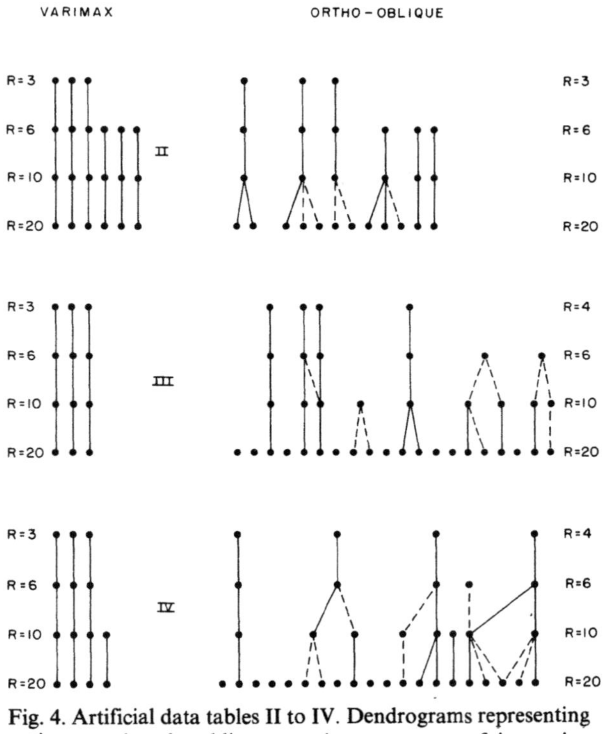

Fig. 4. Artificial data tables II to IV. Dendrograms representing varimax and ortho-oblique rotation sequences of increasing numbers of principal components ($R = n$ as in Fig. 3). Strength of similarity is indicated by two levels of the conjunction coefficient, C (see text), between unipolar component pairs at adjacent levels of rotation. Solid line = $C > .85$ and broken line = $.85 > C > .7$.

Boreal forest understorey data

Table 2 includes summary information for PCA of the boreal dataset. Strictly these figures are not comparable with the artificial data table results because total variation is spread among 93 components rather than 30. However, by scaling the first 6 components to a similar sum of variance, 97.0 % of the total instead of 70.89 %, these figures may be compared (Table 2). With only the first principal component emerging as a unipolar axis, the boreal data appears to fall between Tables 2-II and 2-III in terms of overall structure, although the details are somewhat different.

Rotation at various levels led to the construction of dendrograms (Fig. 5) as above. The pattern of orthogonal and oblique dendrograms are similar, with a comparable number of unipolar components emerging from $R = 4$ to $R = 14$. Only at $R = 23$ do significantly more unipolar

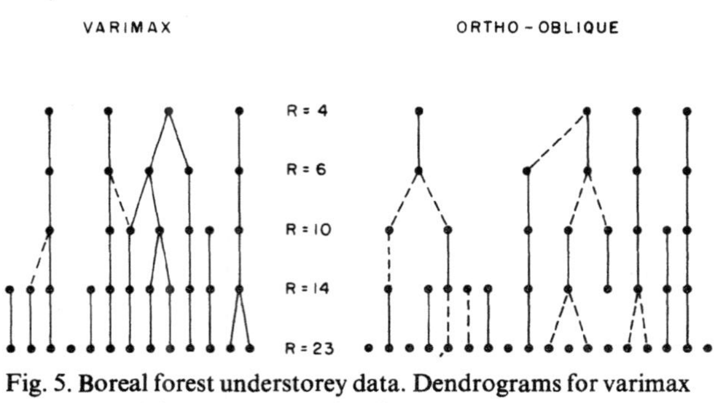

Fig. 5. Boreal forest understorey data. Dendrograms for varimax and ortho-oblique rotation as in Fig. 4.

axes appear with the oblique technique. In each case rotation of the first 4 components describes the major phases of the vegetation complex: lowland bog forest, dry conifer forest, upland deciduous forest and mixed deciduous-coniferous forest. Rotation at higher levels represents division of these phases, leading to a predominantly hierarchic dendrogram structure. If the additional information held in more principal components described new groups, a more vertical structure would be seen and if group composition changed at each level of rotation a more reticulate dendrogram would result. The basically hierarchic scheme, with a small degree of reticulation and a few additional new groups at $R = 23$ in the oblique dendrogram, accords with a data structure intermediate between Tables 2-II and 2-III (cf. Fig. 4).

A detailed consideration of the boreal vegetation data, by unipolar varimax component analysis, is presented elsewhere (Carleton 1978, Carleton & Maycock 1980). in general the oblique solution produces a more interpretable grouping of stands on the terminal portion of each unipolar component. While at $R = 4$ the orthogonal and oblique solutions are very similar, they diverge at higher rotational levels. Even at $R = 14$, however, similarities are strong. For example, one of a pair of general white spruce-balsam fir understorey types, described by uniÄolar components in the varimax analysis, is refined more by the ortho-oblique analysis into a very old white spruce-balsam fir forest type. As with the artificial data analysis, the ortho-oblique method is superior to varimax rotation in cluster seeking by unipolar components.

In looking for group stability, the close agreement between both methods up to, but not beyond, $R = 14$ would suggest this as the natural end point upon which to concentrate attention (Fig. 5). From the artificial data analyses it is clear that varimax rotation can be used as a guide to the number of 'independent' groups which can reasonably be extracted from a dataset. However, it is the oblique unipolar components which form the best descriptors of such groups.

Discussion

As a form of factor analysis, derived, unipolar component analysis is unique to vegetation ecology. Noy-Meir (1971) equates each unipolar varimax component with a vegetational nodum. He defines a nodum as an entity in which a group of species and a group of stands coincide and where neither the groups nor the nodum are necessarily discretely bounded from, or hierarchically related to, other groups or noda (cf. Poore 1956, Lambert & Williams 1962). The duality of R and Q approaches inherent in component analysis, (Gower 1966, Orlóci 1966, 1967) coincides with the above description of the nodum. This has resulted in non-centred PCA followed by varimax rotation being termed 'nodal component analysis' (Noy-Meir & Whittaker 1977). However, there seems to be little reason why the term cannot be extended to the use of any informative simple structure rotation such as the ortho-oblique method. Results on both the artificial tables and the boreal forest dataset indicate that substantially the same overlapping clusters are identified but that oblique rotation resolves the central core of each cluster more accurately (i.e. the true nodum according to the definition supplied above). The use of both varimax rotation and oblique rotation, as complementary aspects of the same analysis, can be regarded as evidence for the existence and size of noda. Each, therefore, should comprise a part of nodal component analysis.

The concerns expressed by Noy-Meir & Whittaker (1977) that oblique rotation will result in too many unipolar axes being assigned to one nodum are not wholly justified. Oblique rotation of the boreal forest data does not result in an excessively large number of unipolar components, even at $R = 23$. More, apparently fortuitous, unipolar components emerge when rotating a large number of principal components from the artificial data tables. However, these can be ascribed to the crudeness of the artificial tables compared with real vegetation data. Fluctuations in magnitude due to the selection of random numbers over a wide range have determined many single species/stand clusters at $R = 20$. Fortunately, varimax rotation has not shown a similar sensitivity and hence its use as a guide to the appropriate level of oblique rotation is advocated.

To many researchers in vegetation ecology, nodal component analysis will seem a convoluted way of achieving unremarkable results in comparison with more conventional computer-based methods of stand classification. However, there are several features of significance to consider.

Firstly, nodal component analysis is a polythetic divisive technique for exploring inherent structure within a set of data. Most agglomerative methods force the fusion of objects (stands) into groups, whether the data is inherently discontinuous or smoothly continuous.

Secondly, nodal component analysis can detect overlapping species/stand clusters and this conforms to the reasonable precept that a given species may contribute to more than one vegetation type. As a corollary, some stands may not typify any noda and are consequently assigned a low score on all unipolar components. Most clustering techniques form exclusive groups and all stands are forced to become group members. Comparisons of this type are more fully discussed by Noy-Meir & Whittaker (1977).

Thirdly, nodal component analysis is a truly interchangeable R/Q technique. Factoring (eigenanalysis) plus subsequent rotation can be carried out either on a species x species or stand x stand resemblance matrix, whichever is the smaller. This has strong implications for the feasibility of analysing very large bodies of stand data.

Most polythetic clustering methods operate only in the Q-mode. To effect a stand classification it is necessary to compute and work from a stand x stand resemblance matrix. A large dataset of, for example, 300 species records in 2000 stands would be beyond the capacity of most computers when classifying in the Q-mode. However, when coupled with objective ranking criteria, to further reduce the species list to the most informative members (Orlóci 1973, 1976, Orlóci and Mukattu 1973, Grigal & Ohmann 1975), such classification would be a routine and economical operation in the R-mode. Once the most informative solution of rotated species loadings is obtained, the stand scores are calculated by simple vector x matrix multiplication and this can be performed one component at a time to conserve computer storage.

A final advantage is that most computer facilities have routines implemented to perform PCA and rotation techniques. No specialized programme is necessary to carry out a nodal component analysis. Against this must be set the disadvantages of more intimate human involvement with each step in the analysis and the longer time taken to complete all phases. However, when exploring the structure of large stand data collections this additional effort is minor compared to the resources expended in gathering such raw data.

Summary

Non-centred principal components analyses followed by varimax rotation and ortho-oblique rotation are applied to 4 test data tables of known structure. These methods are also applied to a set of boreal forest understorey vegetation data. The effectiveness of varimax and ortho-oblique rotation techniques at generating unipolar components is considered and the ortho-oblique method with $\gamma = 0.5$ appears optimal. Efficiency of cluster seeking, through the emergence of unipolar components, is assessed in terms of test data table recapture. The ortho-oblique method is superior at all levels of data structure considered.

Stability of clusters at different levels of component rotation is examined. Beyond a level of fundamental structure, oblique-roatation produces fortuitous unipolar components whereas varimax rotation is very conservative. The use of both rotation techniques is advocated as complementary aspects of nodal component analysis. The advantages of nodal component analysis in vegetation study are briefly discussed, especially with respect to large collections of stand data.

References

Austin, M.P. 1976. Performance of four ordination techniques assuming three different non linear species response models. Vegetatio 33: 43–49.

Carleton, T.J. 1978. A phytosociological analysis of boreal forests in the region south of James Bay. Ph. D. Thesis, University of Toronto, Toronto, 348 pp.

Carleton, T.J. 1979. Floristic variation and zonation in the boreal forest south of James Bay: a cluster seeking approach. Vegetatio 39: 147–160.

Carleton, T.J. & P.F. Maycock. 1980. Vegetation of the boreal forests south of James Bay: non-centered component analysis of the vascular flora. Ecology (In Press).

Fasham, M.J.R. 1977. A comparison of nonmetric multidimensional scaling, principal components and reciprocal averaging for the ordination of simulated coenoclines and coenoplanes. Ecology 58: 551–561.

Feoli, E. 1977. On the resolving power of principal component analysis in plant community ordination. Vegetatio 33: 119–125.

Fisher, R.A. & F. Yates. 1963. Statistical tables for biological, agricultural and medical research. Oliver & Boyd, Edinburgh and London, 146 pp.

Gauch, H.G., Jr., R.H. Whittaker & T.R. Wentworth. 1976. A comparative study of reciprocal averaging and other ordination techniques. J. Ecol. 65: 157–174.

Gittins, R. 1969. The application of ordination techniques. In: Rorison, I.H. (ed.), Ecological aspects of the mineral nutrition of plants. Symp. Brit. Ecol. Soc. 1968, 9: 37–66.

Gower, J.C. 1966. Some distance properties of latent root and vector methods used in multivariate analysis. Biometrika 53: 325–338.

Goodall, D.W. 1954. Objective methods for the classification of vegetation. III. An essay in the use of factor analysis. Aust. J. Bot. 2: 304–324.

Grigal, D.F. & L.F. Ohmann. 1975. Classification, description and dynamics of upland plant communities within a Minnesota wilderness area. Ecol. Monogr. 45: 389–407.

Harman, H. 1976. Modern Factor Analysis. University of Chicago Press, Chicago, 487 pp.

Harris, C.W. & H.F. Kaiser. 1964. Oblique factor analytical solutions by orthogonal transformations. Psychometrika 29: 347–362.
Hinneri, S. 1972. An ecological monograph on eutrophic deciduous woods in the South-west archipelago of Finland. Annal. Univers. Turku, Series A II, 50, 131 pp.
Ivimey-Cook, R.B. & M.C.F. Proctor. 1967. Factor analysis of data from an East Devon heath: a comparison of principal component and rotated solutions. J. Ecol. 55: 405–413.
Kaiser, H.F. 1958. The varimax criterion for analytic rotation in factor analysis. Psychometrika 23: 187–200.
Lambert, J.M. & W.T. Williams. 1962. Multivariate methods in plant ecology. IV. Nodal analysis. J. Ecol. 50: 775–802.
Nichols, S. 1977. On the interpretation of principal components analysis in ecological contexts. Vegetatio 34: 191–197.
Noy-Meir, I. 1971. Multivariate analysis of the semi-arid vegetation in southeastern Australia: nodal ordination by component analysis. Proc. Ecol. Soc. Aust. 6: 159–193.
Noy-Meir, I. 1973a. Data transformations in ecological ordination. I. Some advantages of non-centering. J. Ecol. 61: 329–341.
Noy-Meir, I. 1973b. Divisive polythetic classification of vegetation by optimized division on ordination components. J. Ecol. 61: 753–760.
Noy-Meir, I. & R.H. Whittaker. 1977. Continuous multivariate methods in community analysis: some problems and developments. Vegetatio 33: 79–98.
Orlóci, L. 1966. Gemetric models in ecology. I. The theory and application of some ordination methods. J. Ecol. 54: 193–215.
Orlóci, L. 1967. Data centering: a review and evaluation with reference to component analysis. Syst. Zool. 16: 208–212.
Orlóci, L. 1973. Ranking characters by a dispersion criterion. Nature 244: 371–373.
Orlóci, L. 1976. Ranking species by an information criterion. J. Ecol. 64: 417–419.
Orlóci, L. & M.M. Mukkattu. 1973. The effect of species number and type of data on the resemblance structure of a phytosociological collection. J. Ecol. 61: 37–46.
Poore, M.E.D. 1956. The use of phytosociological methods in ecological investigations. IV. General discussion of phytosociological problems. J. Ecol. 44: 28–50.
Prentice, I.C. 1977. Non-metric ordination methods in ecology. J. Ecol. 65: 85–94.
Rohlf, F.J., J. Kishpaugh & D. Kirk. 1974. NT-SYS. Numerical taxonomy system of multivariate statistical programs. State Univ. N.Y., Stony Brook, & Quantra Devt. Corp.
Seligman, N.G. 1973. A quantitative geobotanical analysis of the vegetation of the Golan. Ph. D. Thesis, Hebrew University, Jerusalem. (Hebrew with English summary).
Shmida, A. 1972. The vegetation of Jebel Maghara, Sinai. M. Sc. Thesis, Hebrew University, Jerusalem. (Hebrew with English summary).
Thurstone, L.L. 1947. Multiple-Factor Analysis. University of Chicago Press, Chicago. 535 pp.

Accepted 4 December 1979

SOME APPLICATIONS OF PRINCIPAL COMPONENTS ANALYSIS IN VEGETATION: ECOLOGICAL RESEARCH OF AQUATIC COMMUNITIES*

Gerhard WIEGLEB

Fachbereich IV der Universität, Ammerländer Heerstr. 67–99, 2900 Oldenburg, West-Germany

Keywords: Association matrix, Classification, Data reduction, Ordination, PCA, Water plant communities

Introduction

Several recently published studies (Nichols 1977, Feoli 1977, Noy-Meir & Whittaker 1977) have made a fundamental contribution to clarify certain misunderstandings that have been linked to the application of PCA and the interpretation of its results. It can be concluded that with certain options such as data transformation, standardization of variables, different similarity measures, and axis rotation, PCA is ecologically appropriate. Nichols (1977) discussed three important aspects of PCA.

1. Variables reduction. This is a general intention of all multivariate methods. This aspect will also be of great importance for the treatment of ecological data. (See Symader 1977 for an important application in hydrology). This kind of data reduction may be very useful for the selection of predictor variables in regression and discriminant analysis. Also the substitution of ecological or vegetational data by principal components and the analysis of the reciprocal correlations (Orlóci 1978) may possibly help to solve the problem of connecting such different data sets.

2. Analysis of association matrices. Complex species association matrices are hard to analyse using other numerical methods. Satisfactory results are only obtained with trivial matrices, which could also be solved intuitively, or by simple methods like plexus diagrams and contingency tables. Noy-Meir & Whittaker (1977) consider PCA with subsequent varimax rotation of the components (nodal analysis) to be an appropriate technique. As defined by Noy-Meir (1973), this technique can be regarded as a divisive polythetic classification technique that partly produces overlapping species groups. The ecological validity of such groups have to be tested.

3. Graphical representation of the eigenvector coefficients. Such graphical representation (ordination sensu latu) is efficient only if the first two axes explain a maximum of variance of the matrix. This condition is sometimes neglected. It is formally required that the graphical representation offers an improvement of understanding to the viewer by

a. showing gradients, not necessarily of linear shape, or
b. showing discontinuities, or
c. being interpretable ecologically in some other sense, for example succession.

Complex diagrams, the meaning of which could only be visualized by intricate isolines, have to be considered of doubtful value.

In simple matrices the claim of ordination sensu latu is identical to that of ordination sensu strictu. The representation is mostly interpretable in the sense of an ecological gradient. In most studies indirect gradient analysis is carried out on the stands. From the mathematical point of view the solutions of the stands and species ordination (Q- and R-technique) may be identical. However, from the pragmatical point of view the information referring to species is often more interesting than that referring to stands that are unknown to the reader.

All such applications should less aim at the canonical structure of data but should rather be considered as tools for hypothesis generation. In the present paper, an example will be given for each of the possibilities and interpreted ecologically.

For computations the program package SPSS, subroutine FACTOR, was used (description in Nie et al. 1975). When raw data are used the calculation of principal components is based on a correlation matrix. In addition calculations were performed with previously calculated similarity

*Nomenclature follows F. Ehrendorfer. Liste der Gefässplanzen Mitteleuropas. 2 Aufl. Fischer, Stuttgart 1973.

matrices. The computations were carried out on the TR 440 of the Regionales Rechenzentrum Oldenburg-Osnabrück der Universität Oldenburg.

Data reduction

Data and preparation

Vegetation ecological investigations were carried out in 79 stagnant waters in Southern Lower Saxony (Wiegleb 1978a). The 33 most frequent species (estimated in a tripartite dominance-abundance scale) as well as some hydrochemical parameters were compiled into a matrix. Firstly only the vegetation data were taken into account. On the stands a principal components analysis (Q-technique) was carried out. The representation of the unrotated component scores along the first and second principal axes produced no new information with respect to the phytosociological grouping of the stands. Additionally, the principal components were varimax-rotated.
The first five principal components, which explain a high amount of the variance (48.8 %), were chosen for further analysis. Within the initial matrix, the dominance-abundance values of the species were replaced by those of the principal components of the Q-analysis. Subsequently, the product moment correlation between the ecological parameters and the rotated principal components was calculated.

Results

No ecologically interpretable correlations could be found between the unrotated principal components and the hydrochemical data. On the other hand, several significant correlations were found between the varimax rotated components and the hydrochemical data (see Table 1).

The first principal component of the vegetation data is significantly correlated (d = 0.05) to hydrogen carbonate, calcium, conductivity, and nitrate maxima, whilst the second is negatively correlated to pH-minimum, pH-maximum, hydrogen carbonate, calcium, conductivity, and sodium. In both cases, the hydrogen carbonate shows the highest simple linear correlation to the principal components. This fact stresses the great importance of the hydrogen carbonate content for the composition of vegetation in stagnant waters (cf. Pietsch 1972, Wiegleb 1978a). The calculated correlations reflect the intercorrelations between the parameters in a certain way. This first principal component represents a general nutrient factor (nitrate!), the second one reflects parameters associated to the calcium hydrogencarbonate system.

The third principal component shows a high correlation to both nitrogen fractions, especially ammonia. The differentiating effect of the ammonia content was pointed out before (Wiegleb 1978a). The fourth principal component is correlated to the phosphate maxima. The simultaneous correlation to the pH-maxima indicates a more indirect ecological mechanism. The fifth principal component emphazises the importance of the calcium in water ecosystems.
The results produced by this method correspond to those obtained using other methods, like analysis of variance and direct gradient analysis (Wiegleb 1978a). They can be regarded as a valuable completion.

Table 1. Significant linear correlations between hydrochemical parameters and varimax-rotated principal components of a Q-analysis ($\alpha = 0.05$).

	QR 1	QR 2	QR 3	QR 4	QR 5
pH-minimum		-0.39			
pH-maximum		-0.34		0.36	0.32
Hydrogen carbonate	0.36	-0.42			
Carbon dioxide					
Calcium	0.34	-0.33			0.36
Conductivity	0.36	-0.38			
Sodium		-0.34			
Chloride					
Potassium					
Nitrate maxima	0.32		0.34		
Ammonia			0.54		
Permanganate value					
Phosphate maxima				0.33	

Analysis of an association matrix

Data and preparation

PCA was applied to a matrix, which had already been analysed by means of matrix and plexus techniques (Wiegleb 1978b). This matrix shows the phytosociological similarities between the 47 most frequent macrophytes in Central Europe. The data were taken from publications. The Hamming coefficient was used as similarity measure; it can be calculated from the contingency coefficient ϕ by simple transformation (Haeupler 1974).

Results

By means of PCA 16 principal components (PC) with an

eigenvalue > 1 were extracted, which explain a total of 68.6 % of the variance. This indicates the great complexity of the matrix. The first two principal components explain only 10.5 resp. 7.6 % of the variance.

The varimax-rotated eigenvectors of the 16 principal components are presented in Table 2. Here, only those scores which exceed 0.30 were taken into account allowing each species to appear at least once.

PC 1 is mainly determined by pleustophytic species. The highest loading is shown by *Hydrocharis morsus-ranae* (+ 0.84), followed by *Lemna minor*, *Lemna trisulca*, *Spirodela polyrhiza*, and *Stratiotes aloides*, *Elodea canadensis* and *Hottonia palustris*, which occur frequently together with pleustophytes, also belong to this group. PC 2 is loaded by *Callitriche hamulata* (+0.80) and *Ranunculus peltatus*. This species group constitutes the vegetation of rhitral, non-calcareous oligotrophic running waters in Central Europe. The same applies especially to *Myriophyllum alterniflorum*. The exclusive occurrence of *Callitriche platycarpa* here may be caused by wrong identification or neglection in several relevés which were taken into analysis. *Sparganium emersum* and *Elodea canadensis* are only remotely connected to this group and also belong to other groups.

The competitive floating leave species *Nymphaea alba*, *Nuphar lutea*, and *Potamogeton natans* load PC 3, completed by *Myriophyllum verticillatum*. *Utricularia vulgaris* is only losely associated. PC 4 is loaded by species of rhitral, calcareous eutrophic running waters. The very high loading of *Callitriche obtusangula* (+ 0.88) may also be caused by the lack of sufficient exact data material. Certainly, the ecological affinity to the other species of this group (*Ranunculus fluitans*, *Ranunculus trichophyllus*, *Zannichellia palustris*, *Hippuris vulgaris*, and *Groenlandia densa*) is undisputed.

Potamogeton pectinatus shows the highest loading on PC 5. This group consists of species of a parvopotamid life form only which occur together especially in running waters and ponds. These are *Potamogeton pusillus*, *Potamogeton crispus*, *Potamogeton friesii*, and *Zannichellia palustris*. The předominant mesotrophic occurrence of *Najas marina* is not separated because certain forms of *Potamogeton pectinatus*, *Potamogeton pusillus*, and *Zannichellia palustris* often are associated with this species. The taxonomical uncertainty in this field causes a mixture of two ecological groups. PC 6 comprises competitive elodeids such as *Myriophyllum spicatum* (+ 0.77), *Ranunculus circinatus*, *Potamogeton perfoliatus*, and *Potamogeton lucens*; also *Ceratophyllum demersum* and *Elodea canadensis*, which are frequent in magnopotamid stands.

PC 7 is again loaded by a group of pleustophytic species.

Table 2. Macrophytes grouped according to their loadings on 16 principal components (denoted F1, F2 etc.).

F 1		F2		F 3		F 4	
Hydrocharis m.-r.	0.84	Callitriche hamul.	0.80	Nymphaea alba	0.79	Callitriche obtus.	0.88
Lemma minor	0.80	Ranunculus pelt.	0.77	Myriophyllum vert.	0.73	Ranunculus fluitans	0.76
Lemna trisulca	0.80	Callitriche plat.	0.66	Nuphar lutea	0.57	Ranunculus trich.	0.51
Spirodela polyrhiza	0.70	Myriophyllum alt.	0.47	Potamogeton natans	0.43	Zannichellia pal.	0.42
Stratiotes aloides	0.64	Sparganium emersum	0.40	Utricularia vulg.	0.34	Hippuris vulgaris	0.39
Elodea canadensis	0.49	Elodea canadensis	0.30			Groenlandia densa	0.37
Hottonia palustris	0.48						
F 5		**F 6**		**F 7**		**F 8**	
Potamogeton pect.	0.73	Myriophyllum spic.	0.77	Riccia fluitans	0.79	Lemna gibba	0.83
Potamogeton pusill.	0.67	Ranunculus circin.	0.70	Ricciocarpus natans	0.79	Potamogeton friesii	0.58
Potamogeton crispus	0.47	Potamogeton perfol.	0.57	Utricularia vulg.	0.45	Ceratophyllum dem.	0.43
Potamogeton friesii	0.37	Potamogeton lucens	0.45	Salvinia natans	0.32	Spirodela polyrhiza	0.39
Zannichellia pal.	0.33	Ceratophyllum dem.	0.44			Lemna minor	0.37
Najas marina	0.31	Elodea canadensis	0.30			Potamogeton pusill.	0.31
F 9		**F 10**		**F 11**		**F 12**	
Sagittaria sagit.	0.84	Trapa natans	0.74	Potamogeton obtus.	0.83	Potamogeton gramin.	0.78
Sparganium emersum	0.74	Nymphoides peltata	0.67	Potamogeton alpinus	0.60	Myriophyllum alt.	0.63
Nuphar lutea	0.55	Salvinia natans	0.66	Hottonia palustris	0.39		
Potamogeton natans	0.39	Najas marina	0.37				
Potamogeton crisp.	0.38	Ceratophyllum dem.	0.30				
Potamogeton alpinus	0.34						
F 13		**F 14**		**F 15**		**F 16**	
Potamogeton color.	0.83	Potamogeton acutif.	0.81	Polygonum amphibium	0.78	Hippuris vulgaris	0.54
Groenlandia densa	0.68	Potamogeton trich.	0.56	Potamogeton trich.	0.30	Potamogeton lucens	0.50
Ranunculus trich.	0.33					Zannichellia pal.	−0.40

Here, some mesotrophic species show the highest loadings (*Riccia fluitans* and *Ricciocarpus natans*, + 0.79), also *Utricularia vulgaris* and *Salvinia natans*. The species that load PC 8 belong to two life forms, namely the pleustophytes (*Lemna gibba*, *Lemna minor*, *Spirodela polyrhiza*, *Ceratophyllum demersum*) and the parvopotamids (*Potamogeton pusillus*, *Potamogeton friesii*). These species form a pronounced eutrophic group, occurring frequently together, especially in small waters.

PC 9 integrates the species of the potamal running waters (*Sagittaria sagittifolia*, *Sparganium emersum*, *Nuphar lutea*, *Potamogeton natans*). *Potamogeton crispus* is also associated. The assignment of *Potamogeton alpinus* to this group can be regarded as surprising since one would rather assume a connection to the *Callitriche hamulata* group (cf. Wiegleb 1979a, Kohler & Zeltner 1974). PC 10 comprises a group of species which prefer relatively warm conditions. These are *Trapa natans*, *Nymphoides peltata*, *Salvinia natans*, and *Najas marina*. The group preference for eutrophic waters is indicated by *Ceratophyllum demersum*.

PC 11 includes *Potamogeton obtusifolius*, *Potamogeton alpinus* and *Hottonia palustris*, some mesotrophic elements the assignment of which is not definitive. Two other species of moderately nutrient rich stagnant waters (*Potamogeton gramineus*, *Myriophyllum alterniflorum*) load PC12. *Potamogeton coloratus* dominates the loadings of PC 13. By the loadings of *Groenlandia densa* and *Ranunculus trichophyllus* this group is related to the fourth one. Both groups colonize calcareous waters. However, the species of the 13th group prefer a lower nutrient level.

PC 14 is loaded by parvopotamid species that show less relations elsewhere (*Potamogeton acutifolius*, *Potamogeton trichoides*). These are rare resp. overlooked species. *Polygonum amphibium* loads a principal component of its own (15). The last and 16th principal component is loaded by *Potamogeton lucens* and *Hippuris vulgaris*. This is the the first bipolar component with *Zannichellia palustris* showing a high negative loading.

The first 15 components are unipolar, which indicates that the species form several disjunct clusters (Noy-Meir & Whittaker 1977). The first 10 components seem to be well interpretable ecologically. Within these groups species of similar ecological claims are united. With certain restriction, this is also true for the components 12 and 13, whilst the components 11, 14, and 15 are loaded extremely high by single isolated species. The bipolar group 16 may be excluded from interpretation because of the minimal account of extracted variance (2.2 %).

Altogether, the results are satisfactory. Some bimodal distributions are discovered, for example of *Myriophyllum alterniflorum* and *Zannichellia palustris*. Some other indications for the sensitivity of the technique can be found, for instance the assignment of *Nuphar lutea* to the groups 3 and 9, or that of *Sparganium emersum* to the groups 2 and 9. Especially the composition of the groups 1 to 6 and also 9 clearly points out the good quality of the results.

Ordination

Date material and preparation

The third example refers to the floristic composition of 14 river drainage areas in Lower Saxony (Wiegleb 1979b). In this case, the frequency of the species was counted and standardized in a 5-step scale (1 = in 0 to 20 % of the plots, 2 = in 20 to 40 % of the plots, etc.). The forty most frequent species were taken into analysis.

Results

The result of the analysis is presented in Fig. 1. The unrotated eigenvectors were plotted along the first two axes. Using the eigenvalue criterion ($\lambda > 1$), no further axes were extracted. Principal component 1 explains 60.7 % of the variance, principal component 2 explains 16.0 %.

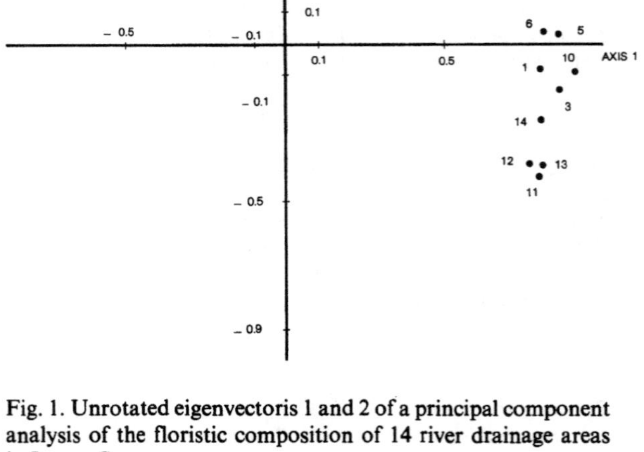

Fig. 1. Unrotated eigenvectoris 1 and 2 of a principal component analysis of the floristic composition of 14 river drainage areas in Lower Saxony.

The representation of the unrotated eigenvectors seems to be efficient.

The result meets the requirement of indirect gradient analysis (ordination sensu strictu). The drainage areas are ordered along a slightly curved line in a sequence that is ecologically interpretable. There is a strong polarity between rhitral-influenced drainage areas (8 – Upper Leine Area, and 9 – Upper Weser Area, both situated in the Southern Lower Saxony Mountain Area) and potamal drainage areas (11 – Hunte Area, 12 – coastal areas, 13 – Leda-Jümme Area, and 14 – Hase Area). The latter are situated in western Lower Saxonian lowland areas. Also the areas 4 – Northern Aller tributaries, 2 – Eastern Elbe tributaries, 6 Souther Aller tributaries excl. River Leine, 5 – Upper Aller and tributaries, 1 – Western Elbe tributaries, 10 – Ochtum and Middle Weser Area, and 3 – Hamme-Wümme Area incline to the potamal areas, whilst the area 7 – Middle and Lower Leine Area occupies a mid-position. The polarization between the areas 3 (Hamme-Wümme Area) and 4 (Northern Aller tributaries) is of special interest. The first is an area poor in ground water with mainly potamal characteristics, while the running waters in the adjacent area 4 get a certain rhitral-potamal mixed character from the strong inflow of ground water. The strong influence of this ecological gradient (rhitral – potamal) which is caused by geographical conditions is clearly shown. On the other hand, it is amazing that anthropogeneous interference has no greater influence on these circumstances.

Consequently the hypothesis was developed that this ecological factor complex is of great importance, not only for the distribution of macrophytes in large areas but also for small-scale distribution in individual river systems, superimposing very much the influence of hydrochemical factors. This hypothesis has been confirmed by the work of Worbes (1979). Worbes (1979) studied the macrophyte distribution in the river Fulda (Northern Hessen). He divided the river into socalled 'phytosociological river sections (vegetationskundliche Flußzonen)' by using occurrence as well as dominance of certain species as criteria for division. Within these relatively homogeneous river segments, he ordered the single stands according to a 'rhitrality index' which depends on average water depth and current velocity of the stands. He finds that certain species occur in all river sections in rhitral-influenced stands. These are the batrachid species like *Ranunculus fluitans*, *Ranunculus peltatus*, *Ranunculus aquatilis*, and *Callitriche hamulata*. Other species, mostly nymphaeids like *Sparganium emersum*, *Nuphar lutea*, *Sagittaria sagittifolia*, and *Potamogeton nodosus* dominate in potamal stands. A similar pattern of distribution had been observed earlier (Wiegleb 1979a, see also Krause 1979), but could not be verified by exact measurements. Elodeid species (*Myriophyllum spicatum*, *Potamogeton pectinatus*, *Potamogeton perfoliatus*) are rather indifferent within this polarity, but in heavier polluted waters they prefer stands with a stronger current velocity (cf. the statements of Weber 1976 on the interaction of current velocity and water pollution, especially suspended load). All these observations are supported by vegetational and ecological data from different running waters in Western Lower Saxony, which are investigated at present (unpublished data).

Conclusions

Each researcher who is going to use any multivariate method should carefully consider the following aspects before starting any computation:

1. What do I really want to know? Or rather: What do I want to know that cannot be shown by means of simpler techniques?.
2. Which is the appropriate technique?

To answer these questions knowledge of the structure of the own data is required.

In the first example the purpose of reduction of the vegetational data is to connect these with the hydrochemical parameters in a linear model. It is safe to suppose that the measured parameters don't have a linear effect on the occurrence of a single species. The distribution of the single species was investigated by a non-linear approach regarding the frequency of the species within certain threshold values (Wiegleb 1978c). However, a hypothesis can be generated to find out which of the parameters measured is correlated highest with the observed variance in the vegetation.

An adequate technique to find such correlations is canonical correlation analysis (CCA). A complete solution including the reciprocal variance explanation of both matrices is necessary (Gaensslen & Schubö 1975). The program available in Oldenburg (CANCORR) does not achieve this. Programs like MANOVA (Beutel et al. 1978) are not yet available in Oldenburg. Hitherto, CCA has only been used for ordination purpose in vegetation ecology, for example Gauch & Wentworth (1976).. All eigenvalue methods that produçe such kind of axes, can be used here (factor analysis, principal component analysis, multiple discriminant analysis, canonical correlation analysis, Noy-Meir & Whittaker 1977). The statistical elements of these techniques are not much used (cf. Feoli 1977 on PCA). In the present example the statistical test refers to the product-moment correlation coefficients between the rotated components of Q-analysis and the hydrochemical parameters. This has nothing to do with actual PCA.

In the case of the association matrix it was difficult to find the appropriate question. In the first step, ordination techniques sensu latu were used (Wiegleb 1978b). These can be optimized by numerical techniques (plexus diagrams – multidimensional scaling or related techniques, cf. Matthews 1978; matrix arrangement programs, e.g. COLPER, Späth 1975). However, the actual interest concerns species groups, which means that a classification technique is required. A classification with overlapping groups seems to be efficient since especially in water plants several species are disjunctly distributed with regard to hydrochemical parameters.

Clustering algorithms that assign each object to a cluster (for example KMEANS and related algorithms, that are based on the variance criterion, Späth 1975) would produce artificial assignments. Agglomerative methods, like single linkage and related techniques Orlóci 1975), are unsuitable for the analysis of such complex matrices because they suppress to much information.

The threshold value of the eigenvectors (> 0.30) for inclusion into the clusters is arbitrary, but all these techniques include arbitrary elements of this kind. The choice of this value depends on the distribution of the scores of the eigenvectors.

The object of the third example is to show that an ecological polarity, the importance of which can be assumed a priori for the distribution of vegetation, is not superimposed or distorted by other possible polarities (polluted – unpolluted). For this purpose, an ordination technique is well appropriate. Because of the low dimensionality of the matrix, this can be easily done. In this example ordination is subsequently used for hypothesis supporting and hypothesis generation. Appreciation of PCA as an ordination technique mainly depends on the personal attitude of the researcher. An objective test for the best ordination is not available, and that is why comparisons might give very different results (cf. Bouxin 1976, Robertson 1978, Gauch et al. 1977, Gauch & Whittaker 1972, Kessel & Whittaker 1976).

Taking into consideration the examples as a whole two general remarks can be made:

1. Common to all examples is the relatively inexact scaling of the data. In the second example (association matrix) only presence-absence data were used. Nevertheless, a satisfactory result was obtained. In other cases, presence-absence data will not produce such good results (van der Maarel 1979). In the other examples, 3- or 5-step estimation scales were used, that do not refer to cover, importance, or frequency values within a single homogeneous stand, but to the dominance-abundance in partly large stagnant waters (example 1) or to the frequency in a whole river drainage area (example 3). In both cases, the existing variation regarding the ecological parameters is levelled by the estimation method. Unreflected application of PCA on a matrix of single relevés with whatever sampling and transformation methods seems to be dubious since random numerical effects may influence the results a great deal (cf. Nichols 1977).

2. In two cases (examples 1 and 2) the rotated solutions of PCA prove to be superior with regard to ecological interpretability. The importance of rotated solutions was pointed out already long before. (Ivimey-Cook & Proctor 1967, Noy-Meir 1973). These possibilities have often been neglected when judging PCA, since they do not aim directly at ordination. The rotated solutions seem to be all the better, the more complex the matrices are (example 2). Simple matrices can also be interpreted with unrotated solutions in a satisfactory way.

Summary

Three examples for the application of PCA to vegetation-ecological data are given.

1. Vegetation data concerning 33 species from 79 stagnant waters were replaced by the varimax rotated principal components of a Q-analysis and subsequently correlated to measured hydrochemical parameters according to a linear model. The correlations were calculated and statistically tested. The hydrogen carbonate proved to be a very important ecological parameter for the composition of the vegetation.

2. 16 principal components were extracted from an association matrix of the 47 most frequent macrophytes of Central Europe (according to 3000 relevés taken from publications) and varimax rotated. Only eigenvectors > 0.3 were taken into account for interpretation. The method produces a classification of overlapping species groups. The groups related to the first 10 principal components are very well interpretable ecologically and

correspond largely to former classification attempts.
3. The frequency of the 40 most frequent water plant species in 14 drainage areas was estimated in a 5-step scale. The result of the ordination was presented as a scatter diagram of the eigenvectors along the first and second unrotated principal axes. This showed that a gradient influences the composition of vegetation, which can be explained by the contrast between rhitral and potamal running waters.

Finally, two other items are emphasized:
1. An appropriate scaling should be used.
2. Varimax rotated solutions are often superior to unrotated ones, particularly in cases of a complex matrix.

References

Beutel, P., H. Küffner, E. Röck & W. Schubö. 1978. SPSS 7. Statistikprogrammsystem für die Sozialwissenschaften. 2. Aufl. Fischer, Stuttgart.

Bouxin, G. 1976. Ordination and classification in the upland Rugege forest (Rwanda, Central Africa). Vegetatio 32: 97–115.

Feoli, E. 1977. On the resolving power of principal component analysis in plant community ordination. Vegetatio 33: 119–125.

Gaensslen, H. & W. Schubö 1976. Einfache und komplexe statistische Analyse. 2. Aufl. UTB 274. Ulmer, Stuttgart.

Gauch, H.G. & T.R. Wentworth 1976. Canonical correlation analysis as an ordination technique. Vegetatio 33: 17–22.

Gauch, H.G. & R.H. Whittaker. 1972. Comparison of ordination techniques. Ecology 53: 446–451.

Gauch, H.G., R.H. Whittaker & T.R. Wentworth. 1977. A comparative study of reciprocal averaging and other ordination techniques. J. Ecol. 65: 157–174.

Haeupler, H. 1974. Statistische Auswertung von Punktrasterkarten der Gefäßpflanzenflora Süd-Niedersachsens. Scripta Geobot. 8: 1–141.

Ivimey-Cook, R.B. & M.C.F. Proctor. 1967. Factor analysis of data from east Devon heath: A comparison of principal component and rotated solutions. J. Ecol. 55: 405–413.

Kessel, S.R. & R.H. Whittaker. 1976. Comparisons of three ordination techniques. Vegetatio 32: 21–29.

Kohler, A. & G.H. Zeltner. 1974. Verbreitung und Ökologie von Makrophyten in Weichwasserflüssen des Oberpfälzer Waldes. Hoppea (Regensburg) 33: 171–232.

Krause, A. 1979. Zur Kenntnis des Wasserpflanzenbesatzes der westdeutschen Mittelgebirgsflüsse Fulda, Ahr, Sieg und Saar. Decheniana, Bonn. 132: 15–28.

Maarel, E. van der. 1979. Transformation of cover-abundance values in phytosociology and its effects on community similarity. Vegetatio 39: 97–114.

Matthews, J.A. 1978. An application of non-metric multidimensional scaling to the construction of an improved species plexus. J. Ecol. 66: 157–173.

Nichols, S. 1977. On the interpretation of principal component analysis in ecological contexts. Vegetatio 34: 191–197.

Nie, N.H., C.H. Hull., J.G. Jenkins., K. Steinbrenner & D.H. Bent. 1975. SPSS. Statistical Package for Social Sciences. New York.

Noy-Meir, I. 1973. Divisive polythetic classification of vegetation data by optimized division on ordination components. J. Ecol. 61: 753–760.

Noy-Meir, I. & R.H. Whittaker. 1977. Continous multivariate methods in community analysis: Some problems and developments. Vegetatio 33: 79–98.

Orlóci, L. 1978. Multivariate analysis in vegetation research, 2nd ed. Junk, The Hague, IX + 451 pp.

Pietsch, W. 1972. Ausgewählte Beispiele für Indikatoreigenschaften höherer Wasserpflanzen. Arch. Naturschutz Landschaftsforsch. 12: 121–151.

Robertson, P.A. 1978. Comparison of techniques for ordinating and classifying old growth floodplain forests in southern Illinois. Vegetatio 37: 43–51.

Späth, H. 1975. Cluster-Analyse-Algorithmen zur Objektklassifizierung und Datenreduktion. Oldenbourg, München/Wien, 217 pp.

Symader, W. 1976. Multivariate Nährstoffuntersuchungen zu Vorhersagezwecken in Fließgewässern am Nordrand der Eifel. Kölner Geographische Arbeiten 34.

Weber, H.E. 1976. Die Vegetation der Hase von der Quelle bis Quakenbrück. Osnabrücker Naturw. Mitt. 4: 131–190.

Weber-Oldedop, D.W. 1977. Fließgewässertypologie in Niedersachsen auf floristisch-soziologischer Grundlage. Göttinger Flor. Rundbr. 10: 73–80.

Wiegleb, G. 1978a. Untersuchungen über den Zusammenhang zwischen hydrochemischen Umweltfaktoren und Makrophytenvegetation in stehenden Gewässern. Arch. Hydrobiol. 83: 443–484.

Wiegleb, G. 1978b. Vergleich ökologischer und soziologischer Artengruppen von Makrophyten des Süßwassers. Verh. Ges. Ökol., Kiel 1977: 243–249.

Wiegleb, G. 1978c. Der soziologische Konnex der 47 häufigsten Gefäßmakrophyten Mitteleuropas. Vegetatio 38: 165–174.

Wiegleb, G. 1979a. Vorläufige Übersicht über die Pflanzengesellschaften der niedersächsischen Fließgewässer. Naturschutz und Landschaftpflege in Niedersachsen 10: 85–116.

Wiegleb, G. 1979b. Struktur, Verbreitung und Bewertung von Makrophytengesellschaften niedersächsischer Fließgewässer. Limnologica, Berlin. 13, in press.

Worbes, M. 1979. Die Makrophytenvegetation der Fulda. Dipl. Arbeit. Göttingen. Mskr.

Accepted 4 December 1979

ON SELECTING INDIRECT ORDINATION METHODS*

Roger DEL MORAL**

Department of Botany (AJ-10), University of Washington, Seattle, Washington 98195, USA

Keywords:
Direct ordination, Multidimensional scaling, Ordination comparisons, Polar ordination, Principal components analysis, Reciprocal averaging

Introduction

The purpose of this paper is to compare some indirect ordination methods to determine which can best extract information from real phytosociological data. There are several comparative studies of such techniques, but few employ field data. I have distilled from the literature several hypotheses concerning the relative merits of polar ordination (PO), principal components analysis (PCA), reciprocal averaging (RA), and multidimensional scaling (MDS). Each method was applied to data from an alpine fell-field to test the following hypotheses.

H_1. RA produces an efficient ordination in one dimension, but subsequent axes may reflect nonlinearities and noise that are not ecologically meaningful.

H_2. PO can produce ecologically meaningful ordinations in several dimensions.

H_3. Standardized PCA can produce efficient ordinations up to about five half-changes (HC).

H_4. Standardized centered PCA yields less distorted results than merely centered PCA.

H_5. Outliers affect PCA and RA more than they affect PO.

H_6. MDS produces distortion-free ordinations that reflect stand interrelationships better than metric methods.

Whittaker & Gauch (1978) and Robertson (1978) provide recent evaluations of ordination methods. Noy-Meir & Whittaker (1977, 1978) discuss continuous multivariate methods and suggest procedures to improve ordination efficiency. The appropriateness of any method is influenced by properties of the data such as beta diversity, dimensionality, outliers, and sampling error. Therefore, no definitive test of specific hypotheses is possible, but results can be extended to data with similar properties. However, criteria which tend to confirm hypotheses developed from simulation studies are likely to be more useful in the analysis of field data than criteria that refute such hypotheses.

Comparison of methods

Several studies (Gauch & Whittaker 1972, Kessell & Whittaker 1976, Cottam et al. 1978) show PO with % distance to be superior to PCA. Gauch et al. (1977) found PO to be inferior to RA in unidimensional data, but PO is less sensitive to the effects of outliers and discontinuities (unless an outlier is a reference stand) than are PCA and RA. MDS appears to improve ordinations resulting from other methods (Fasham 1977). Austin (1976) found RA and MDS preferable to PCA for coenocline ordinations, but cautions that RA is affected by the usual non-linearity of species response curves. Centered PCA (PCA-C) can result in strongly involuted coenoclines and coenoplanes (Austin & Noy-Meir 1971), while centered and standardized PCA (PCA-ST) reduces this distortion.

PO is thought to have these advantages. It is mathematically direct and relatively free from distortions at moderate beta diversity. It is robust against the effects of clusters, outliers, and noise. It often disentangles combined gradients. The disadvantages are those inherent in stand selection. Inappropriate reference stands seriously compromise the interpretability of the results.

* Nomenclature follows that of Hitchcock & Cronquist (1973)

** Funds provided by the University of Washington Graduate School. Field assistance was provided by J.E. Canfield, M.J. Cushman, and A.F. Watson. L.C. Anderson provided helpful comments on the manuscript.

PCA is useful only where noise and beta diversity are low and outliers or clusters are lacking. Standardization may mitigate the involution and distortion.

RA is thought to be the best general purpose method now available (Austin 1976, Fasham 1977, Gauch et al. 1977, Robertson 1978). It resists the effects of noise, outliers, discontinuities, and clusters better than PCA. Second and higher axes may not be interpretable due to the combined effects of noise and distortion.

MDS makes no metric assumptions and optimizes local, rather than global, relationships. However the method may fail if a local stress minimum is reached (Kruscal 1964). This circumstance can be avoided by initiating the ordination with a configuration derived from an initial metric ordination. MDS alleviates distortions and is insensitive to disjunctions. The major disadvantage is that results differ depending upon the maximum number of dimensions calculated and best results are obtained only if the proper number of dimensions is used.

Evaluations of methods from field data

A few studies have evaluated indirect ordination methods using field data. This is because there is no objective 'true' result. Whittaker & Gauch (1978) review these studies. Prior to 1971 investigators preferred PCA to PO, but more recent studies appear to reverse the preference. RA has been favored over both methods since shortly after its introduction (Chardy et al. 1976, Gauch & Wentworth 1976, Peet 1975, and Robertson 1978). Del Moral and Watson (1978) found that results varied with complexity of the data, but PCA never produced results better than RA or PO in 12 sets of data. Second axes of RA were often interpretable.

Methods

The study site

The study area is located on Mt. Stuart, a granitic batholith in the Wenatchee Mountains (Lat. N. 47 °29′, Long. W. 120 °48′). 10 × 10 m sample plots were located subjectively in a small alpine basin at elevations between 2260 and 2560 m. Within plots, ten 1-m^2 quadrats were arrayed in a predetermined pattern and used to determine vegetative cover. (See further del Moral 1979). Soils are derived from granodiorite and are moderately acid. Texture is coarse sand to gravel at the surface. Slopes greater than 15 ° are usually rocky, but more gentle slopes may have relatively deep soil. Flats and depressions accumulate moisture and have relatively high organic content.

The following habitat data were collected: elevation, slope, aspect, soil organic matter, and soil texture. These data are summarized in Table 1.

Data set characteristics

Forty plots were selected to reflect the range of fell-field conditions from xeric to mesic sites. The plots were classified using MINFO (Goldstein & Grigal 1972), an agglomerative information metric based method, to produce several clusters. Table 2 lists 39 species found in nine or more stands and their relative cover in each group.

Prior to any analysis, the data were treated as follows. Species with less than 10 % occurrence were deleted, resulting in the loss of 10 of 58 species. Logarithmic transformations were applied to all values to reduce the effects of dominant species and improve normality (Clifford & Stephenson 1975). When PS is used the log transformation of cover values improves PO. The species maxima were standardized to 100 and sample totals are relativized. This procedure effects the Wisconsin double standardization which has generally favorable or neutral effects on PO (Gauch 1972) and RA (Noy-Meir & Whittaker 1977). No study indicates that this standardization is counter productive.

Stand and coenocline descriptors

Due to the standardizations and transformations employed, alpha diversity and absolute cover variations are not major sources of distortion, but beta diversity is. Richness, Shannon diversity (H′), and absolute cover are reported in Table 1. Beta diversity was calculated as half-changes (HC) by the method of Whittaker (1960). The coefficient of similarity between replicate samples (α) was estimated by extrapolation of the curve relating similarity to coenocline separation. Axis I of RA provided the ordination used to make this calculation. Similarity between extreme samples (z) was estimated by linear extrapolation from the midpoint of the coenocline. $HC = (\log \alpha - \log z) / \log 2$.

Ordination methods

PO was calculated with a percent distance (PD) resem-

Table 1. Characteristics of fell-field plots.

Plot	Elevation (m)	Slope (°)	Aspect	Organic Percent	Cover (%)	Richness	H'
I-1	2350	8	S	4.0	65	22	3.56
I-2	2450	10	SE	4.0	15	21	3.69
I-3	2520	10	WSW	3.6	25	20	3.66
I-4	2360	22	S	3.9	3	14	2.14
I-5	2500	22	SE	3.2	10	15	3.20
Means	2436	14.4		3.7	24.6	18.4	3.23
± S.D.	78	6.9		2.3	23.9	3.6	0.63
II-6	2300	12	SE	4.3	40	25	3.67
II-7	2385	0	--	6.5	35	19	3.05
II-8	2300	10	NW	2.0	14	17	2.90
II-9	2375	36	NE	1.5	10	23	3.19
II-10	2440	32	NE	2.3	55	24	3.15
Means	2359	17.8		3.1	30.8	21.6	3.19
± S.D.	60	15.0		1.8	18.7	3.4	0.29
III-11	2400	31	SSE	3.6	20	35	4.12
III-12	2450	14	SE	4.7	20	20	3.73
III-13	2490	8	SE	10.5	65	19	3.08
III-14	2410	26	SE	3.6	50	25	3.90
III-15	2450	8	SW	5.2	40	26	3.78
III-16	2340	6	S	3.4	40	22	3.03
Means	2424	15.5		5.2	39.1	24.5	3.60
± S.D.	52	10.5		2.7	17.4	5.8	0.45
IV-17	2485	14	SE	11.2	80	23	3.32
IV-18	2480	15	SW	4.6	65	19	2.51
IV-19	2430	8	N	2.9	15	21	4.10
IV-20	2560	34	NE	1.1	19	21	3.14
IV-21	2510	16	NE	2.4	60	16	2.36
Means	2493	15.4		4.2	47.8	20.0	3.08
± S.D.	59	11.4		3.9	29.0	2.6	0.69
V-22	2400	16	NW	2.7	21	16	3.43
V-23	2470	6	SE	3.2	16	23	3.59
V-24	2500	20	W	2.4	20	23	3.41
V-25	2380	20	W	1.1	2	13	2.86
V-26	2370	6	ESE	1.2	12	35	4.27
Means	2424	13.6		2.1	14.2	22.0	3.51
± S.D.	58	7.1		0.9	7.7	8.4	0.50
VI-27	2260	25	SW	3.2	2	22	3.45
VI-28	2410	34	S	4.2	5	26	3.59
VI-29	2325	22	S	6.4	15	20	3.02
VI-30	2320	18	SE	5.6	5	16	3.05
VI-31	2500	6	S	2.0	10	15	3.43
Means	2364	21.0		4.3	7.4	21.0	3.27
± S.D.	92	10.2		1.8	5.1	4.2	0.29
VII-32	2260	28	W	2.0	10	21	3.63
VII-33	2322	25	E	2.1	16	30	3.81
VII-34	2320	3	E	1.0	8	36	4.02
VII-35	2320	6	ESE	1.3	20	25	2.06
VII-36	2300	40	W	1.4	12	29	3.84
VII-37	2340	22	SW	0.7	18	26	3.49
VII-38	2370	15	SW	0.8	6	24	3.17
VII-39	2320	18	W	1.2	2	20	3.76
VII-40	2315	8	ESE	0.9	5	28	3.45
Means	2317	18.0		1.2	10.8	25.4	3.46
± S.D.	30	12.2		0.6	6.2	5.9	0.55

blance matrix (Gauch 1973). Reference stands were selected in these ways: 1) subjectively, using a number of pairs suggested by environmental data, 2) automatically based first on the least similar pair of stands, followed by a second axis defined by that pair of least similar stands found in the middle of the first axis, and 3) automatically using reference stands determined from each eigenvector method. Cottam et al. (1978) provide the details of this method.

Despite its flaws, numerous workers have used variants of PCA to good advantage (Austin 1968, del Moral 1974, van der Maarel 1979). Two variants are used here. PCA-C translates the origin to the species centroid and extracts eigenvalues and eigenvectors from the variance-covariance matrix. PCA-ST is similar except that species are standardized by unit variance after centering.

RA was calculated from the cross-products matrix of Hill's (1973) Appendix I. Each ordination method mentioned above was calculated with the Cornell Ecology Package ORDIFLEX (Gauch 1977).

MDS (Kruscal 1964) was employed because it appears to be the best currently available nonmetric method. Parametric mapping (Noy-Meir 1974) was applied in a preliminary study but is too slow. Given any resemblance matrix, here the PD matrix, MDS seeks to arrange stands in a space of specified dimensionality so that a monotonic relationship between the dissimilarity values and distance between objects is produced. Stress measures the departure from the monotonic relationship, but Fasham (1977) indicated that stress is not a good measure of ordination success. The iterative procedure terminates when stress is minimized. By using an initial configuration based on RA, local minima are avoided and the result converges quickly.

Evaluation methods

An RA ordination using elevation, slope, aspect, soil texture, organic matter, and plant cover was used to produce a crude direct ordination. The indirect ordinations were compared in several ways. Two-tailed Spearman rank correlations were calculated between the direct ordination and each indirect ordination. The distribution of plots in two dimensions was compared to the MINFO classification. The degree of agreement between each ordination and the classification was estimated in terms of the number of compact clusters recovered and the number of plots 'erroneously' included with the wrong cluster.

The shape of each two dimensional ordination was inspected for distortion and disjunction. A disjunction plot is separated by at least 20 % of the axis length from its nearest neighbor. Skewness was determined by comparing mean ordination scores of each ordination to the mean for all ordinations. The discriminant ability of each method was judged by the standard deviation of ordination scores. Large deviations imply better spread over the axis than small scores. Ordinations were interpreted for environmental trends by inspection of species distributions and habitat data. The relevance of these interpretations was compared to the habitat ordination and to each other. The degree to which the species by stand table of Axis I approached a diagonal was estimated for each method.

Each of these criteria was used to assign a score of from 1 (good) to 3 (poor). Mean values were calculated to provide an efficiency estimate for each method.

Results and discussion

Community types and data characterization

Tables 1 and 2 provide information about composition and habitat of the seven vegetation groups recognized. Plot numbers are those shown in the ordination figures.

Table 2. Mean relative cover of species present in more than 20 % of the stands in each of seven groups.

SPECIES	GROUP (Sample Size) I (5)	II (5)	III (6)	IV (5)	V (5)	VI (5)	VII (9)
Antennaria alpina	8.2	5.4	8.6	8.4	2.0	9.9	4.3
Arabis lyallii	1.4	2.2	.3	0.1	0.5	-	0.9
Arenaria capillaris	5.7	.4	11.8	0.1	0.1	-	0.3
Arenaria obtusiloba	1.7	.8	1.4	3.7	7.2	3.3	2.4
Artemisia trifurcata	.1	2.0	1.8	9.3	2.9	1.6	0.6
Campanula scabrella	-	-	.5	0.4	0.1	1.6	-
Carex nardina	-	.1	1.1	-	2.3	-	-
Carex nigricans	-	-	.5	1.0	-	0.3	-
Carex proposita	7.0	2.0	6.0	.6	15.5	6.2	21.9
Carex rossii	-	.1	.1	-	-	1.7	2.9
Carex spectabilis	6.3	14.7	16.2	20.2	0.3	11.7	0.3
Draba paysonii	-	0.3	0.1	0.6	1.4	0.1	0.4
Epilobium latifolium	-	-	-	-	-	-	4.3
Erigeron aureus	1.9	2.6	5.4	3.8	4.5	3.3	2.6
Erigeron compositus	-	-	-	.2	7.4	0.3	3.6
Erigeron peregrinus	2.8	1.6	1.0	1.0	-	0.1	-
Eriogonum ovalifolium	-	-	0.1	0.1	0.6	2.2	0.5
Eriogonum pyrolifolium	4.2	0.3	4.6	0.4	0.5	0.1	0.1
Festuca ovina	-	0.7	-	1.9	2.4	0.4	1.8
Juncus parryi	5.1	0.8	0.9	0.1	0.1	13.2	4.4
Lupinus lepidus	8.5	16.4	14.2	32.0	22.7	7.4	29.6
Luzula campestris	-	.8	.7	3.0	1.4	1.3	.3
Oxyria digyna	-	-	-	0.4	-	-	2.9
Penstemon davidsonii	3.8	-	0.1	0.2	1.6	9.9	5.8
Penstemon procerus	5.6	1.6	8.0	0.8	1.3	3.1	1.2
Phacelia sericea	-	-	-	-	0.1	-	2.4
Phlox diffusa	21.3	0.3	8.2	0.4	0.1	7.6	-
Phlox pulvinata	-	0.2	1.5	1.0	12.3	1.2	0.1
Phyllodoce empetriformis	3.5	4.5	0.1	-	-	0.7	-
Phyllodoce glanduliflora	0.1	2.6	1.6	0.1	-	0.3	0.1
Poa cusickii	8.0	6.7	1.6	1.6	2.4	4.4	4.2
Poa gracillima	-	-	-	0.1	1.8	-	0.3
Salix cascadensis	-	26.5	-	4.1	-	-	4.0
Saxifraga bronchealis	-	1.3	-	0.2	0.5	-	0.3
Saxifraga caespitosa	-	0.1	-	0.2	-	2.0	0.6
Sedum divergens	1.4	2.7	2.0	1.7	1.1	0.7	1.4
Senecio canus	-	0.7	1.3	0.7	0.9	0.4	0.8
Trisetum alpinum	-	-	-	0.6	0.5	0.3	1.2
Veronica cusickii	2.0	2.1	0.2	0.1	-	0.1	-

Vegetative cover is the best single index of site severity. Low cover implies xeric conditions, late snow melt, or extreme winds, while high cover values imply relatively good growth conditions. Species richness varies from 13 to 36 and peaks in plots of intermediate stress. In very stressful plots, only a few species tolerate extremes of cold or drought, while in favorable plots, a few species achieve competitive dominance.

Group I is characterized by *Eriogonum pyrolifolium* and *Phlox diffusa*. Cover is moderate, except for plot 1, dominated by *Phyllodoce empetriformis*, which is sometimes identified as an extremely moist stand. Plot 4 is sometimes identified as an extremely dry stand. Inspection of species composition suggests that both interpretations are valid. The classification is based on absolute cover and unifies these stands on the basis of similar dominants, despite significant differences in subordinate species.

Group II is characterized by *Salix cascadensis* and dominated by *Lupinus lepidus* and *Carex spectabilis*. Several stands are identified as extremely moist ones by various ordinations and plots 5 and 10 are suggested as cool extremes in some ordinations.

Group III is characterized by *Arenaria capillaris* and *Artemisia trifurcata* and dominated by *L. lepidus*, *Penstemon procerus*, and *Phlox diffusa*. Plots 14 and 16 are recognized as mesic by PCA.

Group IV is characterized by *Antennaria alpina*, *Arenaria obtusiloba*, and *Erigeron aureus*, but is dominated by *L. lepidus* and *C. spectabilis*. Several of these plots are sometimes recognized as an axis extreme.

Group V is characterized by *Phlox pulvinata* and dominated by *L. lepidus* and *Carex proposita*. Plot 25 is identified as a xeric site in PCA-C, while plots 22, 23, and 24 are identified as cool sites by PCA and RA. These data suggest that plot 25 is misclassified and could be put with group VII.

Group VI is characterized by *Antennaria alpina*, *Phlox diffusa*, and *Poa cusickii*, but lacks a strong dominance. Plot 27 is an extreme possibly warm site in most of the ordinations.

Group VII is dominated by *L. lepidus* and *Carex proposita*, with *A. alpina*, *Phacelia sericea*, and *Oxyria digyna* characteristic species. Various members of this group are recognized as dry extremes in several ordinations.

Beta diversity, calculated from the order in Axis I of RA is about 2.1. This is a low value and suggests that severe distortions and involutions are not to be expected.

Rank order comparisons

The available habitat data are suboptimal, but they provide an independent means for direct ordination without recourse to floristic data. RA accounted for 47 % of the variance in Axis I and produced a better distribution of the stands than did PCA. This axis reflects moisture conditions. Plot order varies from steep to gentle slopes, from low to high cover, and from low to high soil organic matter. The second axis reflects a temperature gradient.

Table 3. Spearman rank correlation between ordinations. ($0.31 \leqq r < 0.40$, $P = 0.05$; $0.40 \leqq r < 0.51$, $P = 0.01$; $r \geqq 0.51$, $P = 0.001$. Blanks are not significant).

METHOD	Number	RANK CORRELATION (x 100)																				
		1	2	3	4	5	6	7	8	9	10	11	12	13	14	15	16	17	18	19	20	21
Habitat RA-I	1	—	-74		74		-74		62		-53	33	34	52		38	-67		72		-68	
PCA + C-I	2		—		-98		.97		-90	40	80	-42	-42	-85	-32	-58	97		91	46	92	
PCA + C-II	3			—		90		94	-39	-40	48	67	58	-43	-84			69		59		67
PCA + ST-C-I	4				—		-97		94	-36	-85	36	41	87		-51	-96		-88	-53	-91	
PCA + ST-C-II	5					—		96	-34	-35	45	68	68	-36	82			65		84		63
RA-I	6						—		-88	34	75	-38	-48	-82		-55	94		-89	44	93	
RA-II	7							—	-50		59	59	58	-50	73			64		61		62
MDS(3)-I	8								—		-96			-92		45	-89	-33	73	-65	-81	
MDS(3)-II	9									—		.71	-43		-66		42		-51		51	
MDS(2)-I	10										—			-88			80	40	-63	60	71	38
MDS(2)-II	11											—	74		86		-43		57		48	
PO-Auto-I	12												—		72	34	-42		-63		-53	
PO-Auto-II	13													—		44	-83	-37	68	-62		-35
PO-Subj-I	14														—		-34	50	54		-42	48
PO-Subj-II	15															—	56	49	60		-51	52
PO-PCA+C-I	16																—		-.90	45	.91	
PO-PCA+C-II	17																	—		78		99
PO-PCA+ST-C-I	18																		—		-95	
PO-PCA+ST-C-II	19																			—		63
PO-RA-I	20																				—	
PO-RA-II	21																					—

Table 3 is a matrix of Spearman rank correlations between ordinations. Axis I of PCA-C, PCA-ST, and RA have the highest rank correlations to the habitat ordination, but the correlation of MDS(3)-I is significantly reduced. POs derived from PCA or RA also have high correlation. These results indicate that either PCA or RA of floristic data provide more meaningful environmental interpretations than the other methods.

The three eigenvector methods produce similar ordinations in each axis. PCA-ST-I has the largest and RA-I the lowest correlation to MDS(3)-I. The correlations of each factor method to MDS(2)-I are lower than those to MDS (3)-I, which suggests that MDS (3) is a more realistic interpretation of the data structure than any of lower dimensions.

Correlations of habitat scores and of the factor methods to automatic and subjective PO analyses are low. Axis II of PO-Auto appears to represent first axes identified by eigenvector methods. This indicates that the floristically most extreme stands, which define axis I, do not define the most relevant axis identified by most other methods.

The rank order analysis suggests that any of the factor analyses provided results similar to the habitat analysis. In the absence of definitive data, however, these observations are viewed as preliminary and must be confirmed by more intensive study.

Interpretation

Each axis was interpreted by inspecting environmental characteristics and species composition of terminal stands. Moisture is influenced by aspect, steepness, and exposure. Plant cover is correlated to effective moisture and was used to assist interpretation. Species indicators were determined from an inverse analysis coupled with habitat data.

Table 4 shows that in most cases axis I is interpretable as a moisture gradient and that axis II can be interpreted as a temperature or growing season gradient. Exceptions are noted in the table.

Species ordinations were obtained only from the eigenvector methods and POs derived from them. These

Table 4. Characteristics and evaluation of indirect ordination methods.

CHARACTERISTIC	METHOD									
		Polar Ordinations				Eigenstructure			Nonmetric	
	SUBJ.	AUTO	PCA-C	PCA-ST	RA	PCA-C	PCA-ST	RA	MDS(3)	MDS(2)
Rank Correlation	3	3	2	1	2	1	1	1	2	3
Disjunction I	2	2	1	2	2	1	1	1	2	2
II	3	2	3	2	3	3	1	1	1	1
Skewness* I	62.3	40.8	41.1	66.7	33.9	39.3	61.6	32.0	50.0	50.0
score	2	2	2	3	3	2	2	3	1	1
II	72.3	59.0	61.9	70.2	61.9	60.9	44.2	42.2	50.0	50.0
score	3	1	2	3	2	2	1	1	1	1
Distortion	1	1	1	1	1	2	2	2	1	1
Distribution* I	22.3	19.4	24.5	22.6	22.2	29.7	28.7	26.8	24.3	23.5
score	2	3	2	2	2	1	1	1	2	2
II	18.2	20.0	15.3	18.0	15.3	22.1	23.0	22.0	20.0	18.8
score	3	3	3	3	3	2	2	2	3	3
Diagonal	-	1	1	2	1	2	2	1	-	-
Compaction*	4(1)	3(1)	5(1)	5(1)	3(2)	5(2)	4(1)	5(2)	5(1)	4(2)
score	2	3	1	1	3	1	2	1	1	2
Inclusions*	10	21	11	11	8	5	8	4	9	8
score	2	3	2	2	2	1	2	1	2	2
Interpretation*I-Stand	M	T	M	M	M	M	M	M	M	M
I-species	-	T	M	M?	M	M	M	M	-	-
II-Stands	T	M	T	?	T	T	T	T	T	T
II-species	-	M	T	T+M	T	T?	T	T	-	-
score	2	3	1	2	1	1	1	1	1	1
Objectivity	3	1	1	1	1	1	1	1	2	2
Mean Score	2.33	2.15	1.69	1.92	2.00	1.54	1.46	1.30	1.46	1.61

*Notes: Skewness values are mean ordination scores; distribution values are standard deviations of ordination scores; compaction measured by the number of plot groups clearly together, with additional, loosely recognized groups in parentheses; inclusion values are the number of plots isolated from other members of their group; interpretation: M indicates a moisture gradient, T indicates a temperature gradient.

ordinations reveal similar gradients. PO-PCA-ST-II is a combined moisture-temperature gradient and axis I is not a clearly defined moisture gradient. PCA-C-II is probably a temperature gradient, but the interpretation is not definitive.

Shape

Fig. 1 (A to J) shows the two-dimensional shape of each ordination. There are different degrees of disjunction, skewness, distortion and plot distribution. These parameters are quantified in Table 4, using methods described earlier.

PCA-ST and RA result in the least disjunction, while PO-SUBJ and PO-RA the most. The mean displacement from an axis mid-point is 12.8 units for axis I and 14.2 units in axis II. All ordinations are moderately skewed, a result that is due more to plot selection than ordination method. Departures less than 8 units are considered little skewed and those over 16 units greatly skewed. RA-I, PO-RA-I, and PO-PCA-ST-I are all strongly skewed. In axis II, means within 10 units of 50 are not considered skewed. RA-II, PCA-ST-II and PO-Auto are relatively unskewed. The method of calculation for MDS precludes skewness.

Distribution, a measure of discriminant ability, was measured by standard deviation of the ordination scores. Values larger than 25 units indicate good distribution, those less than 20 units, poor. On this basis, Axis I of each eigenvector method has good discrimination. The PO's tend to bunch samples at some distance from either reference stand and hence have lower standard deviations.

MDS does no better than PO methods. In each case, second axis distribution is worse than the respective first axis distribution.

Distortion was estimated by determining the degree of curvature in two dimensions. MDS and PO do not ordinarily produce humps, while each of the eigenvector methods do. PCA-C produces less distortion than the others, but in no case is distortion pronounced.

Species ordinations were produced in RA, PCA, and automatic POs. The diagonal distribution of species values was inspected for the species by stand tables produced. RA produced the best diagonal, e.g. the curve connecting the midpoints of species' distribution was the smoothest. The POs, except for PO-ST, have good diagonals, while the PCA diagonals are only fair.

The distribution of the seven communities in each ordination was inspected to determine the degree of group cohesiveness. This analysis is subjective, but the same criteria are applied to each diagram. No method completely recovers the original classification, though both PCA-C and RA have five compact groups (of a possible seven) with the remaining two groups recognizable. Likewise, these two methods have the fewest included stands, which indicates reasonably good separation. MDS (3) is reasonably good, though there is some confusion in group VI.

The objectivity of each method was rated. Once decisions common to all methods are made, the eigenvector methods and their derived PO analyses are the most objective. MDS requires that dimensionality be determined a priori. PO is the most subjective method employed.

Conclusions

By the criteria used, all PO methods are inferior to the others. On a scale of 1 (good) to 3 (poor), RA scores a mean of 1.30. There is little to choose between the two PCA methods. MDS (3) produces a reasonable ordination, marred by poor separation in the second axis. It might improve were more dimensions to be used.

That RA produces an efficient ordination in one dimension is confirmed (H_1). In this instance, the second dimension is also interpretable. It is suggested that such will usually be the case where beta diversity is low. POs usually produce ecologically meaningful ordinations in two or more dimensions (H_2), but the results are susceptible to the effects of outliers selected as reference stands. Ordinations were usually less well related to the habitat ordination than even the eigenvector ordination from which it derived.

PCA-ST produces efficient ordinations with low beta diversity, confirming H_3. Both PCA methods are similarly distorted, but PCA-ST is less skewed and displays little disjunction relative to PCA-C. Thus, H_4 is partially confirmed. With low beta diversity, outlier stands appear to have less effect on RA, PCA, and MDS than on most PO methods. This contradicts H_5. However, unlike the other methods, the effect of outliers on PO can be avoided by alternative, subjective stand selection.

MDS produces essentially distortion-free ordinations, confirming H_6. However, it does not separate closely related stands well and requires that the dimensionality of the data be known a priori for best results.

This study suggests that these principles be followed. Data with unknown properties should be subjected first to a direct ordination based on habitat data or species indicators. Beta diversity should be calculated from this ordination. If beta is low (less than 3 HC), then RA or PCA-ST should produce reliably interpretable results. The latter is preferred beyond two dimensions. Should refined results be required, MDS, using relatively high dimensions should be used. Though not tested in this study, it is suggested that the suitable dimensionality in MDS is related to beta diversity. PO's may be used once other methods have revealed outliers.

Should beta be between 3 and 10 HC, then RA, followed by several POs should be attempted. Beyond 10 HC, the data should be divided by a numerical classification method and the subsets analyzed separately.

It bears repeating that no ordination method yet devised can produce an accurate, undistorted picture of stand relationships. The results of ordinations are not facts, only hypotheses, and the goal of ordination should be to develop hypotheses that are both interesting and testable.

Summary

A set of forty stands from an alpine fell-field in the Washington Cascades was used to test these ordination methods: polar ordination, principal components analysis, reciprocal averaging, and multidimensional scaling. Because no objective 'true' ordination can be known for such data, a series of evaluative methods were employed to characterize each method. These criteria included the rank correlation to a direct ordination based on habitat

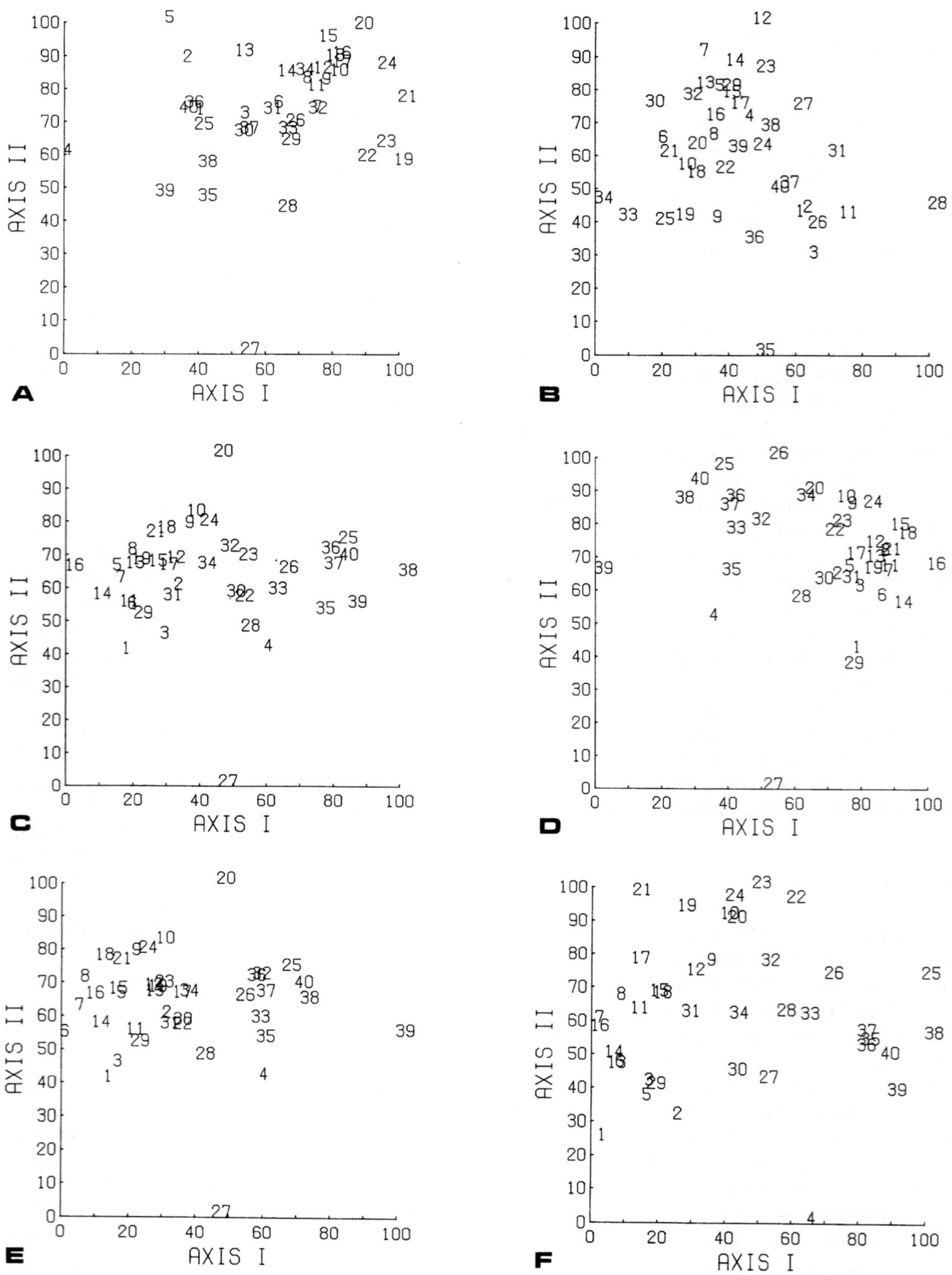
A
B
C
D
E
F
AXIS I
AXIS II

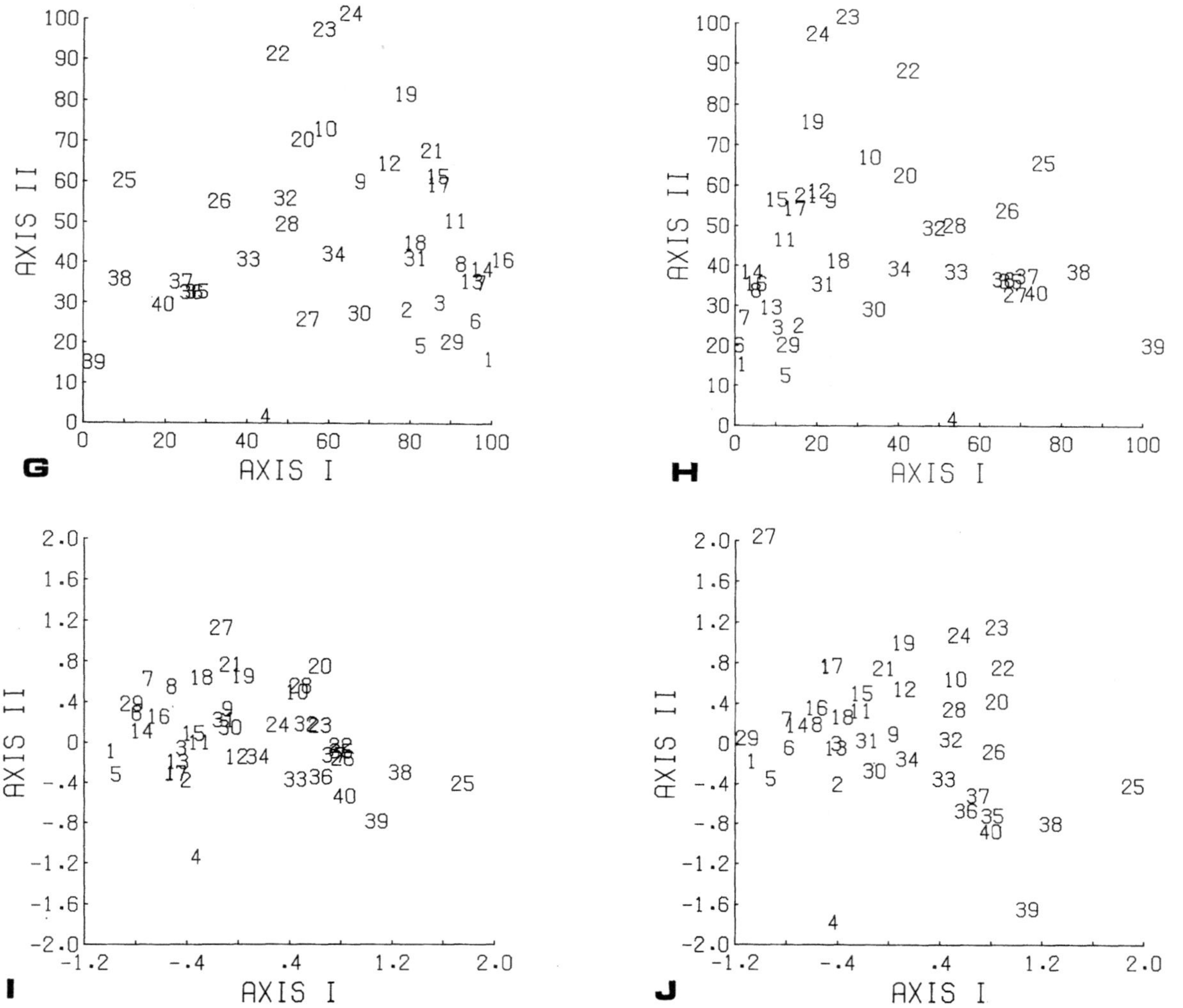

Fig. 1. Two dimensional ordinations. Numbers are stand sequence number in each case. A. PO-Subjective; B. PO-Automatic; C. PO-PCA-C; D. PO-PCA-ST; E. PO-RA; F. PCA-C, axis I = 17.7 % and axis II = 13.2 % of the variance; G. PCA-ST, axis I = 15.3 % and axis II = 13.2 % of the variance; H. RA, axis I = 14.8 % and axis II = 12.3 % of the variance; I. MDS(3), stress = 0.333; and J. MDS(2), stress = 0.414.

data, the degree of distortion, skewness, and disjunction revealed, the ability to separate stands, the efficiency of species ordinations, ease of interpretation, objectivity, and how well the method recovers a classification of the same data.

Based on these criteria, reciprocal averaging performs best, with principal components analysis and multidimensional scaling not much worse. Most polar ordinations produce significant disjunctions and fail to separate stands well in one or more dimensions. Both of these properties result from the selection of partially disjunct stands as end points. These conclusions apply only to data with low beta diversity.

References

Austin, M.P. 1968. An ordination study of a chalk grassland community. J. Ecol. 56: 739–758.

Austin, M.P. 1976. Performance of four ordination techniques assuming three different non-linear species response models. Vegetatio 33: 43–49.

Austin, M.P. & I. Noy-Meir. 1971. The problem of non-linearity in ordination: experiments with two-gradient models. J. Ecol. 59: 763–773.
Cottam, G., F.G. Goff & R.H. Whittaker. 1978. Wisconsin comparative ordination. In R.H. Whittaker (ed.), Ordination of plant communities. p. 185–215. Junk, The Hague.
Clifford, H.T. & W. Stephenson. 1975. Introduction to numerical classification. Academic Press, New York, 229 pp.
Fasham, M.J.R. 1977. A comparison of nonmetric and multidimensional scaling: principal components and reciprocal averaging for the ordination of simulated coenoclines. Ecology 58: 551–561.
Gauch, H.B., Jr. 1973. A quantitative evaluation of the Bray-Curtis ordination. Ecology 54: 829–836.
Gauch, H.B., Jr. 1977. ORDIFLEX. Ecology and Systematics, Cornell University, Ithaca, New York, 123 pp.
Gauch, H.B., Jr. & T.R. Wentworth. 1976. Canonical correlation analysis as an ordination technique. Vegetatio 33: 17–22.
Gauch, H.B., Jr. & R.H. Whittaker. 1972. Comparison of ordination techniques. Ecology 53: 868–875.
Gauch, H.B., Jr., R.H. Whittaker & T.R. Wentworth. 1977. A comparative study of reciprocal averaging and other ordination techniques. J. Ecol. 65: 157–174.
Goldstein, R.A. & D.F. Grigal. 1972. Computer programs for the ordination and classification of ecosystems. Oak Ridge Natl. Lab. ORNL-IBP-71-10.
Hitchcock, C.L. & A. Cronquist. 1973. Flora of the Pacific Northwest. Univ. Washington Press, Seattle. 730 pp.
Hill, M.O. 1973. Reciprocal averaging: an eigenvector method of ordination. J. Ecol. 61: 237–249.
Kessell, S.R. & R.H. Whittaker. 1976. Comparison of three ordination techniques. Vegetatio 32: 21–29.
Kruscal, J.B. 1964. Multidimensional scaling by optimizing goodness of fit to a nonmetric hypothesis. Psychometrika 29: 1–27.
Maarel, E. van der. 1979. Multivariate methods in phytosociology, with reference to the Netherlands. In: M.J.A. Werger (ed.), The study of vegetation, p. 161–227. Junk, The Hague.
Moral, R. del. 1974. Species patterns in the Upper North Fork Teanaway River drainage, Wenatchee Mountains, Washington. Syesis 7: 13–30.
Moral, R. del. 1979. High elevation vegetation of the Enchantment Lakes basin, Washington. Canad J. Bot. 57: 1111–1130.
Moral, R. del & A.F. Watson. 1978. Gradient structure of forest vegetation in the central Washington Cascades. Vegetatio 38: 29–48.
Noy-Meir, I. 1974. Catenation: quantitative methods for definition of coenoclines. Vegetatio 29: 89–99.
Noy-Meir, I. & R.H. Whittaker. 1977. Continuous multivariate methods in community analysis: some problems and developments. Vegetatio 33: 79–98.
Noy-Meir, I. & R.H. Whittaker. 1978. Recent developments in continuous multivariate techniques. In: R.H. Whittaker (ed.), Ordination of plant communities, p. 337–379. Junk, The Hague.
Peet, R. 1978. Forest vegetation of the Colorado Front Range: patterns of species diversity. Vegetatio 37: 65–78.
Robertson, P.A. 1978. Comparisons of techniques for ordinating and classifying old-growth floodplain forests in southern Illinois. Vegetatio 37: 43–51.
Whittaker, R.H. 1960. Vegetation of the Siskiyou Mountains, Oregon and Washington. Ecol. Monogr. 30: 279–338.
Whittaker, R.H. & H.B. Gauch, Jr. 1978. Evaluation of ordination techniques. In R.H. Whittaker (ed.). Ordination of plant communities, p. 277–337. Junk, The Hague.

Accepted 5 November 1979

INFORMATION EFFICIENCY AND REGIONAL CONSTELLATION OF ENVIRONMENTAL VARIABLES*

J. LEPART[1] & M. DEBUSSCHE[2,**]

[1] Ecothèque méditerranéenne, C.N.R.S., B.P. 5051, 34033 Montpellier Cedex, France
[2] Centre d'études phytosociologiques et écologiques Louis Emberger, C.N.R.S., B.P. 5051, 34033 Montpellier Cedex, France

Keywords:
Ecological variable, Efficiency, Information, Languedoc, Mediterranean, *Quercus lanuginosa* communities, Regional organization

Introduction

One of the most important aims of plant ecology is the understanding of the relations between species distribution and the main ecological variables. The study area usually has a size between some hundred and some ten thousand ha, thus falling within an ecological sector or an ecological region (Manil 1963). The relations between species and ecological variables depend on the size and the choice of the studied area (Thiébaut et al., 1978); climatic variables are more important at the level of the ecological region, lithological or topographical variables have more significant influence at the level of the ecological sector (cf. van der Maarel 1976). These differences may be strengthened by the linkages between variables on a local level. Hence a generalization can be biased. A first idea of these inter-relations can be obtained in a study of how the ecological variables fit in the main vegetation pattern.

Material and methods

Quercus lanuginosa communities have been studied throughout an area of 120 000 ha (Fig. 1) including the 'Les Garrigues' region, the southern part of the Causse du Larzac and the La Fage mountain. This mainly calcareous

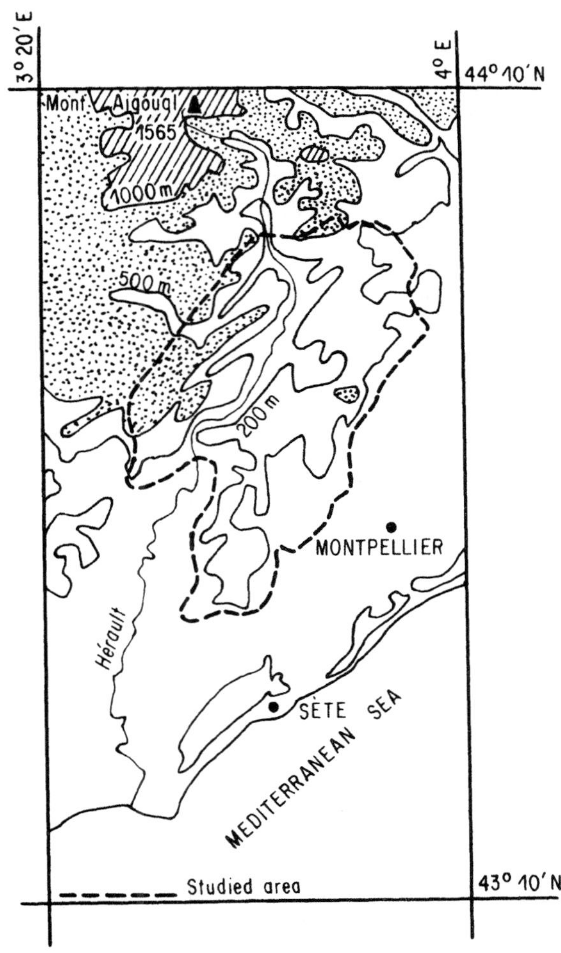

Fig. 1. Map showing the limits of the studied area.

* Plant nomenclature follows Fournier (1961).

** This research was supported in part by DGRST grants N 567-N 633. We are very much indebted to M. Gautier and Ch. Rimbault for their contributions to the translation of this article.

area is situated between 150 m and 980 m above sea level; the average yearly precipitation varies from 800 mm in the south-eastern to 1500 mm in the north-western part of the area. Northwards, *Quercus lanuginosa* is dominant, southwards it is found only in a few appropriate places.

Quercus lanuginosa communities were studied in part of our area with phytosociological methods by Braun-Blanquet (1935, 1951), Kielhauser (1939) and Blondel (1941).

The sampling has been stratified on 35 sub-regions; the division is based essentially on oro-topographic criteria. In each sub-region, we tried to distribute the relevés according to the observed variability of the communities. For a great number of sub-regions, this led us to make at least one relevé in each *Quercus lanuginosa* stand. In fact, it was difficult to obtain a stratified sampling in this area, where one of the most important variables, man's impact, cannot be estimated with accuracy. A sampling at random does not give a good representation of the variability of the studied area because the often grazed and burnt stands of Quercus in the hard limestone part of the area are widespread.

Finally, we made 184 relevés and found 483 species, while 50 variables were studied.

General efficiency of a variable

Methodology

In order to discriminate which variables control species distribution, Godron (1966) proposed to calculate 'the general activity of a variable'; in accordance with Daget (1978), we adopt the expression 'general efficiency of a variable'.

The general efficiency of a variable is measured by the ratio between the mean mutual information variable – species and the entropy of the variable.

The entropy of the variable $H(L)$ is the measure of the uncertainty about the state of the variable in a given relevé:

$$H(L) = \sum_{K=1}^{NK} \frac{R(K)}{NR} log_2 \frac{NR}{R(K)} \qquad (1)$$

with NK: number of states of the variable; NR: whole number of relevés; $R(K)$: number of relevés in the state K.

The mutual information $I(L; E)$ between species E and variable L is the difference between two levels of uncertainty: the entropy of the variable (1) minus the uncertainty about the state of the variable in a given relevé, when the presence or the absence of the species is known. It is also the information given by the presence or absence of the species in a relevé, about the state of the variable in this relevé.

$$I(L, E): \sum_{K=1}^{NK} \frac{U(K)}{NR} log_2 \frac{U(K)}{R(K)} \times \frac{NR}{U(E)}$$

$$+ \frac{V(K)}{R(K)} log_2 \frac{V(K)}{R(K)} \times \frac{NR}{V(E)} \qquad (2)$$

with: $U(K)$: number of relevés of state K with species E; $V(K)$: number of relevés of state K without species E; $U(E)$: number of relevés with species E; $V(E)$: number of relevés without species E.

The mean mutual information $I(L)$ is calculated for the n species with the highest content of information; here (1974).

For further details, see Abramson (1963), Godron (1966, 1968), Daget et al. (1972), Morris & Guillerm (1974).

Application

The general efficiency of the variable is represented by the slope of the straight line going through the origin in a graph in which mean mutual information is plotted against entropy (Fig. 2). The character of the ecological sub-region, the dominant herbaceous species and the cover of the trees higher than 8 m appear as variables with a good efficiency; the species distribution in the studied area is essentially linked to these variables. To the contrary bare ground cover, topography, and aspect are

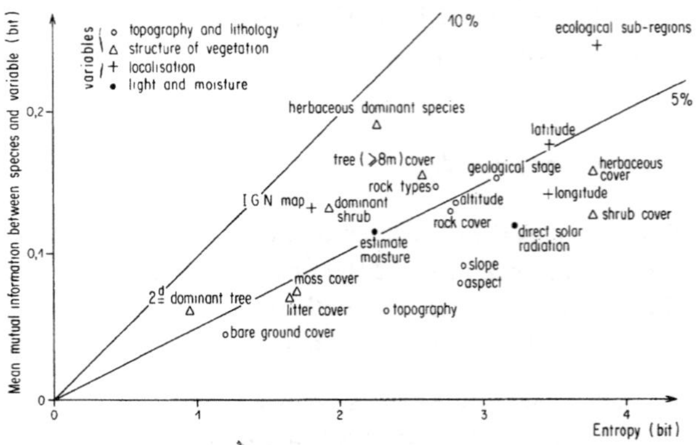

Fig. 2. Relation between mean mutual information for the species with the highest information content and the entropy of some ecological variables.

variables with a poor efficiency; the species distribution in the studied area is only slightly connected with these variables. So, at the level of this study it seems that the localization (more or less linked to the climate and to the biogeographical processes) and the variables related to the structure of the vegetation, have a prominent role.

In order to proceed, we have now to analyze the efficiency related to the states of a variable.

Efficiency profile of an environmental variable

Methodology

We can associate each state of a variable with a value measuring its efficiency to describe the distribution of the species. The set of these values constitutes the efficiency profile of the variable. Calculations may be carried out in two ways which allow to define two profiles: the efficiency profile of the variable and the contribution profile related to the efficiency of the variable.

The efficiency profile of a variable

First, we compare the state of a variable with the set of the other states and we compare the frequency of the species inside and outside this state. This leads us to calculating the entropy and the mean mutual information as if there were only two states '*K*' and 'non *K*' (see formula 1 and 2).
Let: *H(K)* be the entropy associated to the state *K* of variable *L*, and *IE(K)* be the mean mutual information species – state *K*.

IE(K) is an estimation of the mean linkage of the species with the presence or the absence of state *K*. It is at best equal to *I(L)*. The equality means that the two states '*K*' and 'non-*K*' describe the distribution species as well as all the states of the variable.

The efficiency *E(K)* of the state *K* of the variable is given by the ratio of its mean mutual information and its entropy. It measures the value of the partition of the relevés in two sub-sets '*K*' and 'non-*K*'. It will be the greater as the species frequencies in each of these two sub-sets are more different.

The values *E(K)* for *K* varying from 1 to *NK* constitute the efficiency profile of the variable.

The information gain and the contribution profile

The second way to approach the problem is both more direct and natural because it avoids the introduction of an artificial state 'non-*K*'. It allows us to answer the question: what information do we obtain on the presence and on the absence of a species in a relevé if we know that the variable presents state *K* in this relevé?

This information gain can be measured as the difference between the information given by knowing that the relevé is in state *K* and the information given by the same fact if we know that species *E* is present or absent. The procedure is described hereafter.

There are *NR* possibilities to choose a relevé without any precision on its characteristics; the corresponding uncertainty is equal to $\log_2 NR$. If we specify that the variable *L* presents the state *K*, there are only *R(K)* possibilities to choose it; then, the corresponding incertainty is equal to $\log_2 R(K)$. The knowledge that the relevé is found in the state *K* provides an information equal to $\log_2 NR\text{-}\log_2 R(K) = \log_2 NR/R(K)$.

If we do not consider the whole set of relevés, but are only interested in the sub-set where species *E* is present, there are *UE* possibilities to choose a relevé. The corresponding uncertainty is equal to $\log_2 U(E)$. If we precise that variable *L* presents state *K*, the uncertainty is equal to $\log_2 U(K)$. The knowledge that the relevé is found in state *K* brings up an information equal to $\log_2 U(E)/U(K)$, the frequency $U(K)/R(K)$ of species *E* in state *K* is associated with this information.

In the case of the absence of species *E*, the knowledge of the state *K* in which the variable is found provides an information equal to $\log_2 V(E)/V(K)$ with the associate frequency $V(K)/R(K)$.

The information provided by the knowledge of the state *K* in which we find the variable *L* regarding presence and absence of species *E* is then equal to:

$$I(K\|E) = \log_2\frac{NR}{R(K)} - \frac{U(K)}{R(K)}\log_2\frac{U(E)}{U(K)} - \frac{V(K)}{R(K)}\log_2\frac{V(E)}{V(K)}$$

so (3)

$$I(K\|E) = \frac{U(K)}{R(K)}\log_2\frac{U(K)}{R(K)}\times\frac{NR}{U(E)} + \frac{V(K)}{R(K)}\log_2\frac{V(K)}{R(K)}\times\frac{NR}{V(E)}$$

This information is maximal and equal to $\log_2 NR/R(K)$ if the species is always present in state *K* and absent in the other states or vice versa. It is equal to zero if the frequencies of the species in state *K* and in the whole set of relevés are equal. This information has been defined by Rényi (1966) as 'information gain'; it corresponds to the 'minimum discrimination information statistic' of Kullback divided by $2\times NR$ (1959, in Orlóci 1975).

The mean information gain $GM(K)$ is obtained for the n species with the highest mutual information.

The contribution to the efficiency of the variable $CE(K)$ will be the ratio between the mean information gain and the maximum gain which is $\log_2 NR/R(K)$:

$$CE(K) = \frac{GM(K)}{\log_2 \dfrac{NR}{R(K)}}$$

The word 'contribution' is used because of the relation between $I(L)$ and $GM(K)$; we have in fact:

$$I(L) = \sum_{K=1}^{NK} \frac{R(K)}{NR} GM(K) \qquad (4)$$

The set of $CE(K)$ for all the states of the variable constitutes the contribution profile related to the efficiency of the variable.

Application

We shall now analyse the elements we have to take into account in the study of the profile of a variable, and especially study the meaning of the values and the shape of the profiles. We shall specifically use the contribution profile and refer to the efficiency profile only in comparison with the first one. Finally, we shall try to emphasize the interest of a comparison of the profiles of several variables for the understanding of the regional constellation of environmental variables.

The values of the contribution profile

The mean value of the profile is linked to the efficiency of the variable in a more or less complex way. This stems from relation (4). Thus the profile of a variable with a good efficiency, such as the ecological sub-region (Fig. 3), has generally higher values than those of the profile of a variable with a poor efficiency, such as bare ground cover (Fig. 4).

The profile value is an index of the linkage between the corresponding state and the distribution of a species. A high value shows that the species frequencies in the corresponding state and in the whole set of relevés are very different; in other words, the state of the variable evidences a set of floristically well individualized relevés. This is the case for the following states: ***Bromus erectus*** as dominant herbaceous species (Fig. 3), low or high cover of trees (Fig. 4), highest altitude (Fig. 4), highest cover of moss (Fig. 4), . . . To the contrary, a low value of the

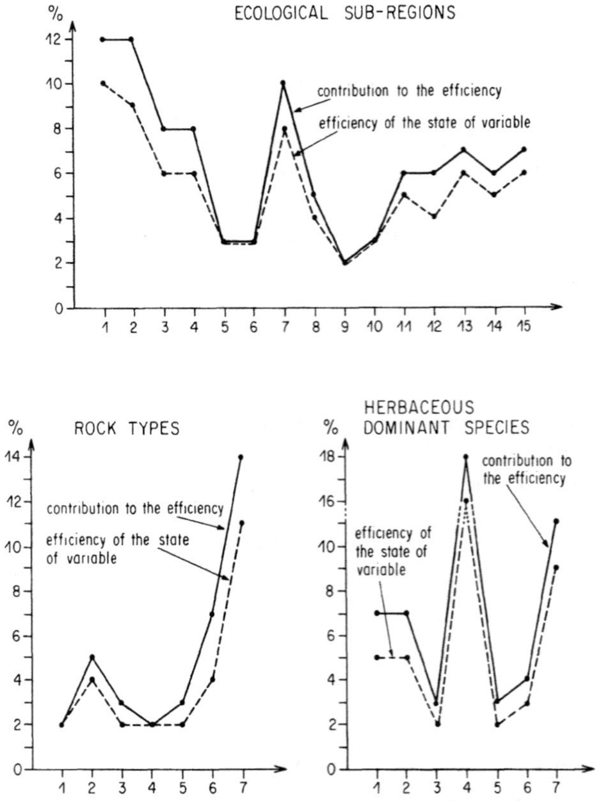

ECOLOGICAL SUB-REGIONS

1. Murviel Hills
2. Aumelas Plateau
3. Montarnaud Hills
4. Causse de Viols
5. Causse de la Selle
6. St Martin Basin
7. Causse de l'Hortus
8. Ganges Basin
9. Brissac Hills
10. Southern side of Serranne Mountain
11. Pompignan Basin
12. Arboras Slopes
13. La Fage Mountain
14. Vis Middle-Valley
15. Causse du Coulet

ROCK TYPES

1. Magnesian limestone
2. "Hard" limestone
3. "Soft" limestone
4. Marly limestone
5. Marl
6. Sandstone
7. Flint limestone

HERBACEOUS DOMINANT SPECIES

1. *Festuca heterophylla*
2. *Melica uniflora*
3. *Brachypodium pinnatum*
4. *Bromus erectus*
5. *Carex glauca*
6. *Carex halleriana*
7. *Carex humilis*

Fig. 3. Profile of efficiency and profile of contribution to the efficiency of qualitative variables.

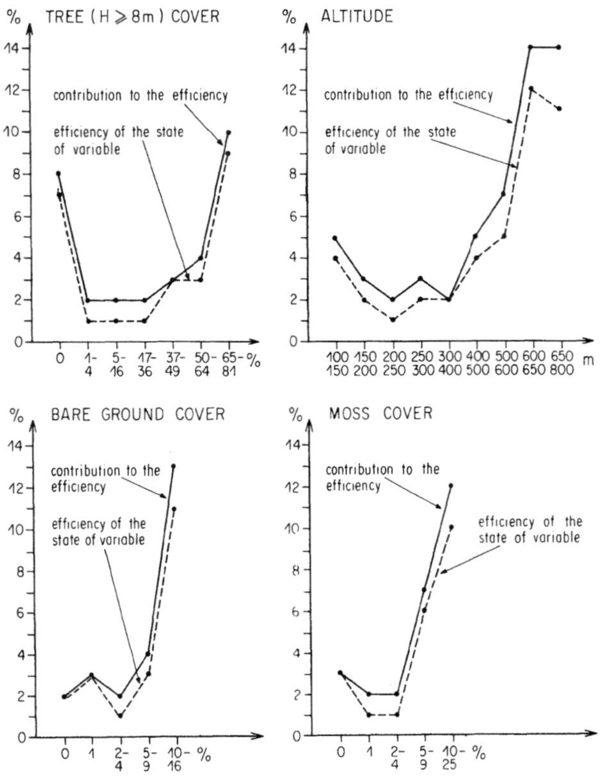

Fig. 4. Profile of efficiency and profile of contribution to the efficiency of scale variables.

profile shows that the state corresponds to a floristically heterogeneous set of relevés. It is very noticeable in the ecological sub-region Brissac Hills (Fig. 3) where we can find clear or dense *Quercus lanuginosa* communities on basic substrate or on acid substrate. It is also the case with *Brachypodium pinnatum* as dominant herbaceous species which due to its plasticity, is the dominant herbaceous species throughout a large panel of environmental conditions.

The profile shape

The profile shape is only interesting in the case of a variable with states that can be ordered; generally, the extreme states have the highest contribution values, whereas the states situated in the middle part of the gradient have the lowest values. This is clearly the case with 'cover of trees higher than 8 m' (Fig. 4), but not with bare ground cover (Fig. 4) and moss cover (Fig. 4), which have a really asymetric profile. Two explanations for this shape may be suggested. (1) The variable acts directly on the species, and in this case it has been noticed that species growing at the extreme parts of the resource gradient have a narrow ecological amplitude, whereas species growing in the middle part of the resource gradient have a broad one (Whittaker 1965, Odum 1971). (In fact, our variables correspond rather seldom to a clear gradient of resource). (2) The extreme states of a variable correspond to a more precise definition of the ecological conditions than the middle states of the variable; for example, an altitude lower than 150 m implies a temperature and precipitation regime which favours the growth of many mediterranean species and, similarly, an altitude higher than 600 m implies a colder and wetter climate which favours the growth of medio-european species. Between these two states of altitude, the two species sets may be found together in various proportions according to aspect, rock types, etc.

Difference between contribution profile and efficiency profile

Contribution profile and efficiency profile are generally of a similar shape. In the first case, the relevés of a state of variable are compared with all relevés. In the other case, the comparison is made with the relevés of the other states. The values of the two profiles are confounded if the contribution of state 'non-*K*' is equal to the contribution of the state '*K*', i.e. if the set of relevés with state *K* is as well or as badly characterized than the set of the others. This occurs generally for the states which have a very weak contribution; but it also occurs sometimes for the states which have a good contribution, but then the distribution of the species is of a bimodal type.

Attempt towards a general analysis

As a result of this approach we consider for each variable in a first approximation, the states which seem to have a clear linkage with species distribution. There are, for instance, the ecological sub-regions (Fig. 3) such as Murviel Hills, Aumelas Plateau, Causse de l'Hortus, Montarnaud Hills; the rock types (Fig. 3) such as sandstone and flint limestone; the dominant herbaceous species (Fig. 3) such as *Bromus erectus*, *Festuca heterophylla*, *Melica uniflora* and *Carex humilis*. In some cases, the states are related to 'efficient' variables (for instance, the ecological sub-regions (Fig. 3)). In other cases, the variable is rather less efficient and only one of the states has a good contribution (for instance, cover of bare ground (Fig. 4)).

The selection of a small number of states of variables makes the study of the linkages between variables easier; therefore, we only take into account the few significant states of variables. For instance, it appears that the set of species linked with the ecological sub-region Causse de l'Hortus (Fig. 3) and the set of species linked with the dominant herbaceous species *Bromus erectus* (Fig. 3) are very similar: both states provide partly redundant information on the species distribution. This is also the case for the dominant herbaceous species *Festuca heterophylla* and of the rock type: limestone with flint.

These linkages between states of variables which describe processes of different nature, emphasize the significance of the regional organization on the relations between species and variables.

Conclusion

The contribution profile and the efficiency profile provide for each state of a variable a measure of their efficiency in describing the species distribution; this efficiency shows very different values according to the various states of a variable. From the comparison between species groups associated with efficient states of different variables we derive a set of redundancies related to the linkage between states of variables; these redundancies must result from either a biased sampling or from the regional organization itself. In the latter case it is difficult to precise the ecology of the species, and to generalize the pattern of relations between variables and species.

Finally, the use of contribution profiles and efficiency profiles should together with other methods give a better understanding and a better testing of the relations between vegetation and environment.

Summary

In any set of phyto-ecological relevés a monothetic partition is possible according to variation in each environmental variable. The significance of the partition realized in this way may be judged by its relationships with the pattern of the vegetation. The mean mutual information may be used as a good criterium here.

The information efficiency related to each state of a variable (i.e. each class of the partition) can be characterized by the value of mean mutual information and information gain associated to this state. From the study of these values, it appears that:

– The states of a variable have very different efficiencies as to the description of pattern in the vegetation. Generally, the extreme states are more discriminant.

– The efficiency related to the states of a variable depends largely on the regional organization (as described through the sampling), i.e. on the one hand the important contrasts between the types of environment, and on the other hand the set of linkages between variables.

These two remarks lead us to a study of the meaning of ecological variables in their relations with the regional framework of the study.

The *Quercus lanuginosa* communities in 'Les Garrigues' region near Montpellier (South of France) are used here as an example.

References

Abramson, N. 1963. Information theory and coding. Mc Graw Hill, London, 201 pp.

Blondel, R. 1941. La végétation de la région de Saint-Paul près de Montpellier. Comm. SIGMA, N° 79, Montpellier, 307–383.

Braun-Blanquet, J. 1936. La forêt d'Yeuse languedocienne Comm. SIGMA. Montpellier, 147 pp.

Braun-Blanquet, J, N. Roussine & R. Negre. 1951. Les groupements végétaux de la France méditerranéenne. CNRS. Paris, 297 pp. .

Daget, Ph. 1976. Ordination des profils écologiques. Naturalia monspeliensia. Ser. Bot., fasc. 26: 109–128.

Daget, Ph. 1978. Ecologie générale et prairie permanente de la réflexion fondamentale à l'application. Thèse d'état, U.S.T.L., Montpellier, 208 pp.

Daget, Ph., M. Godron, J.L. Guillerm, J. Drdos, M. Ružičková & E. Urvichiarová. 1972. Profils écologiques et information mutuelle entre espèces et facteurs écologiques. 14ème Symposium, Association Internationale de Phytosociologie, Rinteln/Weser. (1970): 120–149, Junk, Den Haag.

Fournier, P. 1961. Les quatre flores de la France, Corse comprise (générale, alpine, méditerranéenne, littorale). Lechevalier, Paris, 1105 pp.

Godron, M. 1966. Application de la théorie de l'information à l'étude de l'homogénéité et de la structure de la végétation. Oecol. Plant. 1: 187–197.

Godron, M. 1968. Quelques applications de la notion de fréquence en écologie végétale (recouvrement, information mutuelle entre espèces et facteurs, échantillonnage). Oecol. Plant., 3: 185–212.

Kielhauser, G.E. 1939. Zur Oekologie des Quercetum galloprovinciale pubescentetosum. SIGMA, 67: 24–42 Montpellier.

Kullback, S. 1959. Information and Statistics. Wiley, New-York. 394 pp.

Maarel, E. van der. 1976. On the establishment of plant community boundaries. Ber. Deutsch. Bot. Ges. 89: 415–443

Manil, G. 1963. Niveaux d'écosystèmes et hiérarchie de facteurs écologiques. Un exemple d'analyse dans les hêtraies arden-

naises de Belgique. Bull. Cl. Sci. Acad. Roy. Belgique, 49, 6: 603–623.
Morris, J.W. & J.L. Guillerm. 1974. The ecological profiles technique applied to data from Lichtenburg, South Africa. Bothalia, 11: 355–364.
Odum, E.P. 1971. Fundamentals of ecology. Saunders, Philadelphia, 3rd ed., 574 pp.
Orlóci, L. 1975. Multivariate analysis in vegetation research Junk, The Hague, 276 pp. 2nd ed.
Rényi, A. 1966. Calcul des probabilités. Dunod. Paris, 620 pp.
Thiebaut, B., J. Lepart & A.M. Bacou. 1978. Etude des hêtraies de l'arc montagneux périméditerranéen, de la vallée du Rhône à celle de l'Ebre. III. Information mutuelle moyenne espèces-variables. Oecol. Plant., 13: 301–319.
Whittaker, R.H. 1965. Dominance and diversity in land plant communities. Science. 147: 250–260.

Accepted 20 December 1979

A SYNTACTIC BASIS OF CLASSIFICATION

M.B. DALE

CSIRO, Division of Tropical Crops and Pastures, Cunningham Laboratory, St. Lucia, 4067, Australia

Keywords:
Classification, Ecological grammar, Semantics, Syntax

Introduction

Numerical classification has been treated somewhat superficially by ecologists: they practice but they do not believe, they plant figures and reap ink. Since, as Shakespear says, 'There is occasions and causes why and wherefore in all things', some explanation of this is needed. The confusion existing over the plethora of similarity measures and the profusion of competing methodologies suggests that a stronger theoretical foundation is required, linked more closely to the objectives of the users, and providing sufficient explication to allow rational choices.

The earliest vegetation classifications were not based on formula or method. Instead you learnt to classify by classifying, by paying attention and doing what one thereby discovers has to be done – an apposite description but originally applied by Huxley to love. Such an approach is, I feel, no longer tolerable – at least for classification. Yet the present alternative which is offered, the statistical decision theoretic method which I too embraced in my salad days when I was green in judgement, is also ineffectual. The difficulty is not that it does not suffice, but that it is bereft of *any* relationship to the aims of the ecologist. The pattern is no longer for the the agent and the search an endless exercise in artifice.

Now as Freud, almost, said something which leaves unsatisfied and drives to rebelliousness so large a number of practitioners has neither the prospect of continual existence nor deserves it. Certainly the statistical path is attractive and if we do not know (or care) where we are going we shall certainly arrive somewhere. I must add of course, that I would not wish to reject all stochastic elements. That would be rather like the architect in Morgenstern's poem, whose mansion was built of the holes between fence slats:

Es war einmal ein Lattenzaum
mit Zwischenraum, hindurchzuschauen.
Ein Architekt der dieses sah
stand eines Abend plötzlich da
und nahm den Zwischenraum hinaus
und baute draus ein grosses Haus.

C. Morgenstern

Of objectives and grammar

If we start by examinig the objectives of ecologists it seems to me – and my mind may of course be late maturing or rotting early – that phytosociology is concerned with the processes which generate the variety of vegetation. If classification is to be useful it must aid in elucidating these processes without dumosity, which means covered in bushes and briars, and will itself receive support and explication from an appropriate theoretical basis. One of the most convenient ways of describing processes, and specifically the kinds of discrete processes apparent in vegetation, is to develop an appropriate grammar. A grammar, talking as a linguist not a logician, is a formal tool for defining infinite or finite varation with finite, albeit recursive, rules. An interesting example is given by Lidov & Gabura (1973) in musical composition. As an aside I find myself ill attuned to deterministic differential equations which smear plants like butter across a landscape as the filling in a biomass cake, or energy sandwich.

Since it is foolish to make a long prologue and be short in the story I shall for the rest of the paper glance briefly over the why and hows of grammars.

While Fu (1977) provides a simple introduction to the kinds of grammars, to demonstrate here I shall use a very simple example due to Knuth (1968). This is a context-free grammar for binary numbers – (Fig. 1a).

$$P = \begin{array}{l} N \rightarrow E.E \\ N \rightarrow E \\ E \rightarrow EB \\ E \rightarrow B \\ B \rightarrow 1 \\ B \rightarrow 0 \end{array} \qquad \begin{array}{l} V_N \equiv \{N, L, B\} \\ V_T \equiv \{0, 1\} \\ S \equiv N \end{array}$$

Structure of 1101.01

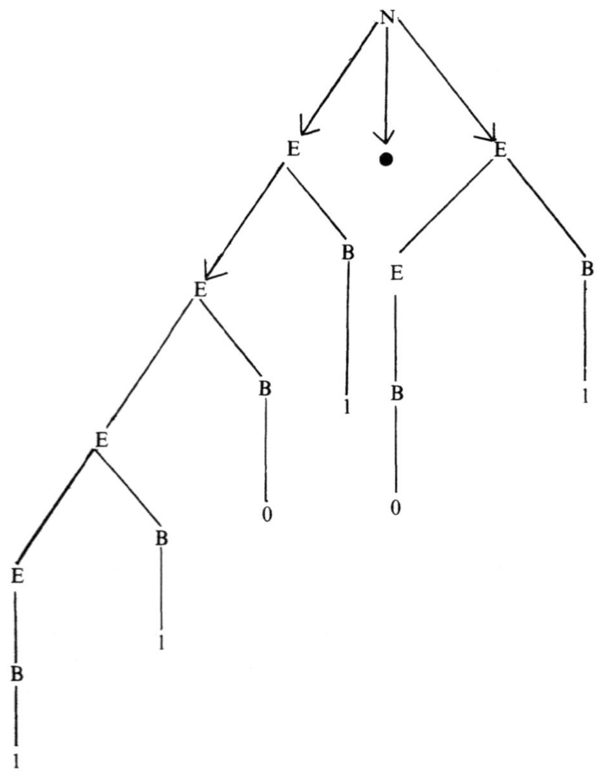

Fig. 1. a. A context free grammar for binary numbers.
b. The structure of a binary number.

The grammar is defined as a 4-tuple $G = (V_N, V_T, S, P)$ where V_N is the set of nonterminals
V_T the set of terminals or primitives
S the start symbol, a single nonterminal
P the set of rules or productions.

I want to emphasise 8 things about this grammar, without going into rigorous definitions.

1. It is a sequential grammar. A rule is selected non-deterministically and applied to one instance of the string.
2. The grammar is context-free. There are no rules of the form $AXL \rightarrow AYL$ where X can be written as Y if surrounded by A and L, and only in that context.
3. The grammar is deterministic in the sense that there are no probabilities attached to the rules to control their selection and use.
4. The only relation between primitives is 'next to'.
5. The grammar assigns a particular tree structure to every number – (Fig. 1b).
6. This grammar shows only how the number is constructed, but does not show what the constructions mean. There are no semantics.
7. The primitives are provided from outside the grammar.
8. The grammar is given. We do not have any means. given to infer such a grammar from examples of binary numbers.

These remarks are not intended to make you treat me like a cuckoo in June, heard but not regarded. They simply point up obvious problems, which I shall now attempt to resolve.

Grammars for ecology

The earliest biological examples of grammars are those of Lindenmayer (1968), whose L-systems were introduced to prescribe morphological development. Their most important feature is that L-systems are parallel grammars, that is a rule applied simultaneously to all the parts of the string which permit it. L-systems have now been generalised to three-dimensional problems (Reusch 1976, Mayoh 1974), and Herman (1970) has discussed the role of environmental variations, though regrettably not from an ecological viewpoint. Perhaps most significant in our present context, Hogeweg & Hesper (1974) showed that numerical classification techniques could retrieve the grammar underlying a set of examples, and that such techniques could be used to aid in identifying processes.

The first ecological grammars are apparently those of Haeffner (1975) who attempted to formalise a niche approach to ecosystems. He required an essentially encyclopaedic description of every species and of its interactions with abiotic and biotic environments; at the risk of rubbing the poor itch of my opinion I feel that this simply demonstrates the irrelevence of the niche concept. Partly Haeffner's problems arise from using a deterministic grammar whose complexity will approach the real system as its adequacy increases, whereas a stochastic grammar need not be so complex (Wharton 1974). Partly it is because he adopts a sequential grammar when a parallel grammar is more appropriate. Significantly he later (1978) attempted to show that no class of grammars less than transformational grammars could be adequate for ecosystem description. Now it is true that every

Given #XXXX#

rules

XXX → XYX	#XX → #cX
XXY → XaY	XX# → Xc#
XYY → XbY	cXc → ccc
aX → aa	bX → bb
Xa → aa	bY → bb
aY → aa	Xb → bb
Sequential application	#aaaa#
Parallel generation	#bbbb#
Parsed parallel	#cccc#

(i.e. reverse the rules)

is an end of string marker.

Fig. 2. A grammar applied in sequence and in parallel.

parallel grammar can be rewritten as a sequential grammar but Rosenfeld (1971) showed that parallel context free grammars in general are equivalent to context sensitive sequential grammars. Certainly the same grammar applied sequentially and in parallel will give divergent results – (Fig. 2). Parallel grammars are almost always simpler than equivalent sequential grammars, as Gips (1974) demonstrates so delightfully with his snowflake grammar.

Gips (loc. cit) also illustrates that grammars can be applied to more complex relationships, in his case to geometric shapes. Indeed there has been a considerable effort given in the last decade to developing grammars for picture recognition, and in this case the rules of the grammar must not only specify changes in the symbols but also specify how the new symbol will be connected to other components of the scene. For complex relationships this is not a trivial task. However, Moayer & Fu (1977) in looking at grammars for fingerprints introduced special rules simply to subdivide the picture. Such subdivision was used to reduce the computation by concentrating attention on interesting regions. Kelly (1971) had earlier made use of this approach with somewhat greater finesse. So classification, in the sense of subdivision of an area, is here used to emphasise interesting areas at the expense of less interesting areas, which surely has some ecological analogue.

Grammars impose structure on examples, and it is the existence of this structure which really distinguishes syntactic from decision theoretic methods. The latter ignore any such structure, regarding it as unfortunate logical dependency between primitives, a nuisance to be avoided or at best tolerated – plunge not the finger of enquiry into the pie of imprudence. The former regard the structure as being of particular importance, as it represents a presumed process of generation to which the idea of classes can be bound. Small changes at early stages can produce massive changes at the terminal level. Fu and Lu (1977) have suggested that similarity can be reasonably defined by using this structure. For two entities *A* and *B* they suggest that the rules for generating *A*, augmented by specific error rules, be used for producing *B*. The similarity of *A* and *B* is then a function of the number and kind of errors which are necessarily introduced. It is interesting to note that Sussman's (1975) program for skill acquisition also emphasises the importance of errors. Further, the use of Manhattan metrics in cladistic studies can be seen as representing the single step nature of mutations, which in a sense are errors.

Semantics

Although grammars may well formally describe patterns, this is a somewhat barren occupation unless interpretations can be given. Knuth (1968) presents one approach for the binary numbers – (Fig. 3) which is to assign to each symbol a finite set of attributes. Each attribute may be synthetic or inherited: the start symbol has no inherited attributes, the primitives no synthetic attributes. To each rule in the grammar further semantic rules

Each B has a 'value' $v(B)$ which is a rational number
Each B has a 'scale' $s(B)$ which is an integer
Each L has a 'value' $v(L)$ which is a rational number
Each L has a 'length' $l(L)$ which is an integer
Each L has a 'scale' $s(L)$ which is an integer
Each N has a 'value' $v(L)$ which is a rational number.

Syntactic Rules	Semantic Rules
B → 0	$v(B) := 0$
B → 1	$v(B) := 2^{s(B)}$
L → B	$v(L) := v(B)$
	$s(B) := s(L)$
	$l(L) := 1$
$L_1 \rightarrow L_2B$	$v(L_1) := v(L_2) + v(B)$
	$s(B) := s(L_1)$
	$s(L_2) := s(L_1) + 1$
	$l(L_1) := l(L_2) + 1$
N → L	$v(N) := v(L)$
	$s(L) := 0$
$N \rightarrow L_1.L_2$	$v(N) := v(L_1) + v(L_2)$
	$s(L_1) := 0$
	$s(L_2) := -l(L_2)$
Synthetic attributes are	$v(B)$, $v(L)$, $l(L)$, $v(N)$
Inherited attributes are	$s(B)$, $s(L)$

Fig. 3. Attribute grammar for semantics of binary numbers.

are attached to show how the attributes are related functionally.

Formal semantics is presently an area of great activity in computer science. Knuth's proposals are related to coordinate grammars, and, I suspect affix grammars. What is important here is not the precise formulation but that the semantic meanings can be coupled to the syntactic rules, leading to the possibility of automatic interpretation of patterns within some other domain. For vegetation the most obvious domain might be the abiotic environment. The separation of synthetic and inherited attributes is, incidentally, for convenience rather than of necessity, and one of the interesting corollaries is that monothetic methods, or something close to them, might make interpretation easier since they relate class formation to specific primitives.

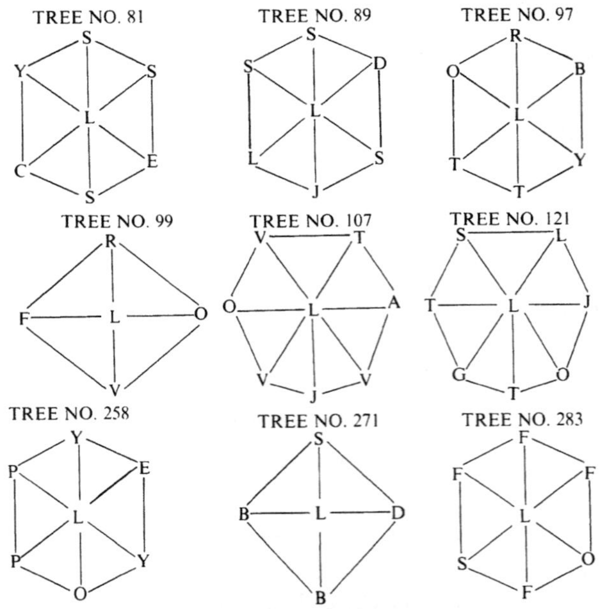

Fig. 4. Initial labelling of 9 trees based on their Delaunay triangulation neighbours.

Inference and primitives

We are left with two trouble spots; can we infer grammars from examples, and what are the ecological primitives. It can be shown (e.g. Fu & Booth 1975, Chou & Fu 1976, Berger & Pair 1978) that inference is possible for some classes of grammars and procedures have been developed to do this. The difficulty as Herman & Walker (1972) point out, is that we have first to infer the hypothesis space, and then the particular system within it, which is not a standard inference problem. In fact, most of the available procedures are computationally intractable, even for quite simple problems, although Cook & Rosenfeld (1976) have made some progress with heuristic methods. Indeed Cook and Rosenfeld's method is remarkably like agglomerative classification. The key difference is that there are 2 functions being examined. One relates to the complexity of the grammar, while the other relates to the adequacy with which it explains the examples. Haeffner (1978) has already indicated that adequacy also involves the existence of rules in the grammar which are not falsifiable at the time, but this has semantic overtones. Klein (1973) has examined inference of semantic rules and and there is much work on this aspect in robot studies.

The selection of primitives forms a distinct problem, the usual solution to which is to adopt the species. I have previously (Dale & Clifford 1976, Dale 1978) suggested the use of other idiotaxonomic categories and other alternatives are of course available. As an experiment I have taken the data of Williams et al. (1969) on rain forest in north Queensland. Each tree is first described by a list of its neighbours, using Delaunay triangulation

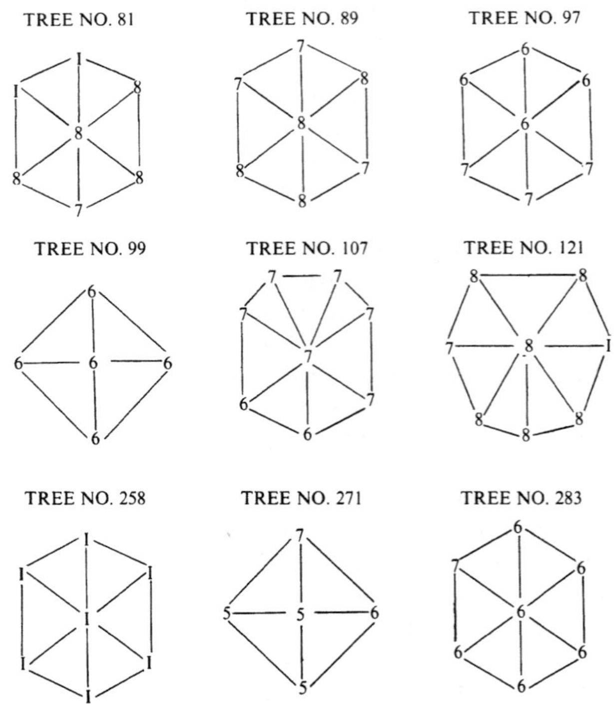

Fig. 5. Relabelling of neighbours after 8 reclassifications.

(Rhynsburger 1973) to find the specific adjoining trees – (Fig. 4). After classifying, each tree is in a specific class, and the new class labels can be used to provide a new description. Of course, this process can be iterated until hopefully there is some convergence. The biggest problem is fixing the number of primitives, that is the number of groups. A number of methods do exist but for this exercise I have subjectively chosen 16 groups. This is less than the original 25 species, and was based on an initial monothetic analysis. The actual classification method used was Indicator Species analysis (Hill, Bunce & Shaw 1975). The results after 8 iterations (Fig. 5) show some convergence. In this particular example, the 9 trees were all of the same species, and the final relabelling suggests that it exists in 5 different environmental neighbourhoods. At the end I shall attempt to obtain a grammar for each species of tree in terms of the class affinities of the neighbours. Knowing the species gives some semantic handles to attach to the rules of the grammar, but this is difficult with the lack of knowledge of rainforest species.

Of course almost any classification of vegetation might be regarded as an attempt at grammatical inference, and hence as an attempt to identify the processes operating in the vegetation. A choice of an hierarchical classification is an attempt to order the processes in terms of their significance and clearly parallels the structure induced by rules of grammars. As an example the work of Noble & Slatyer (1978) is a conscious attempt to identify primitives related to processes of reproduction and regeneration. The usual taxonomic primitives are more strongly related to environmental interpretations, setting spatial rather than temporal boundaries. These two approaches clearly contrast two different views of vegetation, and the resulting organisations of vegetation data will be useful in different contexts. For still other contexts, still other primitives will be required. As examples historical processes may require the use of taxa at other than the species level, while relationship to climate may require the use of structural descriptions.

Conclusion

I hope then that I have clothed grammars in just enough for modesty and no more. Maybe I have committed all the oldest sins in the newest kinds of ways, but I think grammars do provide a useful way of looking at phytosociological processes, and relume the techniques of numerical classification. At least I hope you will no longer treat them as an

egregious professor of Podunk
who once when overly drunk,
remarked I think
I can decline pink
Let me see it goes pink, pank and punk

and I will finish before I rhyme you to death.

References

Berger, J. & C. Pair. 1978. Inference for regular bilanguages. J. Comput. System Sci. 16: 100–122.

Chou, S.M. & K.S. Fu 1976. Inference for transition network grammars. Proc. 3rd Intern. J. Cong. Pattern Recognition pp. 79–84.

Cook, C.M. & A. Rosenfeld. 1976. Some experiments in grammar inference. In: Computer Oriented Learning Programs, J.C. Simon (ed.) Noordhof-Leyden pp. 157–174.

Dale, M.B. 1978. Planning an adaptive numerical classification. Vegetatio 35: 131–136.

Dale, M.B. & M.T. Clifford. 1976. On the effectiveness of higher taxonomic ranks for vegetation analysis. Austral. J. Ecol. 1: 37–62.

Fu, K-S. 1977. Introduction to syntactic pattern recognition. In: Syntactic Pattern Recognition, Applications. K-S. Fu (ed.) Springer N.Y., p. 1–29.

Fu, K-S., & T.L. Booth. 1975. Grammatical Inference: Introduction and Survey. Pt. I & II. I.E.E.E. Trans. System, Man and Cybernetics SMC-5: 95–111, 406–423.

Fu, K-S., & S.Y. Lu. 1977. A clustering procedure for syntactic patterns. I.E.E.E. Trans. Systems, Man and Cybernetics SMC-7: 734–742.

Gips, J. 1974. Shape grammars and their uses. Ph. D. thesis Dept. Computer Science. Stanford University 233 pp.

Haeffner, J.W. 1975. Generative grammars that simulate ecological systems. Simulation Council Proceedings Series 5: 189–211.

Haeffner, J.W. 1978. Ecosystem assembly grammars: generative capacity and empirical adequacy. J. Theoret. Biol. 73: 293–318.

Herman, G.T. 1970. The role of environment in development models. J. Theoret. Biol. 29: 329–341.

Herman, G.T. & A.D. Walker. 1972. The syntactic inference problem applied to biological system. In: Machine Intelligence 7. B. Meltzer & D. Michie (eds) Edinburgh University Press. pp. 341–356.

Hill, M.O., R.G.M. Bunce & M.W. Shaw. 1975. Indicator species analysis: a divisive polythetic method of classification and its application to a survey of native pinewoods in Scotland. J. Ecol. 63: 597–613.

Hogeweg, P. & B. Hesper. 1974. A model study of biomorphological description. Pattern Recognition 6: 165–179.

Kelly, M.D. 1971. Edge detection in pictures using planning. In: Machine Intelligence 6. B. Meltzer & D. Michie (eds) Edinburgh University Press pp. 397–410.

Klein, S. 1973. Automatic inference of semantic deep structure rules in generative semantic grammars. Proc. International

Conf. on Computer Linguistics PISA 27/VIII-1/IX No. 2 pp. 557–574.
Knuth, D.E. 1978. Semantics of context free languages. Mathem. System Theory 2: 127–145.
Lidov, D. & J. Gabura. 1973. A melody writing algorithm using a normal language model. Comput. Studies in the Humanities and Verbal Behaviour. 4: 138–148.
Lindenmayer, A. 1968. Mathematical models for cellular interaction in development 1. Filaments with one-sided inputs. J. Theoret. Biol. 18: 280–289.
Mayoh, B.H. 1974. Multidimensional Lindenmayer organisms. In: Lecture Notes in Computer Science No. 15. B. Groose & J. Hartman (eds) pp. 302–326.
Moayer, B. & K-S Fu. 1977. Fingerprint classification. In: Syntactic Pattern Recognition, Applications. K-S, Fu. (ed.) Springer N.Y. pp. 179–214.
Noble, I.R. & R.O. Slatyer. 1978. The effect of disturbance on plant succession. Proc. Ecol. Soc. Australia 10: in press.
Reusch, P.J.A. 1976. A common approach to retrieval concepts and multidimensional developmental systems based on lattice-like structures. Gesellschaft für Mathematik und Data verarbeitung mBH. Bonn. Ber. Nr. 90 pp. 1–70.
Rhynsburger, D. 1973. Analytical delineation of Thiessen polygons. Geogr. Analysis 5: 133–144.
Rosenfeld, A. 1971. Isotonic grammars, parallel grammars and picture grammars. In: Machine Intelligence 6. B. Meltzer & D. Michie (eds) Edinburgh University Press. pp. 281–296.
Sussman, G. 1975. A computer model of skill acquisition. Elsevier 133 pp.
Wharton, R.M. 1974. Approximate language identification. Information and Control 26: 369–374.
Williams, W.J., G.N. Lance, L.G.J. Webb, J.G. Tracy & J.H. Connell 1969. Studies in the numerical classification of complex rain forest communities IV A method for the elucidation of small scale pattern. J. Ecol. 57: 635–654.

Accepted 31 October 1979

THE MINIMISATION OF RANDOM EVENTS IN THE SEARCH FOR GROUP STRUCTURE

Robert C. JANCEY*

Department of Plant Sciences, The University of Western Ontario, London, Ontario, Canada, N6A 5B7

Keywords:
Agglomerative, Clustering, Group structure, Misclassification, Random

Introduction

Pattern in vegetation exists to the extent that the distribution and abundance of species among relevés is non-random. No matter how distinct the pattern, the distribution of naturally occurring vegetation will always incorporate a random element. It is evident that the recognition and exclusion of this random component will increase the clarity with which pattern can be demonstrated. Species ranking techniques have commonly been used to identify those species with a large non-random component in their distribution, the objective being an economy of data gathering effort. The possibility exists however, of using species ranking as a means of improving the clarity of the revealed group structure. It is entirely possible that the species most highly ranked on the basis of all relevés in a study may, in some sub-set of the whole, be much more lowly ranked given the same ranking criterion. In other words, its distribution in the sub-set has a larger random component. To include such a species when the structure of that sub-set of relevés is being investigated would serve only to reduce the clarity of the structure. This observation is only an extension of the view, commonly held by phytosociologists, that species ranking is a matter of local importance. Given that this attitude is correct, then to the extent that pattern exists, species rankings will differ in different parts of the whole. It should be pointed out that in this communication the word 'random' is used in a particular, restricted sense: It may be argued that if ones scale and system of vegetation measurement were sufficiently precise, the concept of randomness would have little meaning in this context. Practical considerations however, require us to employ data recording techniques where the scores attributed to some species, in some relevés can be considered as contributing little or nothing to the elucidating of vegetational pattern. It is such events that, in the context of this communication, will be described as 'random'.

* This work was supported by a grant from the National Science and Engineering Research Council of Canada.

Procedure

To accommodate the concept of locality of species importance, an iterative approach to species ranking has been developed. While the actual ranking algorithm used by the author will be described elsewhere, it should be pointed out that the iterative approach described below can accommodate any ranking procedure which incorporates an objective cut-off criterion for the species list.

The total species list is first ranked on the basis of species performance in all the relevés used in the study. An agglomerative, hierarchical clustering is then performed, using those species standing above the cut-off point in the first ranking. Two vectors of relevé identifiers are formed; corresponding to those relevés falling on one side of the final clustering fusion as opposed to the other side. These two vectors are stored for the control program and also printed out. The first species ranking is also printed. The ranking is then repeated for all the relevés falling on one side of the final fusion. All the original species are included in this and in subsequent rankings since it is evident that a species which may be of little significance in terms of the total study may well become so in some sub-set of the study. The hierarchical clustering is then repeated, thus giving rise to two more

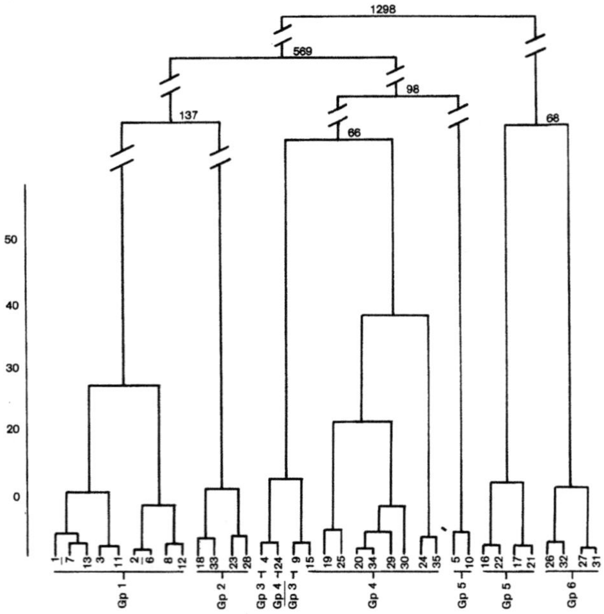

Fig. 1. Hierarchical clustering based on all species, using a sums of squares criterion. Numbers on the horizontal axis identify quadrats, while numbers on the vertical axis represent sums of squares at group fusion.

vectors of relevé identifiers. Again, these are stored and printed out, together with the new species ranking. This process is then repeated until the whole of one side of the final fusion has been explored, the control program keeping track of which sub-branch of the dendogram is being examined. Once one side of the final fusion has been fully explored, the control program turns to the other major branch and repeats the process.

Example

The example given below is based on a set of synthetic data representing 35 'relevés' and 10 'species'. Figure 1 shows a conventional hierarchical clustering, based on a sums of squares criterion (Orlóci 1978). It shows a misclassification of relevés 4 and 9 occurring at the S.S. = 98 fusion level. Figure 2 shows a dendrogram based on the iterative, agglomerative procedure described above, incorporating the same sums of squares clustering procedure. Apart from a much tighter group structure, with lower sums of squares for fusion, it will be seen that the relevés 4 and 9 are now clustered in a way which accords with the original synthetic data structure.

Discussion

Iterative clustering procedures have received little attention in the past since they were found to be very time consuming. Hill, Bunce & Shaw (1975) described a divisive approach, but its objective was not optimisation of the classification but rather the minimisation of computation. It also incorporated a number of subjective steps. The speed and capacity of the newer computers has increased to the extent that we may now start seeking optimality of results rather than computational time saving. The data set used in the above example required under 30 c.p.u. seconds on a P.D.P. 10 computer, with a KL central processor (Copies of the program, written in BASIC, are available from the author). Such an iterative agglomerative approach is thus practicable for use with most sets of real data. Perhaps as interesting as the improved classification is the sequence of species rankings. This allows an insight into the changing forces creating pattern in different sub-sets of the study area.

It must be appreciated that problems arise with very small exemplar data sets. It is hardly meaningful to rank species on their performance in only three or four relevés. In normal sized studies this would not be a problem and in any case, the algorithm could easily be modified to omit the iterative process once relevé vector size fell below some chosen critical value.

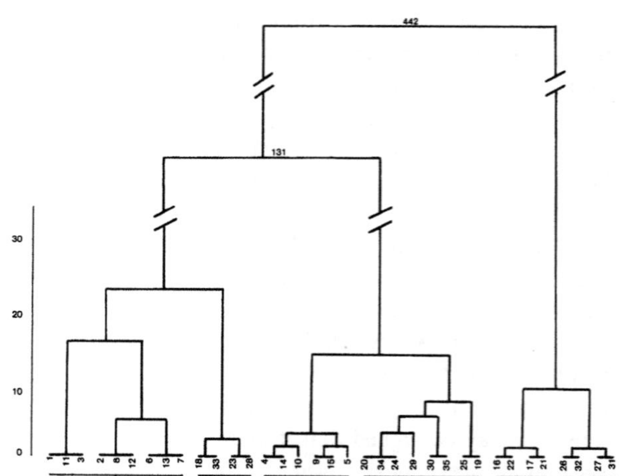

Fig. 2. Hierarchical clustering based on iterative ranking and a sums of squares fusion criterion. Numbers on the horizontal axis identify quadrats, while numbers on the vertical axis represent sums of squares at group fusion.

Summary

An iterative species ranking and agglomerative clustering technique has been described. Its objectives were the minimisation of misclassification due to random species occurrence and the demonstration of the changing role of species in pattern formation.

References

Hill, M.O., R.G.H. Bunce, & M.W. Shaw. 1975. Indicator species analysis, or divisive polythetic method of classification and its application to a survey of native pinewoods in Scotland. J. Ecol. 63: 597–613.

Orlóci, L. 1978. Multivariate analysis in vegetation research. 2nd ed. Junk, The Hague. 451 pp.

Accepted 7 November 1979

RAPID INITIAL CLUSTERING OF LARGE DATA SETS*

Hugh G. GAUCH, Jr.

Ecology and Systematics, Cornell University, Ithaca, New York 14850, USA

Keywords:
Classification, Clustering, Composite clustering, Data processing, Multivariate analysis, Numerical classification, Outliers, Redundancy

Introduction

Multivariate analysis of plant community data primarily involves ordination and classification. The data analyzed are abundance or/and cover values of species recorded for a number of samples, resulting in a species-by-samples matrix of observations. Multivariate analysis of such data has three general aims: (1) to identify similar, redundant samples, (2) to identify outliers (samples very different from all other samples), and (3) to elucidate relationships among samples. (Alternatively analysis may concern species or both species and samples, but for simplicity only samples are discussed here). These three aims correspond to aspects of the data with different underlying biological processes and different requirements for mathematical analysis, and may be characterized as follows.

(1) Redundancy. Replicate vegetation samples may be taken without significant difference in the environment or composition. However, vegetation samples are rarely, if ever, identical; they average 60 % to 90 % similarity (Bray & Curtis 1957, Moore 1972, Janssen 1975). This reduction in similarity from 100 % is caused by historical and stochastic events such as seed dispersal and germination, increase or decrease of resources, disturbance, population fluctuation, and limitations in the sampling process. These matters merit study, but they are not the usual focus of phytosociology. From a phytosociological perspective, field data are considered to be noisy, and samples within a threshold similarity are considered replicates (at a given level of analysis, Janssen 1975).

Most data sets are to some degree redundant. Typically data sets have several to hundreds of replicate samples for each discernibly different vegetation type. Redundancy may or may not be desirable, depending on purposes and practical constraints. Redundancy is necessary in order to produce a vegetation map, or to refine estimates for each vegetation type of the means and variances of its species abundances (Goodall 1978). However, redundancy must be removed in order to describe vegetation economically, or to permit some multivariate analyses that are not practical for large data sets.

(2) Outliers. An outlier is a sample of peculiar composition, having low similarity to all other samples. Outliers come in degrees because (a) the degree of sample singularity is variable, and (b) the concept may be extended to an arbitrary degree to groups of more than one sample that are odd in the same way. Phytosociological data frequently contain outliers because of disturbed, heterogenous or otherwise unusual sites. For various purposes, it may or may not be desirable to include outliers.

(3) Relationships. A major aim of phytosociology is to elucidate the relationships among vegetation samples. Ordination and/or classification of the species-by-samples data matrix are major means of doing so. If environmental or historical data are also collected at each sample site, or if habitat preferences of the species are known, these additional data may be the basis for interpreting relationships.

This view of the data structure relates to the distances in a samples-by-samples distance matrix. Information on redundancy, outliers and relationships is encoded primarily

* This paper benefited at many points from discussions with Mark O. Hill and Robert H. Whittaker. Mark Hill suggested condensed data storage. This work was done under a National Science Foundation grant to Robert Whittaker. I also appreciate technical assistance from Timothy F. Mason and Steven B. Singer.

in the small, large, and intermediate distances respectively.

This data structure has implications also for choice of, and sequence of steps in, mathematical analyses. Redundancy may well be addressed first (unless the entire data set is so small that it can be used intact without problems), since redundant sample groups are relatively uninformative about broader relationships yet cause problems of bulk and tediousness. Outliers should also be considered before relationships because outliers may dominate analyses and confound or obscure other relationships. However, redundancy and outliers can be addressed simultaneously because clustering of similar samples leaves outliers behind. A natural sequence for vegetation analysis is therefore the clustering of similar samples and identification of outliers first, followed by analysis and interpretation of sample relationships. Van der Maarel et al. (1978, Louppen & van der Maarel 1979) reach a similar conclusion saying that the preferred first step of the Braun-Blanquet method is to initiate and homogenize clusters.

The second step in this sequence calls for a variety of ordination and hierarchical clustering methods, but is not the main concern of this paper. The remainder of this paper discusses the first step of this sequence-processing of redundancy and outliers by fast initial clustering. However, the above perspective on the overall multivariate task is the basis of specifications for initial clustering algorithms. Applications to four large data sets are also included in this paper.

Specifications for initial clustering

Computer programs for initial clustering should meet several specifications.

(1) Minimal computer requirements. Because initial clustering involves large data sets, computer time and memory requirements should be minimal.

(2) The basic cluster property to be achieved is within-cluster homogeneity (= community-type homotoneity). Within-cluster homogeneity makes possible inference about a sample's properties based on its cluster membership; this is the aim of general purpose clustering (Sneath 1969, Sneath & Sokal 1973, Gower 1974). Such clustering summarizes redundancy and identifies outliers, and makes sense in terms of the noise inherent in field data (as discussed in the introduction).

The next two specifications concern cluster properties which should be avoided because the implied benefit is small, if not illusory, but the computing cost is considerable.

(3) Algorithmic uniqueness. If the data are naturally strongly clustered, many clustering methods will recover the clusters correctly (although methods may vary greatly in the computer time required). If the data are continuous, there is no natural number of clusters, and there are no natural boundaries to the clusters. Since these features are lacking in continuous data, boundaries must be imposed by the clustering method.

Some clustering methods give unique results, even for continuous data, by means of computing and ranking all the sample pair distances. However, this ranking would be altered by small changes in the data which could arise by sampling errors and which might defy phytosociological interpretation. Also if ties occur in the ranking, they must be broken by some arbitrary procedure. Furthermore, limitations in field data introduce a subjective element in phytosociology which is carried through, rather than eliminated, in subsequent analyses (Moore & O'Sullivan 1970, Moore 1972). Algorithmic uniqueness appears to be attractive, but for these reasons it is an unnecessary if not excessive requirement, and it is dispensed with here.

A related but different question is whether clustering results should be biased by the sequence in which samples are supplied (Louppen & van der Maarel 1979). It seems best to avoid such bias.

(4) Hierarchy. A hierarchy ordinarily is not needed in initial clustering because details of the relationships among essentially replicate samples are not the concern (Williams 1971). Also computation of a hierarchy for a large number of samples is costly (although Bruynooghe 1978 presents a relatively rapid hierarchical clustering method). If desired, initial clusters may be represented by an average (composite) sample for each cluster, and these far fewer clusters may be submitted subsequently to a hierarchical clustering in order to elucidate relationships on a higher level.

Computing requirements

Of the four above specifications, one deserves detailed consideration – computing requirements. These requirements for various clustering algorithms are a function of five numbers:

I number of species
J number of samples
K number of clusters

N average number of species per sample
A amount of stored data $= 2NJ+J$.

The species-by-samples data matrix is usually sparse (mainly zeros), especially for large data sets. Rather than store the entire primary data matrix of IJ values, a condensed storage is preferable, storing only positive data matrix entries. One vector of length NJ stores the positive data entries and another vector of the same length stores the species identification number of each entry; the data are stored by samples sequentially and a third vector of length J indexes the beginning location of each sample's data. Hence the total amount of memory required for the data is $2NJ+J$, which is the amount of data A defined above. For data sets 5 % to 20 % dense, condensed storage is about 10 to 2.5 times as efficient in use of computer memory.

A clustering algorithm requires minimal computer memory if it involves storage of A numbers and little else. Computation and storage of a secondary samples-by-samples (or samples-by-clusters) matrix is to be avoided.

Minimization of computer time involves one or more of the following four devices.

(1) Condensed storage. Condensed storage saves memory, but it also yields a gain in computing speed because (a) only positive data are consulted, and (b) the data are accessed in singly subscripted arrays. This gain is on the order of I/N, times a factor for using single subscripts (say 2). For example, given 1000 species with an average of 40 species per sample, the gain in computing speed is a factor of about $(1000/40) \times 2 = 50$. Furthermore, as vegetation samples are added the number of species I usually increases, but the average number of species per sample N is relatively constant. Hence the ratio I/N tends to increase with data set size, being largest when efficiency is needed most.

(2) Fewer distances. Most clustering techniques compute a samples-by-samples distance matrix (Sneath 1969, Sneath & Sokal 1973). This computation is of order J^2 and becomes prohibitive with large J. Other clustering algorithms, as will be discussed later, instead use sample-by-cluster distances, which are of order JK. This change increases computing speed by a factor of $J^2/(JK) = J/K$. For example, given 2000 samples and 50 clusters, this is a factor of 40. The latter, smaller computation is probably closer to the mental process used in classifying objects, since one tends to compare objects with the classes into which they are placed, rather than comparing each object against all others (especially for large data sets). Furthermore, the number of samples J may increase indefinitely, but the number of clusters K is rarely more than 50 or at most 100. Consequently K may be viewed as a constant with a value of 100 (with deviations from this being on the smaller side). Hence the ratio J/K tends to increase with data set size, which again is convenient because efficiency is needed most with large data sets.

(3) Rapid division. Certain algorithms employing successive approximation with rapid convergence can be made very fast, particularly for sparse data matrices (Hill et al. 1975, Hill 1979).Speed can also be gained by using monothetic rather than polythetic divisions (divisions based on one rather than all species), provided that the criterion for selecting the dividing species involves a modest computation. Monothetic techniques are, however, notorious for misclassification, so this is not an ideal way to reduce computational load. Also a modest gain in speed can be achieved by stopping divisions early, rather than continuing calculations to very fine divisions that are not of interest (Williams 1971).

(4) Data subsets. Ordinarily large data sets are highly redundant, so that repeated clustering of random subsets gives essentially the same pattern (at least for many clustering techniques). This shortcut gives nearly the same result as clustering the entire data set (Williams 1971).

The above four devices for computing speed are independent, and the effects of using two or more are multiplicative. The algorithm described in the next section uses the first two devices. The first device replaces a dependency on I with one for N, and the second replaces one on J for K; the joint effect is to replace a dependancy on IJ with one for NK. In the examples in points 1 and 2 above, these two devices increased computing speed by factors of 50 and 40; the joint effect is a factor of 2000. In this example an algorithm including these two devices accomplishes in 15 seconds what one without them accomplishes in 8.3 hours.

When discussing computing speed it helps to know the theoretical limit for any algorithm which could be considered a clustering algorithm. Comparison of a given algorithm with the theoretical limit then indicates what room is or is not left for improvement.

A theoretical minimal computational load may be derived as follows. (1) Clustering may be viewed as having two steps: cluster definition and assignment of each sample to a cluster. (2) Assignment of samples to (previously defined) clusters is termed identification (Sneath & Sokal 1973). The usual method of identification is comparison of a sample with all clusters, and assignment to the cluster with which it has greatest similarity. Given J samples and K clusters, this requires computation of JK distances. Assuming that all the species are used (a polythetic rather than monothetic distance), each distance requires comparison of I numbers, or preferably about N numbers with efficient programming. Hence the computational task of identification is of order NJK. Since NJ is the same order as the amount of data A, the task is of the order KA. (3) There is no way to specify a minimal algorithm for cluster definition, and hence to declare its computational requirements. The strictest assumption that can be made is to assume that very little computation is required (more exactly, little computation in comparison to that required for identification). (4) Combining the above three points, the task of identification serves as a lower limit for the greater task of clustering; clustering cannot require less computation than identification. This fixes a lower limit for computational load of order $KA + 0 = KA$, where K is the number of clusters (which as noted earlier is effectively a constant), and A is the amount of data.

This argument leads to the conclusion that fast clustering should involve a computational load which is merely a linear function of the amount of data. Also note that for a given average number of species per sample N, this implies a linear dependancy on the number of samples, so that the computation per sample is a constant regardless of the number of samples.

The clustering algorithm presented in the next section has minimal requirements for computer memory and time. Other fast algorithms are cited in the following discussion.

Composite clustering

Fast initial clustering of samples may be done by a technique here termed 'composite clustering.' For this discussion samples are represented by points in a multidimensional Euclidian space where axis 1 is the abundance/cover of species 1, axis 2 is the abundance/cover of species 2, and so on for I species. Composite clustering involves two phases.

Phase 1 picks a sample at random, and clusters samples within a user-specified radius to that sample. This process is repeated until finished, ignoring any sample previously clustered. Because center samples are selected at random, the order in which samples are given introduces no bias in the results.

The first phase has an undesirable property: as spherical clusters are placed, small regions can be left behind in the interstices, resulting in clusters with only one or a few members. This is undesirable because these extra small clusters increase the number of clusters required to account for the data, and obscure the identification of outliers. This problem can be ameliorated rather simply, as explained next.

Phase 2 reassigns samples from small clusters (defined by having fewer than a user-specified number of members) into the nearest large cluster, provided that the sample is within a user-specified radius. The radius used in phase 2 is usually somewhat (but not greatly) larger than that used in phase 1.

For each resulting cluster, a composite sample is produced by averaging the samples it contains. Composite clustering is implemented in Cornell Ecology Program 30 COMPCLUS, written in FORTRAN and executed interactively (Gauch 1979; information on the CEP series available upon request).

COMPCLUS is designed to present a small amount of information at the terminal during execution. This information includes a table of cluster sizes, listing for each size (in rank order from largest to smallest) the size, number of clusters, cumulative number of clusters, and cumulative number of samples. This table aids the user in setting parameters and in assessing the level of clustering. When clustering results appear satisfactory, output options include (1) a printed list of samples in each cluster, (2) a printed list of major species in each cluster, (3) a punched vector giving the cluster membership of each cluster, and (4) punched composite samples. Other computer programs in the CEP series are useful for preparing data matrices for input in the appropriate format, and for editing, manipulating and printing the data matrix in view of the COMPCLUS clustering results.

The user has choices to make of data transformation, distance measure, and clustering radius. These choices are discussed in detail in the documentation for COMPCLUS (Gauch 1979), and are treated here briefly.

COMPCLUS offers five data transformations. Four of these are applied to each matrix entry individually: square root, logarithm (to base 10 of value plus 1), octave scale (Gauch 1977), and presence/absence. Quantitative data without transformation tend to emphasize dominant species, and there is decreasing emphasis on dominants in the above sequence of transformations. The best emphasis depends on the purpose of a given study, but in general it seems that frequently the scaling in the human mental process is a balance between qualitative and quantitative information so that an intermediate transformation (such as the octave scale) gives an emphasis agreeable to the user's expectations (Van der Maarel 1979). The fifth transformation in COMPCLUS is applied to samples: relativization of sample totals to 100. When samples vary greatly in their total abundance/cover values, relativization may put them on a more equal footing, preventing samples with great differences in species contained from being put together merely because they have little content. As with the other transformations, relativization affects emphasis, but the desirability of a given emphasis depends on the purposes of a study.

COMPCLUS offers three distance measures: Euclidean distance *(ED)*, percentage difference *(PD)* and the complemented coefficient of community (*CD*; Gauch 1979). Because *ED* involves squaring numbers, it weights dominants most heavily, whereas *CD* using presence/absence weights dominants least, and *PD* is intermediate. No universal recommendation is possible, but frequently *PD* is best.

Since both the transformation and distance measure affect the emphasis on dominants, the effects of these choices interact. Any transformation that emphasizes dominants will have intensified effect when combined with *ED*, and on the contrary, choice of *CD* removes the effect of different transformations. Likewise, these choices interact with the data set's properties, particularly the numerical range in which abundances are recorded. Transformations and distance measures affect the location of sample points in the multidimensional space in the first place, whereas for clustering radius selection these locations are given and only the nature of the clusters is affected.

The clustering radius controls the level of clustering since a large radius yields few clusters, and a small radius yields many clusters. Three approaches may be used to obtain a first estimate of an appropriate phase 1 radius. (1) The user may want clusters defined by a certain within-cluster similarity (Janssen 1975). (2) The user may know that certain samples are replicates, whereas other sample pairs are not replicates. Then distances between pairs of replicate samples may be computed, and likewise distances between samples which are not replicates, and the radius selected to encompass most replicate pairs but exclude most non-replicate pairs. (3) The user may want a clustering level with about *K* clusters. COMPCLUS includes a feature to estimate the corresponding radius, based on the principle that given *K* clusters of roughly even size about $1/K$ of the distance values are within clusters and the rest are between clusters. Phase 1 results with the first estimate radius may be satisfactory, or some trial and error may be needed with a slightly larger or smaller radius. Finally phase 2 may be applied to eliminate small clusters from interstices between large clusters.

Tests of composite clustering

COMPCLUS was applied to four field data sets in order to test its effectiveness and speed. An IBM 370/168 computer was used with compilation under FORTRAN H extended with optimization level 2.

Two clusterings for a given data set are compared using percent mutual matches *(PMM)*, defined as 200 times the number of mutual matches, divided by the total number of matches. A 'match' is a sample pair assigned to the same cluster in either clustering, and a 'mutual match' is a sample pair assigned together in both clusterings (a mutual match thus counts twice among the matches). PMM ranges from 100 for identical clusterings to 0 for completely different clusterings. Given two random clusterings with *K* clusters of approximately even sizes *PMM* has an expected value of about $100/K$. For comparing a number of clusterings simultaneously, a clustering-by-clustering matrix of *PMM* values subtracted from 100 to yield distances is computed. This matrix is then ordinated using reciprocal averaging (RA, Hill 1973), or in this case the ordination is a 'secondary reciprocal averaging,' since the matrix is symmetric (Gauch et al. 1977). The ordination program used is ORDIFLEX in the Cornell Ecology Program series (Gauch 1977).

Relationships among composite samples from COMPCLUS were analyzed by ordination and hierarchical clustering. The best ordination for this purpose was detrended correspondence analysis (DECORANA, Hill 1979a; 'correspondence analysis' is a synonym of 'reciprocal averaging'; DECORANA is in the Cornell Ecology Programs series). DECORANA corrects the two main faults of RA: it removes the second axis quadratic distortion of the first axis (the 'arch' or 'horseshoe' problem), and it rescales the first axis in order to remove the compression of axis ends relative to the axis middle. DECORANA makes efficient use of memory and has a very fast direct iteration algorithm with only linear computer time requirement with respect to problem size. The ordination axes of DECORANA are scaled in standard deviation units; empirical results show that these units are approximately equivalent to half-change units (HC, where 1 HC indicates a percentage similarity of 50;Gauch et al. 1977) provided that there are no species of much wider distribution than the others. The model used in DECO-. RANA is appropriate for analyzing plant community data, but it is not appropriate for analyzing a matrix of *PMM* values. Hierarchical clustering was done by the NT-SYS package (Rohlf et al. 1972). Two options were used: complete linkage clustering because its production of tight clusters emphasizes homogeneity within related clusters, and UPGMA because Sneath & Sokal (1973) recommend UPGMA as generally the best method.

The first data set uses cover estimates in 469 relevés from the San Jacinto and Santa Rosa Mountains, southern California, with 271 species and an average of 10 species per sample (data supplied by R.H. Whittaker). The samples had been classified subjectively into 36 community-types by physiognomy and dominance (considering recurring combinations of major species, not single-species dominance only). COMPCLUS was applied

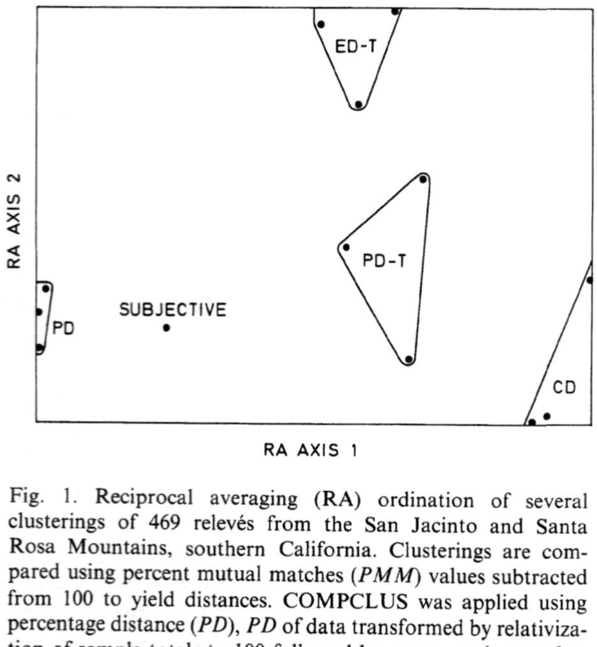

Fig. 1. Reciprocal averaging (RA) ordination of several clusterings of 469 relevés from the San Jacinto and Santa Rosa Mountains, southern California. Clusterings are compared using percent mutual matches (*PMM*) values subtracted from 100 to yield distances. COMPCLUS was applied using percentage distance (*PD*), *PD* of data transformed by relativization of sample totals to 100 followed by octave scale transformation (*PD-T*), complemented coefficient of community (*CD*), and Euclidean distance with this same transformation (*ED-T*). The three replicates of each treatment indicate the variability in COMPCLUS results given this relatively continuous data set. The subjective classification of these data was most similar to COMPCLUS results using *PD*.

using *ED* (Euclidean distance), *PD* (100 minus percentage similarity) and *CD* (100 minus the Sørensen coefficient of community) distance measures, applied to the raw data, or to the data transformed by relativization of sample totals to 100 followed by application of the octave scale (excepting *CD* which is not affected by transformation). The clustering radius was set to result in about 36 clusters. For each of these five treatments three replicate clusterings were obtained by varying the random number initialization. Figure 1 shows a reciprocal averaging arrangement of these classifications by their relative dissimilarities, using the matrix of 100-*PMM* values. The first RA axis places *PD* and *CD* at the axis ends while *PD* with transformed data is intermediate; *ED* with transformed data is displaced along the second RA axis, and *ED* without transformation is excluded here because it was an outlier in an earlier ordination (not shown). This arrangement seems reasonable in view of the properties of these distance measures and transformations. The subjective classification ordinates nearest to the *PD* clusterings; COMPCLUS using *PD* and intuitive definition of dominance-types give similar results. The dispersions within replicate clusterings indicate the inherent variability in COMPCLUS given these data representing an essentially continuous vegetation pattern. The replicate clusterings using the same transformations, distance measure, and cluster radii are, however, quite similar (average *PMM* of 51). They are clearly more similar to one another than clusterings of the same samples with other transformations and distances (average *PMM* of 29), and in this application they produce the same major community-types (with differences in assignments of transitional samples). For ordinary clustering purposes a single clustering of these data using *PD* should usually suffice. For these data, reading the data matrix required about 2 seconds of computer time, and each clustering required about 2 seconds.

The second data set comprised 3270 relevés from a vegetation survey of southeast England (data supplied by H.J.B. Birks). The 876 species occurring in 3 or more relevés were used, and an average sample contained 13 species. COMPCLUS was applied using PD on Domin scale values with square root transformation to reduce the effect of dominants, with the clustering radius set to yield about 40 clusters (and 50 outliers). The clusters were easily interpreted as community-types by M.O. Hill and appear to represent an effective summarization of a large data set.

The clusters were ordinated using DECORANA, RA, and multidimensional scaling (MDS, using the classical nonmetric options of ALSCAL-4, Young & Lewyckyj 1979). The DECORANA first axis (7 standard deviations) went from wall communities to dunes and various grasslands, to dry heath, various woodlands and hedgerow, to alder swamp and wet woodland clearing, to wet heath and sedge fen, then through riverside and emergent communities to duckweed and water weed communities; the trend was clearly one of moisture status. The second axis (4.5 standard deviations) spread the mesic portion of the first gradient in a sequence from oak ash and other woodlands to dry heath and hedgerow to chalk grassland and finally through increasingly weedy grasslands to weed communities; the trend reflected disturbance. MDS gave a similar result, although the computation required several times as much computer time, and MDS could be applied only to the 40 composite samples whereas DECORANA also readily analyzed the 876 species. Unlike DECORANA, RA did not recover an effective second axis. The analysis using COMPCLUS to reduce redundancy and to eliminate outliers, followed by DECORANA to ordinate the composite samples, was an effective approach to summarizing and interpreting this very heterogeneous data set representing the entire range of vegetation of southeast England. COMPCLUS required 18 seconds to read these data and 20 seconds for clustering; DECORANA ordination of the resulting composites required 11 seconds. Complete linkage and UPGMA hierarchical clusterings of the composite samples gave sensible groupings, but did not integrate and display the information as effectively as DECORANA.

The data of the third data set are 2336 relevés from chalk grassland in England (unpublished data supplied by T.C.E. Wells). Species occurring in 3 or more relevés numbered 197, and an average sample contained 28 species. *PD* distances were used, with a clustering radius to obtain 20 clusters (and several outliers from trampled sites, woodland margin, and species poor sites). The first axis of DECORANA (2 standard deviations) went from bare ground and open communities to chalk grassland and *Bromus erectus* communities to *Festuca rubra* dominance to coarse *Arrhenatherum elatius* communities; this sequence indicates increasingly eutrophic conditions. The second axis (1.3 standard deviations) separates the more neutral (less chalky) grasslands on one side and coarse ungrazed *Bromus erectus* on the other side. Reading these data required 22 seconds, and clustering 12 seconds.

For a fourth data set 3270 aquatic macrophyte samples, from a bay of 8 hectares in Otsego Lake, New York, were supplied by F.A. Vertucci. There are 20 species, with an average of 6 species per sample. For these same 3270 sites, 41 environmental parameters were measured or estimated including several major and minor nutrients in the soil and in the pore water, percent available light, percent organic matter, windiness, and percent sand, silt and clay. Initial analyses of these data indicated that the vegetation is treated conveniently as 4 types related to depth and substrate. COMPCLUS was applied to both the vegetational and environmental data matrices with a large *PD* radius to yield 4 clusters, but replicate clusterings with different random number initialization partitioned the data rather differently – the data were too continuous and homogeneous to be treated effectively in this manner. Next a smaller radius was used to obtain 20 clusters for each data matrix, and the resulting composite samples were ordinated using DECORANA and were placed in a hierarchy using complete linkage clustering and UPGMA. The DECORANA analysis of the vegetation showed a primary gradient (3.6 standard deviations) of nutrient status from poor (*Potamogeton amplifolius* and *Chara vulgaris*) to rich (*Ranunculus aquatilis* and *Ceratophyllum demersum*), and a secondary gradient (1.7 standard deviations) primarily indicating depth. The environmental data revealed the same two main gradients, with high organic matter content and N and P supply indicating rich sites, and low organic matter content and high marl content (associated with *Chara vulgaris*) indicating poor sites; depth was expressed weakly in the second axis. The vegetational and environmental patterns were largely congruent. For these data COMPCLUS was effective for initial clustering of highly redundant samples (52 seconds to read both data matrices, each with 3270 samples, and 26 seconds to cluster both). DECORANA was effective for arranging the composite samples in a 2-dimensional pattern which was readily interpretable. The hierarchical clusterings grouped similar composite samples into larger classes generally corresponding to the four types subjectively recognized.

From these tests, it appears that COMPCLUS can give useful initial classifications of large data sets, recognizing community-types generally consistent with those ecologists might define by other criteria, and summarizing in these types extensive, redundant information. The clusters or initial community-types are in turn useful either for hierarchical classification expressing their relationships to both one another and environmental gradients. The combination of initial clustering with ordination of the clusters by reciprocal averaging, or by detrended correspondence analysis, appears to be a powerful approach to extraction of patterns from complex and extensive sets of community data.

Discussion

Sneath & Sokal (1973:483) cite several papers pertaining to rapid clustering. However, these methods are either monothetic or else not particularly fast (not linear with respect to the number of samples).

A computer classification of samples and species into a Braun-Blanquet format is given by Češka & Roemer (1971). A program with similar goals and exceptional economy of computer time and memory is TWINSPAN (Hill et al. 1975, Hill 1979b). TWINSPAN is particularly useful when simultaneous clustering of samples and species is desired. However, arranged tables become awkward for displaying results with more than a few hundred species and samples.

Similar algorithms to COMPCLUS include 'agrégation autour de centres fixes' and 'agrégation autour de centres variables' (Benzécri et al. 1973; also see Diday 1971 and Briane et al. 1977), TABORD, and CLUSLA having a fast initial clustering (Janssen 1975, van der Maarel et al., 1978, Louppen & van der Maarel 1979). These programs differ in details of the algorithms, but in all cases the single criterion applied is within-cluster homogeneity and the results would be in general similar. Some programs require that the user specify a number of clusters rather than a radius to set the level of clustering, but this difference is not important because these two parameters have a straightforward monotonic relationship. All these algorithms have a linear dependency on the number of samples. Of the related algorithms listed in this paragraph COMPCLUS is the fastest, and apparently it is the only program with condensed data storage.

In contrast to initial clustering, the goals in the final steps of hierarchical clustering are numerous, and relate to many properties of the data and of desirable results in a given application. Here a multiplicity of methods giving quite different results seems inevitable. Nevertheless, some standardization of methods is important for meaningful communication (Pielou 1977). Unfortunately a basis for selection of generally preferable hierarchical clusterings in phytosociological applications appears to be lacking at the present.

Although 'clustering,' as one word, might appear to be a single topic, especially in the context of classification versus ordination, it is useful to recognize the fundamental differences between initial and hierarchical clustering. The problem of initial clustering is conceptually simple, and theoretical limitations can be defined against which clustering algorithms may be judged.

Summary

Multivariate analysis of plant community data has three goals: summarization of redundancy, identification of outliers, and elucidation of relationships. The first two are handled conveniently by initial fast clustering, and the third by subsequent ordination and hierarchical clustering, and perhaps table arrangement.

Initial clustering algorithms should achieve within-cluster homogeneity and require minimal computer resources. However, algorithmic uniqueness and a hierarchy are not needed. Computing time should be proportional to the amount of data, with no higher dependencies on the number of samples. A method is presented here meeting these requirements, called composite clustering and implemented in a FORTRAN program called COMPCLUS. The computer time required for COMPCLUS clustering is on the order of the time required merely to read the data, regardless of the number of samples.

Several large field data sets were analyzed effectively by using COMPCLUS to reduce redundancy and identify outliers, and then ordinating the resulting composite clusters by detrended correspondence analysis (DECORANA). Various clusterings of the same data set can be compared using a percent mutual matches *(PMM)* index, and a matrix of such values can be ordinated for simultaneous comparison of a number of clusterings.

References

Benzécri, J.-P. et al. 1973. Algorithmes rapides d'agrégation, pp. 288–319 in: L'Analyse des Données, Vol. I, La Taxinomie (Ed. J.-P. Benzécri et al.). Dunod, Paris. 615 pp.

Bray, J.R. & J.T. Curtis. 1957. An ordination of the upland forest communities of Southern Wisconsin. Ecol. Monogr. 27: 325–349.

Briane, J.-P., J.-J. Lazare & R. Salanon. 1977. Le traitement des très grands ensembles de données en analyse factorielle des correspondances – proposition d'une méthodologie appliquée à la phytosociologie. Departement de Mathematiques, Parc Valrose, 06034 Nice Cedex, France. 38 pp.

Bruynooghe, M. 1978. Classification ascendante hiérarchique des grands ensembles de données: un algorithme rapide fondé sur la construction des voisinages réducibles. Cahiers Analyse Données 3: 7–33.

Češka, A. & H. Roemer. 1971. A computer program for identify-

ing species-relevé groups in vegetation studies. Vegetatio 23: 255–277.
Diday, E. 1971. La méthode des nuées dynamiques et la reconnaissance des formes. Rev. Statist. appl. 19 (2): 19–33.
Gauch, H.G. 1977. ORDIFLEX – A flexible computer program for four ordination techniques – weighted averages, polar ordination, principal components analysis, and reciprocal averaging, Release B. Ecology and Systematics, Cornell University, Ithaca, New York 14850. 185 pp.
Gauch, H.G. 1979. COMPCLUS – A Fortran program for rapid initial clustering of large data sets. Ecology and Systematics, Cornell University, Ithaca, New York 14850. 59 pp.
Gauch, H.G., R.H. Whittaker & T.R. Wentworth. 1977. A comparative study of reciprocal averaging and other ordination techniques. J. Ecol. 65: 157–174.
Goodall, D.W. 1978. Numerical classification, pp. 247–286 in: Classification of Plant Communities. (Ed. R.H. Whittaker). Junk, The Hague. 408 pp.
Gower, J.C. 1974. Maximal predictive classification. Biometrics 30: 643–654.
Hill, M.O. 1973. Reciprocal averaging: an eigenvector method of ordination. J. Ecol. 61: 237–249.
Hill, M.O. 1979a. DECORANA – A Fortran program for detrended correspondence analysis and reciprocal averaging. Ecology and Systematics, Cornell University, Ithaca, New York 14850. 52 pp.
Hill, M.O. 1979b. TWINSPAN – A computer program for arranging multivariate data in an ordered two-way table by classification of the individuals and attributes. Ecology and Systematics, Cornell University, Ithaca, New York 14850. 90 pp.
Hill, M.O., R.G.H. Bunce & M.W. Shaw. 1975. Indicator species analysis, a divisive polythetic method of classification, and its application to a survey of native pinewoods in Scotland. J. Ecol. 63: 597–613.
Janssen, J.G.M. 1975. A simple clustering procedure for preliminary classification of very large sets of phytosociological results. Vegetatio 30: 67–71.
Louppen, J.M.W. & E. van der Maarel. 1979 CLUSLA: a computer program for the clustering of large phytosociological data sets. Vegetatio 40: 107–114.
Maarel, E. van der, 1979. Transformation of cover-abundance values in phytosociology and its effects on community similarity. Vegetatio 39: 97–114.
Maarel, E. van der, J.G.M. Janssen & J.M.W. Louppen. 1978. TABORD, a program for structuring phytosociological tables. Vegetatio 38: 143–156.
Moore, J.J. 1972. An outline of computer-based methods for the analysis of phytosociological data. In: Grundfragen und Methoden in der Pflanzensoziologie (Ed. E. van der Maarel & R. Tüxen). Ber. Symp. Int. Ver. Vegetationskunde, Rinteln 1970: 29–38. Junk, The Hague.
Moore, J.J. & A. O'Sullivan. 1970. A comparison between the results of the Braun-Blanquet method and those of 'cluster analysis'. In: Gesellschaftsmorphologie (Ed. R. Tüxen). Ber. Symp. Int. Ver. Vegetationskunde, Rinteln 1966: 26–29. Junk, The Hague.
Pielou, E.C. 1977. Mathematical Ecology. Wiley-Interscience, New York. 385 pp.
Rohlf, F.J., J. Kishpaugh & D. Kirk. 1972. NT-SYS Numerical taxonomy system of multivariate statistical programs. Department of Ecology and Evolution, The State University of New York, Stony Brook, New York 11790. 74 pp.
Sneath, P.H.A. 1969. Evaluation of clustering methods, pp. 257–271, in: Numerical Taxonomy, Proceedings of the Colloquium in Numerical Taxonomy held in the University of St. Andrews, September 1968 (Ed. A.J. Cole). Academic Press, New York.
Sneath, P.H.A. & R.R. Sokal. 1973. Numerical Taxonomy. W.H. Freeman & Co., San Francisco. 573 pp.
Williams, W.T. 1971. Principles of clustering. Annual Rev. Ecol. Syst. 2: 303–326.
Young, F.W. & R. Lewyckyj. 1979. ALSCAL-4 User's Guide. Data Analysis and Theory Associates, P.O. Box 446, Carrboro, North Carolina 27514. 118 pp.

Accepted 23 October 1979

HIERARCHICAL LEVELS IN SYNTAXONOMY BASED ON INFORMATION FUNCTIONS*

Enrico FEOLI & Duilio LAUSI**

Istituto ed Orto Botanico, Università di Trieste, Via Valerio 30, Trieste, I 34100, Italy

Keywords:
Classification, Dendrogram, Hierarchy, Information, Phytosociology, Syntaxonomy

Numerical taxonomy (formalized cluster analysis) can be defined as a set of procedures which assign objects to groups in such a way that each group reveals some distinct characteristic of the sample population (Orlóci 1978). This means in the classification of vegetation that any cluster of relevés or synoptic tables can be syntaxonomically meaningful only if it can be characterized by a set of species.

The applications of hierarchical numerical classification methods in syntaxonomy produce dendrograms whose characteristics reflect the resemblance functions (Orlóci 1972, Goodall 1978) which control the rearrangements of the data tables and which help to find the characteristic species combination of the clusters. To be useful in the syntanomy of the Braun-Blanquet approach (Westhoff & van der Maarel 1978), it is important for numerical classifications to be capable of producing clusters at different hierarchical levels. This means that the different levels of the syntaxonomy (association, alliance, order and class) must be correlated with the resemblance measure (Kortekaas, van der Maarel & Beeftink 1976, Westhoff & van der Maarel 1978, Lausi & Feoli 1979).

The main problem in the syntaxonomical interpretation of a dendrogram is to find the correspondence between the levels in the hierarchy and the levels of the syntaxa. In the present paper we present a method that can help to solve this problem. The solution is based on the assumption that for each syntaxon maximal correlation between species and clusters exists at a given level in the dendrogram.

* This study was supported by an Italian CNR grant for 'Programma Finalizzato Promozione Qualità dell' Ambiente, Linea di ricerca: Metodologie matematiche e basi di dati'.
** We would like to thank Prof. Dr. L. Orlóci for the revision of the manuscript.

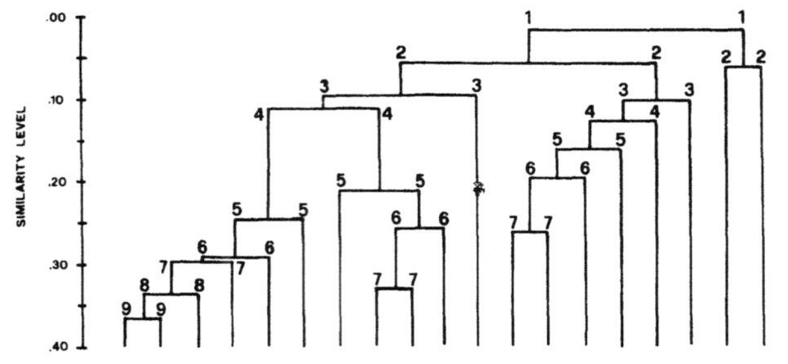

Fig. 1. Cluster analysis of European salt marshes (Lausi & Feoli 1979). The terminal branches in the dendrogram represent the alliance level. Numbers identify dichotomic levels. See explanation in the main text.

A dendrogram of relevés or synoptic tables can serve to identify clusters at different hierarchical levels either on the basis of different thresholds similarities or by considering the dichotomies (see Fig. 1). Each threshold similarity or dichotomy has a set of associated clusters and associated frequency distributions of the species within the clusters. The frequency distributions, condensed in a contingency table, describe the phytosociological classification structure. For each numerical classification based on a dendrogram, a set of contingency tables can be found at each hierarchical level from the lowest to the highest. The correlation between species and clusters can then be measured in the contingency table at any given level in the classification hierarchy.

In a study of the hierarchical classification of European salt marsh vegetation Lausi & Feoli (1979) have proposed a method, based on information functions, to estimate the correlation between species and clusters at different hierarchical levels. The method consists of the following steps:

1. Definition of dichotomic levels and hierarchical levels.

2. Rearrangement of the data into contingency tables corresponding to each hierarchical level. The scores in the tables are the frequencies of the species in the clusters.

3. Measurement of the divergence between the observed frequency distributions (F) and the expected ($F°$).

4. Computation of the correlation between hierarchical levels and the divergence of the corresponding contingency tables.

5. Formation of the optimal hierarchical classification in which the clusters correspond to the peaks in the curve obtained in step 4.

For the levels in the hierarchies we suggest to use dichotomies because in this way the classification is automatic and uniquely definite. Threshold similarities do not have such advantages since no objective basis exist for their definition. The divergence between F and $F°$, that is a measure of correlation between species and clusters of the classification, may be calculated in several ways. For this we have suggested (Lausi & Feoli 1979) the ratio between mutual information and equivocation information. According to Orlóci (1978), these two information functions are defined as follows:

$$I(F_h; F_i) = I(F_h)+I(F_i)-I(F_h, F_i) \text{ (mutual information)} \quad (1)$$

$$E(F_h; F_i) = I(F_h, F_i) - I(F_h; F_i) \text{ (equivocation information)} \quad (2)$$

F_h and F_i are resp. the principal marginal frequency distributions associated with the species and clusters in each contingency table. $I(F_h)$ is the information in the marginal distribution of the species, $I(F_i)$ is the information in the marginal distribution of the clusters, $I(F_h, F_i)$ is the information in the joint frequencies in the body of contingency tables. These quantities are fixed multiples of Shannon's entropy. According to (1) and (2), the ratio expressed by:

$$R(F_h; F_i) = I(F_h; F_i) / E(F_h; F_i) \quad (3)$$

is a relative measure of divergence, it can be compared between contingency tables of different dimensions. In the extreme case of $E(F_h; F_i) = 0$, $R(F_h; F_i)$ is indeterminate. Another simple measure of relative divergence is Rajski's (1961) metric (cf. Orlóci 1978):

$$d(F_h; F_i) = E(F_h; F_i) / I(F_h, F_i)$$
$$= 1 - I(F_h; F_i) / I(F_h, F_i) \quad (4)$$

with limits 0 for maximal and 1 for minimal divergence. It is easy to demonstrate that

$$(1 - d(F_h; F_i)) / d(F_h; F_i) \quad (5)$$

is the ratio $R(F_h; F_i)$. It has been proposed because it is a more sensitive parameter than Rajski's metric divergence. $R(F_h; F_i)$ ranges between zero and infinite. It is zero when the mutual information is equal to zero (minimal divergence), it is infinite when mutual information is equal to joint information (maximal divergence).

In order to limit the range of divergence between 0 and 1 we propose the following function:

$$D(F_h; F_i) = I(F_h; F_i) / (2I(F_h, F_i) - I(F_h; F_i)) \quad (6)$$

Table 1. Values of $R(F_h; F_i)$, $D(F_h; F_i)$ and $d(F_h; F_i)$ corresponding to the dichotomic levels of the dendrogram in Fig. 1. See explanation in the main text.

Dichotomic level	$R(F_h; F_i)$	$D(F_h; F_i)$	$d(F_h; F_i)$
1	0.178	0.081	0.849
2	0.348	0.148	0.742
3	0.490	0.197	0.670
4	0.476	0.192	0.678
5	0.414	0.171	0.707
6	0.420	0.174	0.704
7	0.416	0.172	0.706
8	0.400	0.165	0.713
9	0.380	0.160	0.724

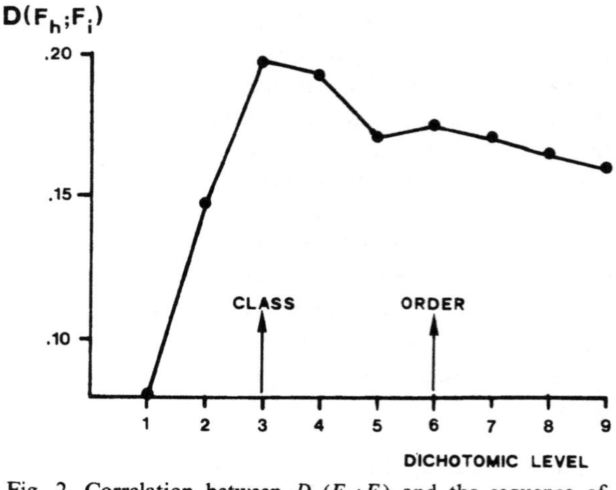

Fig. 2. Correlation between $D(F_h; F_i)$ and the sequence of dichotomic levels in the dendrogram (Fig. 1). Arrows indicate the dichotomic levels at which $D(F_h; F_i)$ reach maxima. See explanation in the main text.

The relation of this function to Rajski's divergence is the following:

$$D(F_h;F_i) = (1-d(F_h;F_i))/(1+d(F_h;F_i)) \quad (7)$$

The greater sensitivity of $R(F_h;F_i)$ and $D(F_h;F_i)$, compared to $d(F_h;F_i)$, is due to the concavity of the functions (5) and (7).

The comparison between $R(F_h;F_i)$, $D(F_h;F_i)$ and $d(F_h;F_i)$ is presented in Table 1. They have been calculated for each contingency table originated at each dichotomic level of the dendrogram in Fig. 1. This dendrogram has been obtained by average linkage clustering (Orlóci 1978) applied to the synoptic tables of some European salt marsh types (Lausi & Feoli 1979). In Fig. 2 only the $D(F_h;F_i)$ values are plotted against each dichotomic level of the dendrogram (Fig. 1) to show the peaks which have been used to establish the order and class level of the European salt marsh syntaxonomy.

The identification of syntaxonomical levels in such a way is consistent with the definition of cluster analysis, since each maximum of divergence identifies a unique classification structure in which clusters are optimally correlated by species. The proposed method is considered as a further extension to cluster analysis in phytosociology.

Summary

A method for defining syntaxa in synsystematics in an objective way is described. The method is based on measures of divergence of contingency tables obtained for the frequency distributions of species in given clusters, defined at different hierarchical levels of the dendrogram. The divergence is measured as an information function. The method is useful to compare different classifications.

References

Goodall, D.W. 1978. Sample similarity and species correlation. In : R.H. Whittaker (ed.) Ordination of plant communities, p. 99–149. Junk, The Hague.

Kortekaas, W.M., E. van der Maarel & W.G. Beeftink. 1976. A numerical classification of European Spartina Communities. Vegetatio 33 : 51–60.

Lausi, D. & E. Feoli. 1979. Hierarchical classification of European salt marsh vegetation based on numerical methods. Vegetatio 39 : 171–184.

Orlóci, L. 1972. On objective functions of phytosociological resemblance. Am. Midl. Nat. 88 : 28–55.

Orlóci, L. 1978. Multivariate analysis in vegetation research. 2nd ed. Junk, The Hague, pp. 451.

Westhoff, V. & E. van der Maarel. 1978. The Braun-Blanquet approach. 2nd ed. In : R.H. Whittaker (ed).), Classification of plant communities, p. 287–399. Junk, The Hague.

Accepted 1 November 1979

NUMERICAL APPROACHES TO LAKE CLASSIFICATION WITH SPECIAL REFERENCE TO MACROPHYTE COMMUNITIES*

Sven JENSÉN[1] & Eddy VAN DER MAAREL[2]

[1] Department of Plant Ecology, University of Lund, Ö. Vallgatan 14, S-223 61 Lund, Sweden
[2] Division of Geobotany, University of Nijmegen, Toernooiveld 1, 6525 ED Nijmegen, The Netherlands

Keywords:
Boundary zone, Classification, Community, Community complex, Lake, Life-form, Macrophyte, Ordination, Synusia, Species performance, Sweden

Introduction

Aquatic vegetation is nearby always markedly zoned with up to 8 zones (sensu Sjörs 1971). Only in a few cases the macrophyte composition changes more or less continuously. In the latter case it is rather difficult to classify the lakes on a plant community basis including all life-forms, notably *helophytes* (*graminids* + *herbs*), *nymphaeids*, *elodeids*, *isoetids* and *lemnids*.

Aquatic vegetation often has a low species diversity with one or a few dominants and a few species of little quantitative importance. In addition there are great differences between oligotrophic and eutrophic lakes as regards the quantitative composition of macrophytes. Barkman (1978) pointed out the great difficulties in classifying lakes on the basis of their macrophytes, because of the partial independence of different lifeform synusiae (cf. Lillieroth 1950, Luther 1951).

Very few applications of numerical classification and ordination of aquatic vegetation are found in the literature (see Seddon 1972, Howard-Williams & Walker 1974 and Jensén 1978, 1979). These applications differ in many respects. For example Seddon (1972) has ordinated lakes only on species presence or absence while using all species together. Jensén (1979) described the floristic composition in a more quantitative way by using cover and frequency data. However, Jensén (op. cit.) classified lakes on the basis of only two life-forms (*helophytes* resp. *nymphaeids*) and one life-form group (*elodeids* + *lemnids* + *isoetids*), where these refer to different parts of the lake. Both the methods used by Seddon (1972) and Jensén (1979) neglect the internal variation in the lake. Alternatively one may classify lakes on the basis of their macrophyte communities (see e.g. Blomgren & Naumann 1925, den Hartog & Segal 1964, Zutshi 1975, Mäkirinta 1978). or more simply on the occurring areas characterized by certain dominants (cf. Jensén 1977).

The aim of this paper is to elucidate some principles of numerical approaches to the classification of aquatic vegetation and of lakes as entire ecosystems. Our considerations are largely based on data from 50 lakes with totally 262 transects and ca. 10.000 squares (1 sq. m. each) (cf. Jensén 1979). All lakes are situated in the southernmost province Skåne (Scania) in Sweden. The size of the lakes varies between 5 and 830 ha. For further information, see Jensén (1977, 1978, 1979). Numerical methods used are described in Jensén (1978, 1979).

Different approaches to lake classification

Some different ways to approach aquatic species are presented in Fig. 1. Any choice of approach must be related to the aim of the investigations. If the aim is e.g. to produce a physiognomical map of the different vegetation areas it is sufficient to estimate the distribution of the dominant species, whilst more detailed studies of the rest of the species can be excluded. But if we want to include interactions between e.g. *isoetids* and *helophytes* or *nymphaeids* in our study much more detailed investigations must be performed.

A main division is that into species or life-form approach and community and/or synusial approach. Another

* Nomenclature follows Lid (1974).

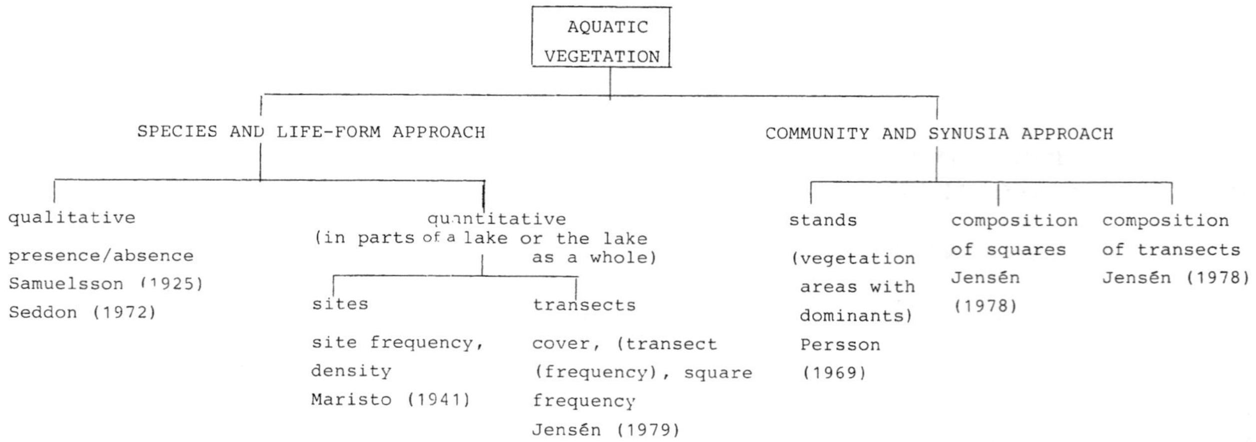

Fig. 1. Different approaches to lake classification.

divergence is related to wether we want to regard the lake as a whole or if we wish to treat the internal variation of a lake. In a regional classification it is often preferable to treat the single lakes as units at least as the first step. But if we want to find a relation between species composition and geomorphology of the area, it can be worthwhile studying different sites or transects in the lakes.

Estimation of species performance

Different measures can be chosen to describe the performance of species, e.g. cover, abundance, frequency, biomass, number of shoots, vitality and fertility. For aquatic vegetation the common phytosociological analyses using a cover estimation or a cover-abundance scale such as those of Domin and Braun-Blanquet are appropriate (see Jensén 1978). However, such values should not be used when comparing life-forms with strongly different sizes, e.g. *Phragmites communis* (2–4 m in height) and *Lobelia dortmanna* (0.05–0.15 m). They can both attain cover values of 100 %, but reach very different biomass values.

Where the indices in the different numerical programmes often are abundance- or cover-weighted, it is useful to make some sort of transformation of the quantitative values, e.g. a logarithmic one of the form (1 + (log cover %)) (Jensén 1978), or to apply an ordinal transformation to combined estimation scales (van der Maarel 1979, 1979a). Finally we can only consider presence or absence of species.

Sampling methods

The simplest analytical method on which the characterization of a lake can be based is the floristic inventory of the lake. This method was used in most of the classical Scandinavian lake studies (see Jensén 1977, 1979 for a review) but also by Seddon (1972). As was elaborated by Jensén (1977) a much more satisfactory method is the sampling of small contiguous quadrats along belt transects running down from the shores of a lake perpendicular to the shoreline. Morever the belt transects can be chosen as to number and location in order to represent the entire vegetational variation in the lake. For each quadrat the species composition is determined, preferably with an estimation of species cover and abundance. In subsequent treatments we can use species parameters such as overall presence, transect frequency (i.e. number of transects in which a species is found), quadrat frequency, overall cover % and characteristic degree of cover (Malmer 1962), i.e. average cover calculated only for the squares in which a species does occur.

The species and life-form approach

Lakes or parts of lakes are characterized only by a list of species, either scored as present or estimated as regards their quantitative occurrence in sites, or in squares arranged along transects (Fig. 1). Because of the considerable ecological differences between the life-forms occurring in lakes a differentiated treatment according to life-form is appropriate. Jensén (1979) showed how Scanian lakes can be classified in a number of *helophyte-*, *nymphaeid-* and *elodeid + isoetid + lemnid-* types and how these types are mutually related.

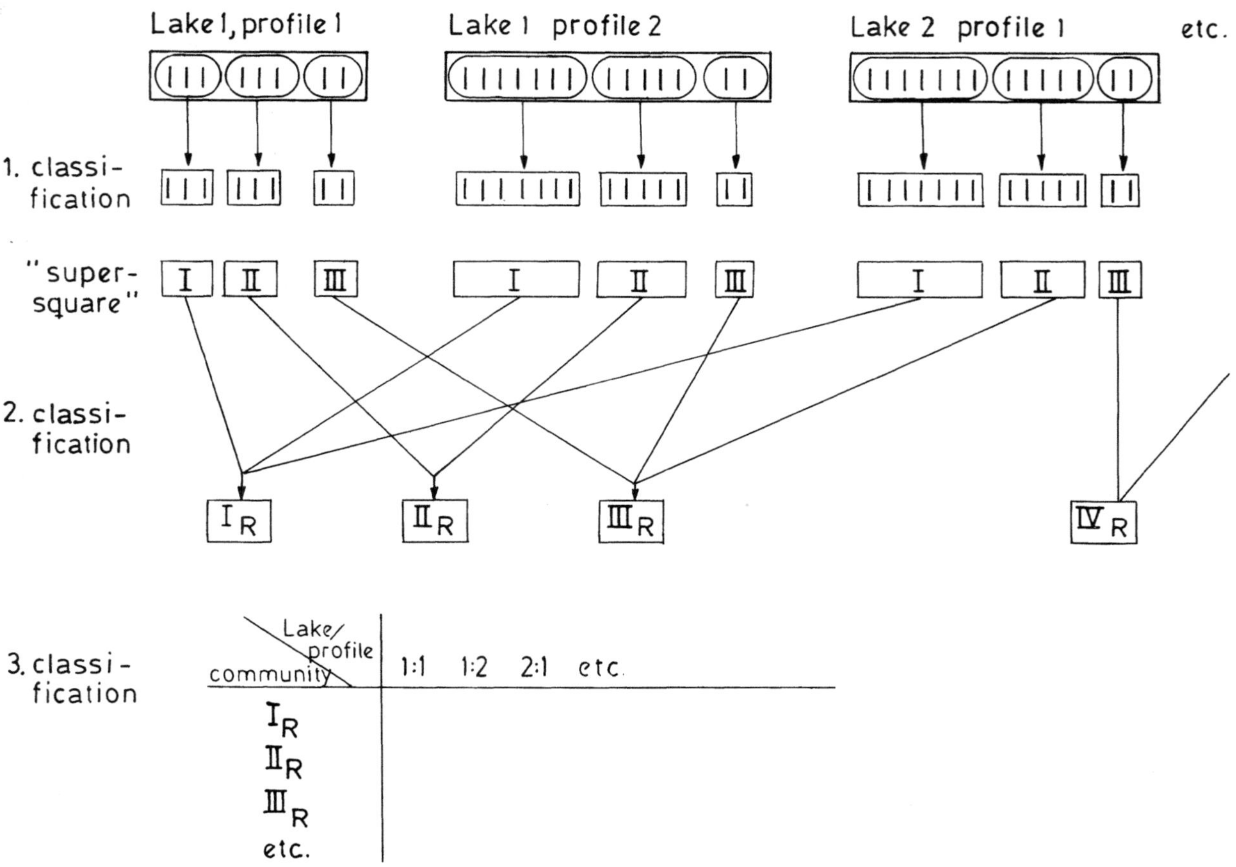

Fig. 2. Stepwise classification of lakes through classification of vegetation in order to reach a final classification of lakes on the basis of macrophyte community composition. Further explanation in text.

The community and synusiae approach

A phytocoenological approach, i.e. a lake classification based on aquatic vegetation, can be performed in three steps (Fig. 2):

1. Primary classification of the squares (sites, stands) per lake (or transect) in order to find local community types, characteristic for each lake (transect).
2. Secondary classification of the clusters arrived in the primary classification to find general community types referring to all lakes involved.
3. Classification of the lakes (or transects) on the basis of their macrophyte communities.

The primary classification is based on cover values, transformed to (1 + log cover %), which gives an acceptable balance between quantity and quality (cf. Jensén 1978, van der Maarel 1979). In the approach presented here the squares have been treated independent of their position in different transects.

In the second classification the primary clusters are used as 'super squares'. It is possible to use both characteristic cover and frequency to indicate the occurrence of species in clusters. Because some clusters are very small in size (2–3 squares) and others are great (20–30 squares) we have chosen characteristic cover as quantitative measure. Some of the secondary clusters are identical to the primary ones, while others arise from fusions of primary clusters.

In the third classification the lakes are classified according to the occurrence of secondary clusters. Here only presence-absence data were used, but of course we could use a quantitative measure as well.

Both the primary and secondary classification can be performed on a synusial basis (one or a few life-forms) as well as a phytocoenological one (all life-forms together).

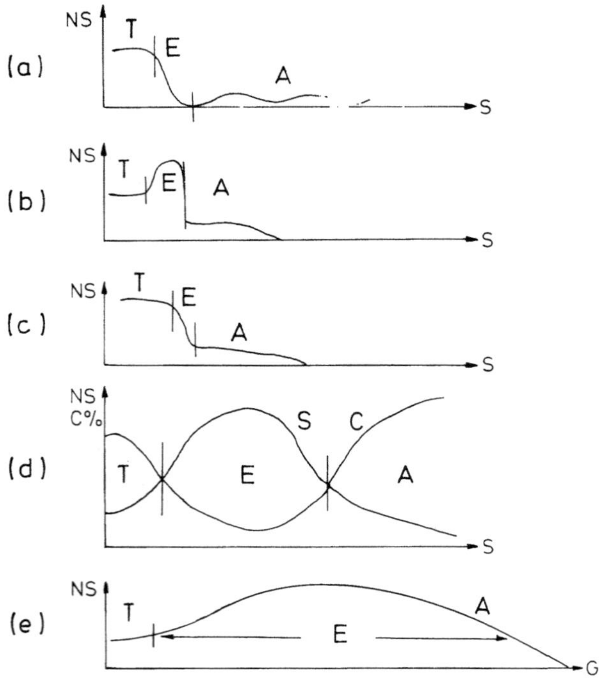

Fig. 3. Different transition zones between land and lake and related profiles of species richness. a. Typical ecotone situation with narrow transition zone in which the species number reaches a local minimum. b. Ibid., but with a first increasing and then decreasing number of species. c. Very narrow ecotone zone, connected with an abrupt fall in the slope. d. Ecocline situation with rapid rise and fall in the number of species and a large number of species as compared with the zones at either side. The pattern of total cover in the same situation shows the opposite course. e. Gradual transition in which the ecocline situation merges into the species-poor deeper aquatic environment. Symbols: A = aquatic, C % = cover in %, E = ecotone, G = ecocline, NS = number of species, S = slope, T = terrestrial.

We present an example based on *helophytes*. When only one life-form is present in the squares compared these two approaches coincide (cf. Fig. 4 squares 10–17). This three-step classification procedure is similar to both the vegetation complex approach as followed in landscape-ecological studies (e.g. Seibert 1968, van Gils & Huits 1978, Stumpel & Kalkhoven 1978) and the sigma-association approach as introduced by Tüxen (1973) and elaborated by Géhu (e.g. 1977, see Tüxen 1978 for application; see also Mériaux 1978, Mériaux & Géhu 1978). In both approaches local plant communities are grouped to spatial complexes. In the vegetation complex approach the complexes are related to more or less uniform landscape-ecological units (ecotopes, geotopes, ecochores). In the sigma-approach the complex units are primarily based on the coherence between local plant communities, which are recorded in loosely defined areas, exactly as plant species are recorded in a relevé. Morever the local communities are all interpreted as syntaxonomical units, while the complexes themselves are also placed in a formal hierarchial system of 'sigma-syntaxa'.

We do not intend to discuss the pros and cons of the latter system, but would still remark that the approach we present here, i.e. a step-wise objective establishment of complex units, is very much needed in sigma-taxonomy.

Plant communities or synusiae

The associations between species are rather loose in aquatic conditions, and different life-form layers may develop independently (Almquist 1929, Lillieroth 1950, Luther 1951). This may lead to a synusial approach (Barkman 1978). For lake vegetation this approach was adopted by Almquist (1929) and also by Mäkirinta (1978), but the latter author described combinations of synusiae (= phytocoenoses). It is thus possible to distinguish three different approaches:

1. Establishment of communities in relation to all layers
2. Establishment of synusiae in relation to one layer, e.g. *lemnids*.
3. Establishment of communities in relation to a combination of layers, e.g. *helophytes* + *isoetids*.

Aquatic vascular species often appear as one-species- or as one-layer-communities over large areas. Then it is difficult to decide to use the synusia or phytocoenosis approach. An environment covered with only one layer may have the potention for other ones. As soon as species interactions or succession are considered the synusial combination is of interest because we may predict future developments along a certain gradient or in the lake as a whole. Thus we adopt the concept of an aquatic phytocoenosis as a synusial combination. If such a combination consists of one layer only the others are considered to have an actual representation of zero (cf. squares 10–17 in Fig. 4).

Types of lake zonation and boundary zones

Lakes can also be characterized by the type of vegetation zonation along the shores. This approach has hardly been developed so far (cf. Jensén 1977). In Fig. 3 we summarize some general types of zonation which we discuss in terms of gradients and boundary zones (cf. Scott 1974).

SQUARES		1	2	3	4	5	6	7	8	9	10	11	12	13	14	15	16	17
STONES (>20 sq.cm) %		20	40	20	70	95	80	50										
WATER AREA %		-	-	-	-	1	50	100										
WATER DEPTH cm		-	-	-	-	-	-	14	30	50	70	80	110	120	131	144	150	164
CAREX NIGRA	(H)	5																
JUNCUS FILIFORMIS	(H)	2																
LYSIMACHIA THYRSIFLORA	(H)	2																
MOLINIA COERULEA	(H)	3																
MYRICA GALE	(S)	60																
GLYCERIA FLUITANS	(H)		2															
JUNCUS BULBOSUS	(I)	1	1	1														
LYSIMACHIA VULGARIS	(H)	2	1	5	4													
MENTHA ARVENSIS	(H)	40	4	10	1													
AGROSTIS STOLONIFERA	(H)	5	2	2	2	2												
CAREX ROSTRATA	(H)	2	3	10	3	20												
HYDROCOTYLE VULGARIS	(H)	30	20	20	20	3												
RANUNCULUS FLAMMULA	(H)	3	2	2	3	1												
LITTORELLA UNIFLORA	(I)	10	50	60	30	3	1											
SCIRPUS PALUSTRIS	(H)	3	4	4	5	3	30	1	2	1								
LOBELIA DORTMANNA	(I)					1	5	2										
EQUISETUM FLUVIATILE	(H)					1	1	2	10	10	4	2	-	1	1	-	1	1
MYRIOPHYLLUM ALTERNIFL.	(E)						1	5	-	1								
ALISMA PLANTAGO - AQUATICA	(H)							1	4									
ISOETES LACUSTRIS	(I)							2	3	1								
NUMBER OF H(+S)		12	8	7	7	6	2	3	3	2	1	1	-	1	1	-	1	1
" OF E+I		2	2	2	1	2	3	3	1	2	0	0	-	0	0	-	0	0
TNVS		14	10	9	8	8	5	6	4	4	1	1	0	1	1	0	1	1
SBCQ:SR			60	90	89	60	44	57	67	60	25	100	0	0	100	0	0	100
DSBCQ:ED			6	1	1	4	5	3	2	2	3	0	1	1	0	1	1	0

Fig. 4. Composition of quadrats in a belt-transect corresponding to the ecotone-type shown in Fig 3c. The lake is of the oligo-mesotrophic type with brownish water (see Jensén 1979). The similarity figures between contiguous quadrats are values X100 of the similarity ratio for presence/absence data. Symbols: S = shrub, H = *helophyte*, E = *elodeid*, I = *isoetid*, TNVP = total number of vascular plants, SBCQ:SR = similarity between contiguous quadrats with index similarity, DSBCQ:ED = dissimilarity between contiguous quadrats with index Euclidean distance.

The boundary between land and water is usually considered an ecotone (see Sjörs 1971) or a limes convergens zone (van Leeuwen 1965, van der Maarel 1976). As van der Maarel (1964, see 1976) emphasized the ecotone concept should be restricted to real tension zones with strongly fluctuating conditions. Such conditions are usually found along lake shores where the fluctuations in water table can cause an instable environment (Fig. 3. a-b), where the species number generally decreases (Fig. 3a). On sheltered shores, however, the opposite may happen, when the number of species increases in the transition zone (Fig. 3b). For gradual transitions from one environmental constellation to another, with no interfering strong fluctuations, the limes divergens concept is appropriate, whereas the term ecocline could be used as a synonym here. The ecotone can become very narrow and more or less disappear on e.g. steep rocky parts of the lakes or on the border between lake and bog (Fig. 3c). It may happen that within an ecotone small 'islands' of aquatic habitat appear with *Lemna minor* and/or *Hydrocharis morsus-ranae*.

The number of species increases within the ecocline zone, which as a whole has more species than the zones at either side. The same pattern is shown by the different life-forms. Apparently the complexity of conditions in this zone of the lake-shore increases. The total cover shows the opposite pattern in relation to the gradient from land to lake (Fig. 3d). Fig. 3e shows the conditions on a very gently sloping shore. Here great overlaps between different species will appear, when the conditions will be favourable for many species.

Fig. 3d presents a typical ecocline profile with a gradual course of the slope and a typical bell-shaped species richness curve as often described for such situations (e.g. van der Maarel & Leertouwer 1967, van der Maarel 1971, see Whittaker 1978). Usually many species are involved but none of them is dominating.

The vegetation zonation related to the type of environmental boundary can be easily demonstrated by means of a phytosociological table in which the quadrats are arranged according to their position along the gradient. Fig. 4 gives an example for a belt transect related to an environmental zonation as presented in Fig. 3c. The number of vascular species decreases along the transect. In squares 1–5 the *helophytes* outnumber the *elodeids* and *isoetids*, in squares 6–9 they balance each other and in sq. 10–17 it is only the *helophyte Equisetum fluviatile* which occurs with scatterd shoots.

The changes in composition along the transect can be followed through the result of a clustering process as described above, (see the lines in Fig. 4). Evidently the clustering causes discontinuities which are not present in the real sequence of quadrats. Square 5 for example forms a cluster of its own but is related to both the neighbouring quadrats 4 and 6. On a presence-absence basis it is more similar to square 4, but as regards shifts in cover, e.g. *Hydrocotyle vulgaris* and *Littorella uniflora*, it is less similar. The compositional changes along a gradient can therefore better be reflected in an ordination diagram (Jensén 1978; see also van der Maarel & Leertouwer 1967 for a discussion of compositional changes).

A method for detecting a compositional profile along environmental gradients which is more direct than numerical classification or ordination is to calculate the resemblance between adjacent quadrats. Fresco (1972), van der Maarel (1976) and Danielsson (1977) have used such 'running similarity' measures. The resulting 'differential profile' (van der Maarel 1976) will not always show clear drops or peaks, but in lake transects they are usually easy to detect (Fig. 4). Because of the dependence of the profile on both qualitative and quantitative shifts and on the total number of species involved, the resemblance may be calculated both as similarity and as distance and for both numerical and presence/absence data.

Species Clustersize	9	6	7	9	12	9	6	6	4	4
Elodeids										
Juncus fluitans	6863323F2	F3331F	....	...F.....	.....F.....F	F....F...	......	F3....	....	.FF.
Myriophyllum altern.	...F.F2..	......	....	....	.........F..	....1....	......	......	....	FF11
Utricularia vulgaris	...F..FF.	F.....	.FF....	....	....	......F..	......	......	....	....
Utricularia intermedia	....	......	1	....	....	.......F.	......	......	....	....
Isoetids										
Isoetes lacustris	....	......	2F1F333	..F......	...F.....FF.	.11FF2F.1	...F.F	...FF.	....	2242
Juncus bulbosus	....	......	....	....	.F.F..2F.F..	2.F.F.1..	F.....	F2....	F.FF	....
Littorella uniflora	....	....F.	..F....	F...1...F	F1.FFFF..FFF	63F23F436	633432	522311	...F	.F..
Lobelia dortmanna	....	F33553	11FF441	432384526	431455625268	32F22F3FF	333324	133355	1411	F...
Nitella sp.	........F	......	....	....	..........F.	....	......	......	....	....
Ranunculus reptans	....	......	.F.....	....	21...11.FF..	....	......	......	....	....
Subularia aquatica	....F...F	......	....	....	....F....FF.	....	......	....F.	....	....
Helophytes										
Carex rostrata	F........	..F...	....	2.FF2.F2F	.F.111FFFF.F	....	......	..F...	...F	....
Equisetum fluviatile	.FF.F....	.F....	F...FF.	F.F.1FFFF	FF.FFFF.FFFF	.F.......	F....F	.F..FF	....	1FFF
Lysimachia thyrsiflora	....F..FF	......	....	....	.F.F.FF.....	....	..F...	......	F1FF	....
Scirpus multicaulis	...F.....	..1FFF	....	....	....	....	......	......	....	....
Scirpus palustris	.....F...	F1FF..	...FF..	1FF...FFF	1FF12FFFF.FF	F1FFF1F22	F.....	......	FFFF	....
Phragmites communis	......F..	......	.FF....	....	..FFFF2FFFF1	....F....	......	FFF1FF	....	....
Scirpus lacustris	....	......	......F	....	.......FF...	....	.F.FFF	...1FF	....	....
Nymphaeids	..F..1.F.	......	....11.	......F.F	....	....	......		F...	....

Fig. 5. TABORD-classification of 72 primary clusters from 19 lakes in S Sweden. Resulting clusters contain *isotids, elodeids* and *helophytes* (and scattered *nymphaeids*). Significant species are underlined. For symbols, see Fig. 7.

Some results of the community and synusia approach to South Swedish lakes

Community classification

Fig. 5 presents an example of a classification of 72 primary clusters from 19 lakes treating all present life-forms together. The program used here is TABORD (van der Maarel et al. 1978, Persson 1977), with the similarity ratio as resemblance function, a fusion level of 0.80 and a threshold value of 0.40. In several cases very heterogeneous clusters result, depending on the zoned conditions and the overlapping distribution of species. In many cases such overlaps occur between different life-forms. Swan (1970) suggested to study only shorter intervals, with monotonic species phases but this is not useful when an extensive material is to be treated.

In the South Swedish lakes taken as an example in this paper widely distributed phytocoena of *helophytes* (pure), *helophytes* + *isoetids* (+ *elodeids*) and *helophytes* + *nymphaeids* have been distinguished. In the associations between helophytes and isoetids (Fig. 5) the *helophytes* usually have low cover values (≦ 20 %) while the *isoetid* cover values varies a lot (10–80 %).

In most cases an open *helophyte* layer occurs with *Carex rostrata, Equisetum fluviatile, Lysimachia thyrsiflora, Phragmites communis* and *Scirpus palustris*. The most common *isoetids* are *Isoetes lacustris* and especially *Littorella uniflora* and *Lobelia dortmanna*. It is obvious from the classification results that the composition of *helophytes* varies with the *isoetid*-combinations.

Occasionally *nymphaeids* and *isoetids* have been found together but the position of the *nymphaeids* is rather variable, so it is very difficult to obtain true associations between the two life-form types (cf. Almquist 1929, Barkman 1978). The *isoetids* are completely lacking in areas with dense *nymphaeids* but in gaps between the *nymphaeid* stands here and there e.g. *Nitella spec., Potamogeton gramineus* and *Subularia aquatica* appear.

The different position of *helophytes* and *nymphaeids* can be shown in a principal components ordination of 19 lakes on the basis of their *isoetid* and *elodeid* communities (Fig. 6). The lakes with pure *Isoetes lacustris*- mats are grouped to the left in the graph (dimension 1). The four quadrants (I-IV indicated on the graph) are characterized in the following way:

I. Lakes having sparse *Lobelia dortmanna* (or locally abundant). *Isoetes lacustris* is absent.

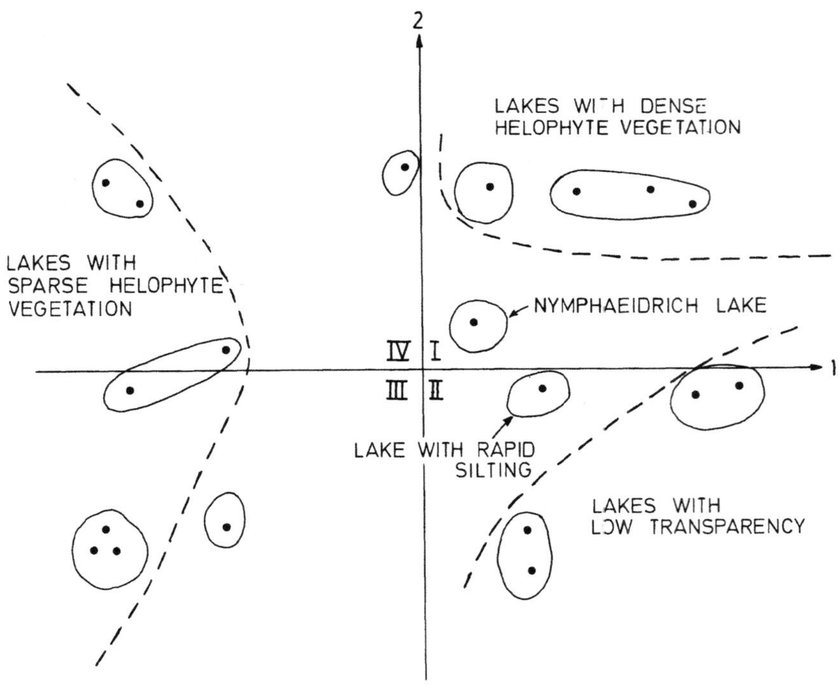

Fig. 6. PCA-ordination of 19 lakes on the basis of their *isoetid* and *elodeid* communities (program ORDINA, see Jensén 1978).

Species Clustersize	17	8	8	10	4	10	10	9
Nymphaeids								
Nuphar luteum	FF1FFFFF21FF12131	FFF1FF13	13161113	.......F..	F.11	....	FF11FF4115	.F....F..
Nymphaea alba	....	....	...F....	F1F3431252	F113	....	....	.....F..3
Potamogeton natans	FFF...F1FF.FFF.F.	.F......	31121432	......F...	FFFF	F111464261	.......F2.	F11F2.312
Helophytes								
Carex rostrata	....	.F......	....	..F.....F.	....	....F.....	....	....
Carex lasiocarpa	....	....	....	2.F1......	....	.......F..	....	....
Equisetum fluviatile	.FFF.F.F.F...1.FF	251FFFF1	431FF11F	FFFFFFFFFF	FF1F	F2FFFFFFFF	..F.FF2FF.	.FFF.F...
Phragmites communis	.....F.....F..F..	.F.....1	....	.........F	2F..	.......FF.	4.F3F.FFF.	F........
Scirpus lacustris	34311531511111111	....	F.......	....	21F2	....	1.....FFF.	1....FFF.
Sparganium simplex	..........F......	.......F	......F.	....	F...	....F....F	....	....
Typha angustifolia	....	....	....	....	....	....	F5235FFFF2	432135411
Typha latifolia	....	....	....	....	....	..1...F...	....	....

Fig. 7. TABORD-classification of 76 primary clusters containing *nymphaeids* and *helophytes* from 48 lakes in S Sweden. Significant species are underlined. Symbols (cover values): F = 1–5%, 1 = 6–14%, 2 = 15–24%, ... 8 = 75–84% and 9 = 85–100%.

II. Limited distribution of *Isoetes lacustris* – communities and *Lobelia dortmanna* – communities (locally both species are abundant).

III. *Isoetes* and *Lobelia* are abundant both associated and in separate communities. In addition the *helophyte* vegetation is sparse-to very sparse.

IV. Lakes dominated by *Littorella uniflora* – and *Lobelia dortmanna*-communities. *Isoetes* is always present but less abundant than in III.

Helophytes and *nymphaeids* are found together in a number of lakes as follows from Fig. 7, showing a TABORD-classification of 48 lakes. The primary clusters from the 48 lakes are classified in a secondary classification at a fusion level of 0.80 and a threshold value of 0.40. Especially in bays the two life-forms (*helophytes* and *nymphaeids*) grow together. It is possible to regard this situation as a successional stage when the *helophytes* can expand into the areas of *nymphaeids* in rather shallow water.

Scirpus lacustris, *Equisetum fluviatile* and *Typha angustifolia* are more often combined with *nymphaeids* (*Nuphar luteum*, *Nymphaea alba* and *Potamogeton natans*) than are the other *helophytes* (e.g. *Carex rostrata*, *Phragmites communis* and *Typha latifolia*). The explanation may be that *Scirpus lacustris* and *Equisetum fluviatile* usually tolerate the greatest water depths and build up the front of the entire helophyte zone where they meet the *nymphaeids* in their innermost parts.

A conclusion to be drawn from the phytocoenological approach is that it is usually possible to trace associations between life-form-types while it is more difficult to find exact species combinations (communities).

Synusial classification of the lakes on the basis of their helophyte communities

In the first classification step the local communities of each of the 50 lakes were distinguished and in the second step the general community-types. The classification of the 50 lakes with the program TABORD on the basis of the general community-types resulted in 12 community spectrum types at a fusion level of 0.80 (Fig. 8). Three groups (1–3) were formed from the 12 types on the basis of the greatest similarity between the types, i.e. for example types I and III are both most similar to type II, and type II is most similar to type III. This only means that a certain type is never more similar to a type outside the group than to the most similar type within the group.

The first three types (group 1) are characterized by the presence of an *Equisetum fluviatile*-community and *Carex*-comm. *Phragmites*- and *Scirpus lacustris*- comm. are more or less lacking. The second group (types IV-VI) is characterized by the presence of *Equisetum*- and *Carex rostrata* – as well as *Phragmites communis* – and *Scirpus lacustris*-comm. The *Typha angustifolia* – comm., which is lacking in group 2, characterizes the third group (types VII-XII), together with *Phragmites communis*- and *Scirpus lacustris*-comm., while here *Equisetum*- and *Carex*-comm. are less frequent. Group 1 and types XI and XII in group 3 have very few communities while group 2 and the rest of the types in group 3 have many more communities. The explanation is probably that group 1 and the lakes belonging to types XI and XII (group 3) are small and have very homogeneous surroundings and shore conditions.

Community: EQ	CR	CL	SP	LY	SM	SL	PC	TA	CE	ST	TL	I	II	III	IV	V	VI	VII	VIII	IX	X	XI	XII
GROUP												1			2			3					
TYPE												I	II	III	IV	V	VI	VII	VIII	IX	X	XI	XII
NL												6	5	3	8	8	2	4	2	2	4	4	2
X	.	.	.	.	.	.	.	.	.	.	.	100	80	67	100	100	100	+	+	+	+	-	-
X	X	.	.	.	.	.	.	.	.	.	.	83	+	-	+	63	100	-	-	+	-	-	-
.	X	.	.	.	.	.	.	.	.	.	.	-	60	100	-	-	-	-	-	-	-	-	-
.	.	X	.	.	.	.	.	.	.	.	.	-	100	-	+	+	-	+	-	-	-	-	-
.	.	.	X	.	.	.	.	.	.	.	.	-	+	+	50	-	100	100	-	-	-	-	-
X				X	.	.	.	.	.	.	.	-	-	67	-	+	-	-	-	-	-	-	-
X			X	.	.	.	.	.	.	.	.	-	-	+	50	-	50	-	-	-	+	-	+
.	.	.	.	.	X	.	.	.	.	.	.	-	-	+	+	-	-	-	-	-	-	-	-
.	X		X	X	.	.	.	.	.	.	.	-	-	-	100	-	-	-	-	-	-	-	-
X	X		X	.	.	.	.	.	.	.	.	-	-	-	63	+	-	-	-	-	-	-	-
X	X					X	.	.	.	.	.				+	75	50						
X						X	.	.	.	.	.				+	63	-						
X	X			X		X	X	.	.	.	.				-	+	50	-	-	-	-	+	-
.	X	X					X	.	.	.	.				+	+	100						
X	X	X					X	.	.	.	.				63	-	50	-	-	-	+	+	-
						X	.	.	.	.	.				63	100	100	100	100	100	100	+	-
						X	X	.	.	.	.				-	+	50	50	100	-	-	75	-
						.	X	.	.	.	.				63	50	-	50	100	100	100	-	100
						.	X	X	.	.	.							75	50	100	100	50	-
						X	X	X	.	.	.							50	100	50	75	-	-
							.	X	.	.	.							-	100	100	-	-	-
							X		X	.	.							-	-	-	50	100	-
										X	.							-	50	-	-	-	100
											X				-	+	-	-	-	-	+	-	+
TOTAL NUMBER OF COMMUNITIES												2	3	3	8	6	10	6	7	5	5	3	2

Fig. 8. Classification of 50 lakes in S Sweden on the basis of the presence or absence of *helophyte* community-types. The lake frequency per type is indicated. The species which characterize the community are indicated by X. Symbols: CE = *Carex elata,* CL = *Carex lasiocarpa,* CR = *Carex rostrata,* EQ = *Equisetum fluviatile,* LY = *Lysimachia thyrsiflora,* PC = *Phragmites communis,* SL = *Scirpus lacustris,* SM = *Scirpus multicaulis,* SP = *Scirpus palustris,* ST = *Scirpus tabernaemontani,* TA = *Typha angustifolia,* TL = *Typha latifolia.* NL = number of lakes. + = single occurrence of a community. These communities are not included in 'Total number of communities' in Fig. 8 below.

Fig. 9 presents the similarity spectrum of the 12 lake-types involved. It is clear that most similarity values are rather low, indicating a rather independent phytocoenological position of the lake-types, which is related to the variation in substrate type, wind exposure, shore inclination, and surroundings (forest, mire, arable land etc.). The result from the classification is effectively checked by a PCA-ordination of the same material, i.e. *helophyte* community spectra in 50 lakes (Fig. 10) (cf. Jensén 1978, 1979). The 12 lake-types and the three main groups are easy to detect.

TYPES	I	II	III	IV	V	VI	VII	VIII	IX	X	XI	XII
I	--	36	23	24	13	11						
II	36	--	35	23	22	18	13					
III	23	35	--	28	34	27						
IV	24	23	28	--	41	34	27	19	12	15		
V	13	22	34	41	--	51	29	30	30	11	14	
VI	11	18	27	34	51	--	34	23	31		11	
VII		13		27	29	34	--	48	35	24	19	5
VIII				19	30	23	48	--	50	34	24	15
IX				12	30	31	35	50	--	42	14	7
X				15	11		24	34	42	--	25	17
XI					14	11	19	24	14	25	--	7
XII							5	15	7	17	7	--

Fig. 9. Similarity matrix for 12 lake groups classified on the basis of the *helophyte* communities. Indicated figures are % values of the similarity ratio. The three groups are shown in squares (group 1 above left, group 2 central and group 3 below right). Only similarities > 10 % are shown, except for within-group similarity.

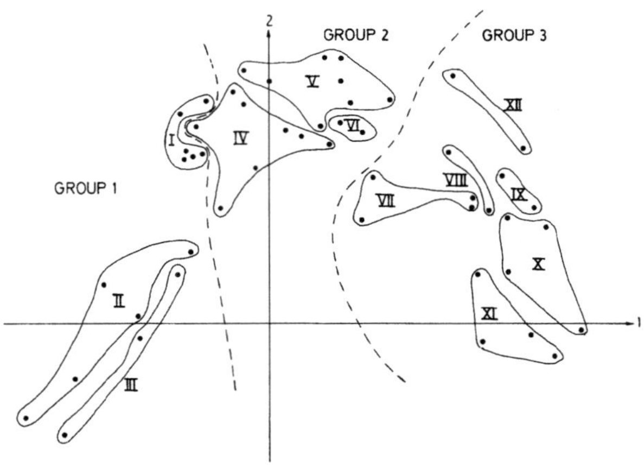

Fig. 10. PCA-ordination of the 50 lakes on the basis of their *helophyte* communities. The 12 distinguished helophyte-community-spectrum types (I-XII) and the 3 groups (1–3) as found with TABORD are indicated.

Finally it is interesting to compare the lake classification on the basis of the species approach (from Jensén 1979) with that based on the community approach as presented here (Fig. 11). The two systems correspond rather well.

Community-based lake-types

Species-based lake-types:		Group 1			Group 2			Group 3						
		I	II	III	IV	V	VI	VII	VIII	IX	X	XI	XII	NL
Group 1	CL	-	1	-										1
	EQ	3	1	-										4
	EQ-CL-CR	3	2	2										7
	SP-PC-EQ	-	1	-	6	1	-							8
Group 2	SL-PC-EQ				1	5	1							7
	SL				-	1	1							2
	PC-SL-CE			1	-	-	-	-	-	-	-	2	-	3
Group 3	PC-SL-TA							3	2	2	-	-	-	7
	PC-TA										3	-	-	3
	PC				1	1	-	1	-	-	1	2	2	8
NL		6	5	3	8	8	2	4	2	2	4	4	2	

		community-based lake-groups 1	2	3
Species-based lake-groups	1	0.76	0.37	-
	2	0.08	0.60	0.13
	3	-	0.11	0.89

Fig. 11. Comparison between species-based lake-types (Jensén 1979) and community-based lake-types presented in Fig. 8. Number of lakes (NL) per type is indicated. The similarity between the groups in the two classifications is shown by means of Sørensen's (1948) similarity coefficient. Symbols: CE = *Carex elata*, CL = *Carex lasiocarpa*, CR = *Carex rostrata*, EQ = *Equisetum fluviatile*, PC = *Phragmites communis*, SL = *Scirpus lacustris*, SP = *Scirpus palustris*, TA = *Typha angustifolia*.

Conclusions

Quadrat size

A variable square size can be used for different life-forms, e.g. ≫ 1 sq.m. for *nymphaeids* and *lemnids* (moving synusia), ≦ 1 sq.m. for *isoetids*, and *elodeids*, and ≧ 1 sq.m. for *helophytes* and *elodeids*. In our investigation all squares are 1 sq.m. in size. This size usually fulfils the minimum area requirement for macrophytes (cf. Dierssen 1975 p. 7). The heterogeneity in the clusters, especially when different life-forms are treated together, must be accepted as long as a constant square size is used. The variation in nature is reflected in the distinguished communities, where constant and significant species often have a constancy of 40–60 % (the accepted constancy value depends on the square size).

Data transformation

As has been shown by e.g. Jensén (1978) the cover values have to be transformed in order to reach an effective classification. This causes a problem, particularly as regards the *helophytes*. The qualitative difference on the basis of *helophytes* and the *helophyte* communities between oligotrophic and eutrophic lakes are sometimes very small or even neglectable, but the quantitative differences are greater (cf. Blomgren & Nauman 1925). After transformation of the cover data and the primary classification of the *helophytes* (community approach) the second classification resulted in secondary clusters, i.e. general community types. These general types are sometimes a mixture of local communities from oligotrophic and eutrophic lakes respectively. The quantitative differences are actually greater than is shown in the tables and could have led to different general community types, separated on a quantitative base. This results in some cases in too small differences between oligotrophic and eutrophic lakes in the third classification. Therefore it is desirable to overcome this weakness in the treatment of *helophyte* communities. The same problems do not exist for *isoetids* and *elodeids* because these life-forms are concentrated in the oligotrophic lakes at least in South Sweden.

Multivariate methods

Numerical classification and ordination programmes have often limitations in the number of squares and/or species treated at the same time. It was often necessary especially in the primary classification, to divide the material into 2–5 parts because otherwise the clusters should become too heterogeneous. The division can be done either by hand or by a classification program such as DIVINF (see Lance & Williams 1968, Jensén 1978). Empirically we have found that when a total information content > 1,000 is found the data matrix must be subdivided. This program DIVINF is also useful because it can be used in the establishment of the initial classification array with which TABORD, as main classification program, here, starts.

The size of clusters is another problem which must be considered in the community approach. When the inner zones (communities) around the lakes are narrower than the outer ones, then the minimum size here must be as small as 1–2 squares. This results in varying cluster sizes. Sometimes it is difficult to be sure of the significance of species occurrences in small clusters. In general the constancy must be higher in small clusters than in larger ones.

Choice of approach

The typology of lakes on the basis of their macrophyte composition is usually approached through the entire species lists of all lakes. As a first improvement the quantitative relations of species can be taken into account, i.e. frequency or/and cover data. Because of the independance of various life-forms and their synusiae it is realistic to compare lakes separately for different life-forms or life-form combinations.

A further elaboration consists of a community-type based classification. The three-step procedure described by us may lead to an objective description of each lake in terms of the spectrum of local plant communities and in an integrated comparison of lakes.

Numerical methods, including both classification and ordination, cannot be missed in this procedure. The reasonable similarity between the species-based and the community-type based lake typologies may suggest that the simpler species-based approach might be preferred. However, this similarity may vanish in case more complex community types with many species are involved. Then we may expect that the community-type based system will be superior. This expectation remains to be tested.

Summary

Various approaches to the classification of S. Swedish lakes on the basis of their macrophyte composition are discussed. A main division is made between species or life-form, and synusia or community approaches. Three steps in the classification are discussed: primary clustering of sites to arrive at a local classification, secondary classification of clusters of sites to arrive at general community types and classification of lakes according to their community composition.

The estimation of species performance is discussed and a logarithmic transformation advocated. The analysis of belt transects is suggested as a satisfactory general approach.

Some problems regarding synusial independence and possible synusial interactions are mentioned. A survey of boundary types in lakes is given.

Examples are given of classifications based on single life-forms and life-form combinations. Helophyte community composition of lakes is effectively shown by a combination of classification and ordination. The idea of aquatic community complexes is discussed with reference to the recent sigma-syntaxon approach in phytosociology.

References

Almquist, E. 1929. Upplands vegetation och flora. Acta Phytogeogr. Suec. 1: 1–624.

Barkman, J.J. 1978. Synusial approaches to classification. 2nd ed. In: R.H. Whittaker (ed.): Classification of plant communities, p. 111–165. Junk, The Hague.

Blomgren, N. & E. Naumann. 1925. Untersuchungen über die höhere Vegetation des Sees Stråken bei Aneboda. Acta Univ. Lund. N.F. Avd. 2. Bd. 21: 6. 51 pp.

Danielsson, M. 1977. Direkta tillämpningar av likhetsindex vid vegetationsundersökningar. Institutet för Vatten- och Lufvårdsforskning (IVL). Rapport B397. 14+VIII pp.

Dierssen, K. 1975. Littorelletea uniflorae. Prodromus der europäischen Pflanzengesellschaften, 2, 149 pp. Cramer, Vaduz.

Fresco, L.M.F. 1972. A direct quantitative analysis of vegetational boundaries and gradients. In: Grundfragen und Methoden in der Pflanzensoziologie, E. van der Maarel & R. Tüxen (eds.), Ber. Symp. Int. Ver. Vegetationskunde, Rinteln 1970: 99–111. Junk, The Hague.

Géhu, J.-M. 1977. Le concept de sigmassociation et son application a l'étude du paysage végetal de falaises atlantiques françaises. Vegetatio 34: 117–125.

Gils, H. van & P. Huits. 1978. Auf zwei Wegen zum Assoziationskomplex. In: R. Tüxen (ed.): Assoziationskomplexe (Sigmeten). Ber. Int. Symp. Rinteln 1977, p. 13–26. Cramer, Vaduz.

Hartog, C. den & S. Segal. 1964. A new classification of the waterplant communities. Acta Bot. Neerl. 13: 367–393.

Howard-Williams, C. & B.H. Walker. 1974. The vegetation of a tropical African lake: Classification and ordination of Lake Chilva (Malawi). J. Ecol. 62: 831–854.

Jensén, S. 1977. An objective method for sampling the macrophyte vegetation in lakes. Vegetatio 33: 107–118.

Jensén, S. 1978. Influences of transformation of cover values on classification and ordination of lake vegetation. Vegetatio 37: 19–31.

Jensén, S. 1979. Classification of lakes in Southern Sweden on the basis of their macrophyte composition by means of multivariate methods. Vegetatio 39: 129–146.

Lance, G.N. & W.T. Williams. 1968. Note on a new information-statistic classification program. Comput. J. 11: 195.

Leeuwen, C.G. 1965. Het verband tussen natuurlijke en anthropogene landschapsvormen, bezien vanuit de betrekkingen in grensmilieu's. (Engl. summary) Gorteria 2: 93–105.

Lid, J. 1974. Norsk og svensk flora. Det Norske Samlaget. Oslo. 808 pp.

Lillieroth, S. 1950. Über Folgen kulturbedingter Wasserstandssenkungen für Makrophyten- und Planktongemeinschaften in seichten Seen des südschwedischen Oligotrophiegebietes. Acta Limnol. 3. 288 pp.

Luther, H. 1951. Verbreitung und Ökologie der höheren Wasserpflanzen im Brackwasser der Ekenäs-Gegend in Südfinnland. I. Allg. Teil. Acta Bot. Fennica 49: 1–332 + app.

Maarel, E. van der. 1964. Review of J.T. Curtis: The vegetation of Wisconsin. Acta Bot. Neerl. 13: 438–440.

Maarel, E. van der. 1971. Plant species diversity in relation to management. In: A.S. Watt and E. Duffey (eds.), The scientific management of plant and animal communities for conservation pp. 45–63. Blackwell, Oxford.

Maarel, E. van der. 1976. On the establishment of plant community boundaries. Ber. Deutsch. Bot. Ges. 89: 415–443.

Maarel, E. van der. 1979. Transformation of cover-abundance values in phytosociology and its effects on community similarity. Vegetatio 39: 97–114.

Maarel, E. van der. 1979a. Multivariate methods in phytosociology, with reference to the Netherlands. In: M.J.A. Werger (ed.): The study of vegetation. pp. 161–225. Junk, The Hague.

Maarel, E. van der, J.G.M. Janssen & J.M.W. Louppen. 1978. TABORD, a program for structuring phytosociological tables. Vegetatio 38: 143–156.

Maarel, E. van der & J. Leertouwer. 1967. Variation in vegetation and species diversity along a local environmental gradient. Acta Bot. Neerl. 16: 211–221.

Mäkirinta, U. 1978. Die Pflanzensoziologische Gliederung der Wasser-vegetation im See Kukkia, Südfinnland. Acta Univ. Ouluensis, Ser. A, Scient. Rer. Natural. 75. Biologica 5.

Malmer, N. 1962. Studies on mire vegetation in the Archean area of southwestern Götaland (South Sweden). I. Vegetation and habitat conditions on the Åkhult mire. Opera Bot. 7(1). 322 pp.

Maristo, L. 1941. Die Seetypen Finnlands auf floristischer und vegetationsfysionomischer Grundlage. Ann. Bot. Soc. Zool. – Bot. Fenn. Vanamo. 15, 5. Helsinki. 314 pp.

Mériaux, J.L. 1978. Etude analytique et comparative de la vegetation aquatique d'etangs et marais du nord de la France (Vallée de la Sensée et Bassin Houillier du Nord-Pas-de-Calais). Documents phytosociologiques. N.S. Vol. II. Lille, Juin 1978.

Mériaux, J.J. & J.M. Géhu. 1978. De l'analyse symphytosociologique des complexes de vegetation a celle des végétation complexes. Example des groupements aquatiques et subaquatiques. In R. Tüxen (ed.): Assoziationskomplexe (Sigmeten), pp. 97–116. (Rinteln 4.–7.4. 1977).

Persson, F. 1969. Makrofytvegetation och litoraltopografi i sjön Trummen. Meddelanden från Forskargruppen för Sjörestaureringar vid Lunds Universitet, 32: 1–19.

Persson, S. 1977. Data program för bearbetning ar vegetationsdata. 1. Klassifikationsprogram-dokumentation och handhavande. Meddn. Avd. Ekol. Bot. Lunds Univ. 33: 1–68.

Samuelsson, G. 1925. Untersuchungen über die höhere Wasserflora von Dalarne. Svenska Växtsociologiska Sällskapets Handlingar IX. 31 pp.

Scott, J.T. 1974. Correlation of vegetation with environment: a test of the continuum and community-type hypothesis. In: B.R. Strain & W.D. Billings (eds.), Handbook of vegetation science, 6: Vegetation and environment, pp. 87–109. Junk, The Hague.

Seddon, B. 1972. Aquatic macrophytes as limnological indicators. Freshwater Biol. 2: 107–130.

Seibert, P. 1968. Gesellschaftsring und Gesellschaftskomplex in der Landschaftsgliederung. In: R. Tüxen (ed.): Pflanzensoziologie und Landschaftsökologie. Stolzenau 1963: 48–60. Junk, The Hague.

Sjörs, H. 1971. Ekologisk botanik. Almqvist & Wiksell, Stockholm, 296 pp.

Sørensen, T. 1948. A method of establishing groups of equal amplitude in plant sociology based on similarity of species content. Danske Vidensk. Selsk. Biol. Skr. 5(4): 1–34.

Stumpel, A.H.P. & J.T.R. Kalkhoven. 1978. A vegetation map of the Netherlands, based on the relationship between ecotopes and types of potential natural vegetation. Vegetatio 37: 163–173.

Swan, J.M.A. 1970. An examination of some ordination problems by use of simulated data. Ecology 51: 89–102.

Tüxen, R. 1973. Vorschlag zur Aufnahme von Gesellschaftkomplexen in potentiellen natürlichen Vegetationsgebieten. Acta Bot. Acad. Sci. Hung. 19: 379–384.

Tüxen, R. 1978. Bemerkungen zu historischen, begrifflichen und methodischen Grundlagen der Synsoziologie. In: R. Tüxen (ed.): Assoziationskomplexe (Sigmeten). Ber. Int. Symp. Rinteln, p. 3–12. Cramer, Vaduz.

Whittaker, R.H. 1978. Direct gradient analysis, 2nd ed. In: R.H. Whittaker (ed.): Ordination of plant communities, p. 7–50. Junk, The Hague.

Zutshi, D.P. 1975. Associations of macrophytic vegetation in Kashmir lakes. Vegetatio 30: 61–66.

Accepted 4 December 1979

PRELIMINARY SURVEY OF THE PEAT-BOG HUMMELL KNOWE MOSS USING VARIOUS NUMERICAL METHODS*

R.S. CLYMO**

Botany Department, Westfield College, London NW3 7ST, England

Keywords:
Mire, Multidimensional scaling, Ordination, Peat-bog, Principal component analysis, Principal coordinate analysis, Reciprocal averaging, Transformation of data

Introduction

The peat-bog Humell Knowe in northern England is unusual in having a nearly complete *Sphagnum* carpet with abundant *Eriophorum vaginatum*, and *E. angustifolium*, but with little differentiation of hummock or hollow topography, This paper reports a primary survey of the vegetation and stratigraphy, made with strictly limited resources, and intended to direct the planning of further work. A variety of numerical aids were used in the vegetation analysis. The usefulness of these aids was judged by the extent to which they both confirmed what was really obvious to the ecologist-on-the spot, and the extent to which they indicated what further work should be done: a method which does not reveal the really obvious must be suspect, but one which does nothing but reinforce prejudice is equally useless.

* Nomenclature follows Clapham, Tutin & Warburg (1962) – vascular plants; Smith (1978) – *Sphagnum*; Watson (1969) – other bryophytes; and Duncan with James (1970) – lichens.

** The following other members of the Mires Research Group contributed to this survey: D. Bale, R. Bhadresa, D.J. Boatman, O. Clymo, M. Crawley, R.E. Daniels, C. Dickinson, W. Fojt, P. Hayward, P.D. Hulme, E.A. Livett, R. Meade, the late M.C. Perason, P.J. Shaw, T. Stevenson, J.H. Tallis, P. Wiltshire. I thank them for allowing me to use their results. The analyses and views expressed in this paper are my own. I also thank Mrs. P. Ratnesar for preparing the results for computing and for drawing the diagrams, T. Clymo for typing the text, and Dr J.H. Tallis and Dr K.E. Clymo for commenting on parts of a draft.

The study area

Hummell Knowe Moss (National Grid reference NY 705715; alt. 230 m; long. 2° 28′ W, lat. 55° 02′ N) is a large blanket mire in the centre of northern England, halfway between Newcastle and Carlisle, and 8 km north of Haltwhistle (Fig. 1). It is one of a group of fine mires known collectively as the Border Mires, which lie in a 30 km band around the central section of the Roman Wall. These include *Butterburn Flow, Broomlee Lough, *Coom Rigg, (Chapman, 1964a, b; 1965), Fozy Moss, Greenlee Lough, *Haining Head Moss, Muckle Moss (Pearson 1960), Robin Rock Flothers and The Wou. (The starred mires

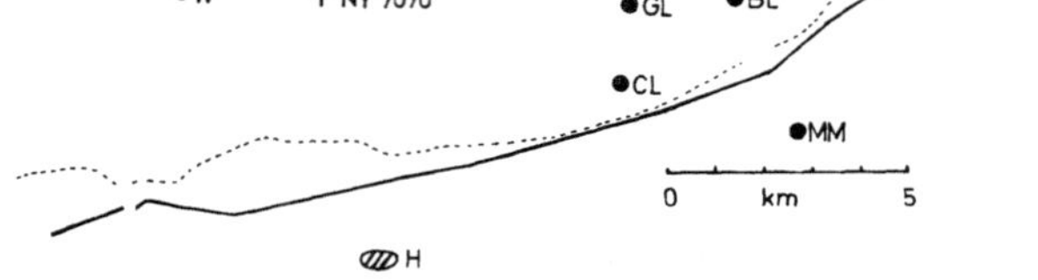

Fig. 1. Location of the Border Mires mentioned in the text. Continuous line, B 6318 road; dashed line, Roman Wall; BF Butterburn Flow; BL Broomlee Lough; CL Crag Lough; CR Coom Rigg; FM Fozy Moss; RRF Robin Rock Flothers; W The Wou. To the south is Haltwhistle (H).

are listed as Grade 1 by Ratcliffe 1977.) Amongst them one may find open water, fen, carr, and bog (much of it apparently undamaged). *Hummell Knowe Moss has a central, slightly domed, bog plain about 0.5 km across (Figs 2, 9) with a nearly complete cover of *Sphagnum magellanicum*, and, in order of abundance, almost constant occurrence, in 25 × 25 cm quadrats, of *Eriophorum vaginatum*, *E. angustifolium*, *Erica tetralix*, *Narthecium ossifragum*, *Andromeda polifolia*, *Vaccinium oxycoccos*, *Drosera rotundifolia* and *Sphagnum tenellum*, with several other species nearly as common. This central area is wet, but with little open water, and is surprisingly free of the hummock and hollow topography, on a scale of about 5 m, which is usually characteristic of such areas. (There are a few hummocks on the moss, not associated with pools, and these may turn out to be of unusual interest.)

Surrounding the central plain (Fig. 2) is vegetation which grades into a variety of marginal communities which border four low (ca. 10–20 m tall) hills to the north-east, northwest and south of the moss. Eastwards and westwards the bog plain slopes downhill, and the surface has obvious drainage channels and eroded peat: *Sphagnum* is less common here, and in the east lichens of the genus *Cladonia* are conspicuous. To the north-west and south are communities with little *Sphagnum*, and dominated by *Molinia caerulea*, *Eriophorum vaginatum* and *Deschampsia flexuosa*. To the south-west is a distinctive region which may be flushed by water running off the low hills behind. This region is recognisable by tussocky *Eriophorum vaginatum* on which grow species such as *Potentilla erecta* which are not found elsewhere on the mire centre, though common on the surrounding mineral ground. Between the tussocks is an impoverished bog flora. At the base of the hills to the north and south is a marginal lagg with *Juncus effusus* and *Sphagnum recurvum* common.

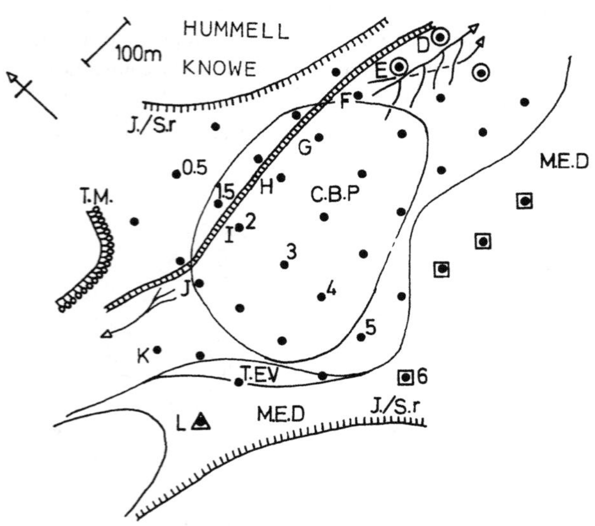

Fig. 2. Isodiametric view of Hummell Knowe bog from the southwest. Filled circles are vegetation sampling sites on a grid (D to L, 0.5, 1.5, 2, 3 . . . . 6). A fence runs E-W across the bog. Erosion at east and west ends is associated with drainage channels. Large circle, triangle and square refer to regions of the ordination diagrams. Subjective vegetation types: C.B.P = central bog plain; M.E.D = *Molinia – Eriophorum vaginatum – Deschampsia flexuosa*; T.E.V = tussocky *E. vaginatum*; T.M = tussocky *Molinia*; J./S.r = *Juncus effusus – Sphagnum recurvum*.

The mire continues for more than 500 m to the east (where it is eroded) and to the north, but these areas were not included in the survey. The surrounding area is grazed by sheep, and the hill to the north, Hummell Knowe itself, and the hills to the north-west, have been ploughed and forested. Some of these features are obvious at a glance, others became clear during detailed vegetation recording.

Field methods

The features surveyed and methods used were a compromise between the desirable and the practicable. The survey was made on 20–21 September 1978 by 18 members of the Mires Research Group. Not all were on the site for both days and not all were equally competent taxonomists. The programme was organised to allow rotation of jobs, but also to ensure that the vegetation records were as reliable as possible. The fieldwork occupied, in total, about 200 h.

Because so many people were involved, so little time was available, and so little was known about the study site, a regular sampling pattern was chosen. A reference grid was established. The base-line ran parallel to and 10 m south of the fence which runs east-west across the mire. The grid interval was 100 paces, except that the first line north of the fence was only 50 paces from the base-line. Pacing was used to save time. The accuracy of this grid is about ± 5 paces in the central row, and about ± 10 paces at the northern and southern limits.

The stratigraphy of a complete profile through the peat deposit was examined at one grid intersection (F2) using a Russian-type peat borer (West 1968). Over the whole area the depth of the basal transition from clay to organic remains was established at alternate grid intersections, those in adjacent rows being staggered. Again, the work was limited by time, and this pattern was chosen to ensure fair cover of most of the site.

The relative height of the mire surface at most grid intersections was measured with a level and staff. A series of closed loops was surveyed. The loop error was usually less than 1 cm.

Vegetation was recorded at most grid intersections in a ladder of 12 contiguous square quadrats, each 25 × 25 cm, on a NE-SW diagonal starting 5 paces SW of the grid point (to avoid trampled areas along grid lines and intersections). A ladder was used for four reasons: so that pattern on a scale up to 3 m would be spanned and might later be revealed by analysis, so that sample site selection was rapid, so that taxonomic advice was readily available, and so that a single recorder could serve as many as four surveyors. The species present in a quadrat were recorded, but no attempt was made to estimate cover, though, as Smartt, Meacock & Lambert (1976) very sensibly point out, there is no virtue in recording a feature simply because it is easily recorded. In the present case fourteen people were concerned at various times in vegetation recording, and different observers of the same quadrat often give grossly different (ten-fold) subjective estimates of cover, especially of fine leaved or small but widely dispersed species, which form an important part of the Hummell Knowe mire vegetation. The objective of consistency of identification was not completely achieved either. It was attempted by having, on the first day, pairs of observers record three adjacent quadrats at each site, with cross reference in any case of uncertain identity. Three of these observers stayed in the vegetation recording party for the whole time and acted as referees. Nevertheless there remain some problems.

1. *Sphagnum magellanicum* could be positively identified when tinged with red, but much of this species, especially where shaded, was green and not consistently distinguishable from *Sphagnum papillosum*. It is likely that *Sphagnum papillosum* is under-recorded and observer biassed.

2. The fruticose *Cladonia* species: *C. tenuis*, *C. arbuscula* and *C. impexa* were not consistently distinguishable, so they have been grouped as *C. arbuscula* agg. In general the quadrats contained either no lichens or more than one species, so this should be unimportant.

3. Small leafy liverworts posed two problems. Firstly, they were usually inconspicuous and could be found only by grab sampling within the quadrat. They were therefore under-recorded. Secondly, it is clear from the records that identification was not consistent: at a series of grid points one group consistently recorded *Mylia* whilst another recorded *Mylia anomala*. Such taxa were grouped as 'sp' or 'spp'.

Table 1. Species recorded from sites at Hummell Knowe Moss. Frequency in 0.75 m^2 at 41 sites.

				Pohlia nutans	3
Andromeda polifolia	37	Erica tetralix	37	Polytrichum alpestre	7
Aulacomnium palustre	8	Eriophorum angustifolium	38	P. commune	8
Barbilophozia sp.	1	E. vaginatum	41	Potentilla erecta	2
Calluna vulgaris	30	Festuca ovina	1	Ptilidium ciliare	1
Calypogeia sp(p?)	25	Galium saxatile	2	Rhynchospora alba	1
Campylopus atrovirens	2	Gymnocolea inflata	6	Sphagnum sp.	1
Carex sp.	2	Hypnum cupressiforme	8	* S. capillifolium	21
C. panicea	3	Lepidozia setacea	20	S. cuspidatum	27
Cephalozia bicuspidata	23	Leucobryum glaucum	4	S. magellanicum	35
Cladonia arbuscula agg.	2	Lophozia sp.	5		
				S. palustre	8
C. furcata	4	Lophocolea sp.	2	S. papillosum	8
C. impexa	5	Mnium hornum	2	S. recurvum	8
C. pyxidata	1	Molinia caerulea	7	* S. rubellum	4
C. tenuis	7	Mylia sp(p?)	32	S. squarrosum	7
C. uncialis	10	Nardia scalaris	1	S. tenellum	34
Deschampsia flexuosa	5	Narthecium ossifragum	36	S. subnitens	2
Dicranella cerviculata	1	Odontoschisma sphagni	34	Splachnum ampullaceum	1
Dicranum sp.	9	Parmelia physodes	2	Trichophorum cespitosum	34
Drosera rotundifolia	32	Plagiothecium undulatum	6	Vaccinium myrtillus	1
Empetrum nigrum	2	Pleurozium scheberi	7	V. oxycoccos	36

* *S. rubellum*, a small form of *S. capillifolium*, is probably not a distinct species (M.O. Hill in Smith 1978) but it was recorded separately.

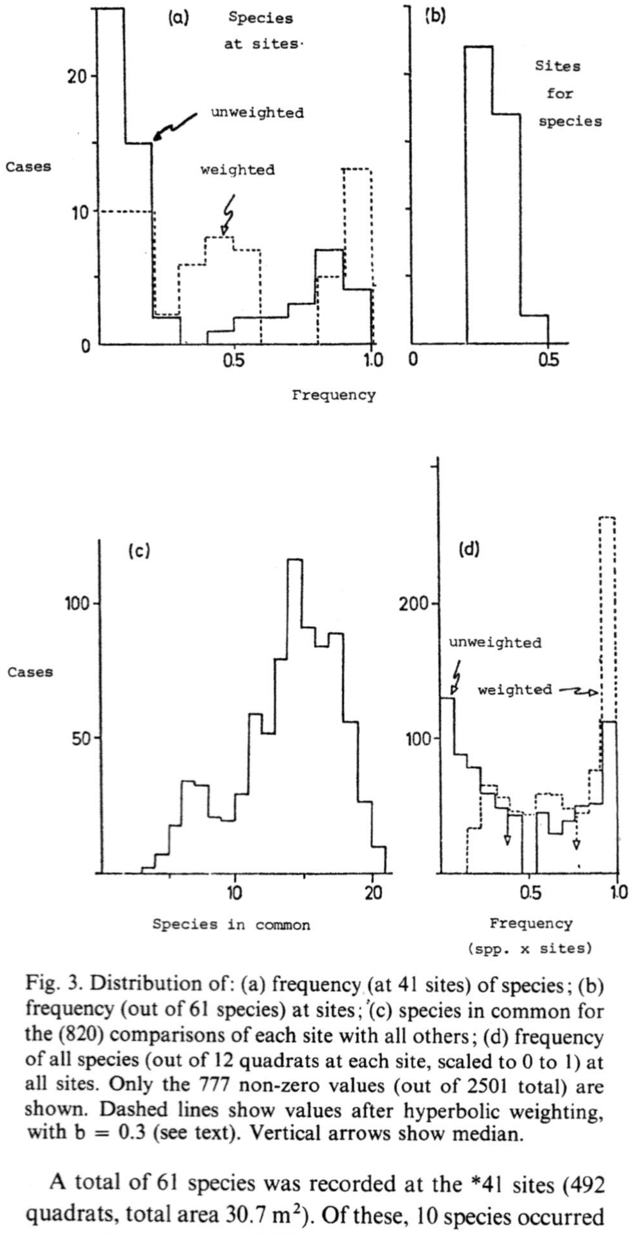

Fig. 3. Distribution of: (a) frequency (at 41 sites) of species; (b) frequency (out of 61 species) at sites; (c) species in common for the (820) comparisons of each site with all others; (d) frequency of all species (out of 12 quadrats at each site, scaled to 0 to 1) at all sites. Only the 777 non-zero values (out of 2501 total) are shown. Dashed lines show values after hyperbolic weighting, with b = 0.3 (see text). Vertical arrows show median.

A total of 61 species was recorded at the *41 sites (492 quadrats, total area 30.7 m^2). Of these, 10 species occurred at one site only (out of 41) and 28 at fewer than five sites (Table 1, Fig. 3). The number of species at a site ranged from 14 to 29, and overall 31 % of the possible species-at-site occurrences were observed, indicating an unusually uniform vegetation; an impression confirmed by the distribution of the number of species common to pairs of sites (Fig. 3).

After the first day of vegetation surveying it was possible to recognise, tentatively and subjectively, the major vegetation types. The approximate position of the boundary of these types was recorded on the second day, and is shown in Fig. 2.

Finally a simple survey of water chemistry was made. Water samples were squeezed from *Sphagnum*, or taken from runnels, and pH and electrical conductivity measured on the spot.

* Only 40 separate sites, but D2 was (accidentally) repeated. The second survey was made by a different group of observers, working in poor light and in a hurry, at the end of the day. The two sets of records have been kept and are both shown in the Figs, to illustrate what is probably the maximum 'recording error'.

Numerical methods

We needed to summarise the results, but had no clear idea of the number of dimensions which would be needed, nor of the most useful proximity measure. The number of sites (41) imposed no serious restrictions, but we did want to experiment with means for giving greater or less weight to local site-site relationships. For these reasons we chose ordination rather than classification.

Transformation

The analyses may be considered in stages (Fig. 4). Transformation is used here to describe any manipulation which changes the field records before the analysis proper. A transformation which can be applied to a datum in isolation is here called a weighting (e.g. log, square-root). One which is applied in the context of the values in a row or column is called a standardisation (e.g. to rank, or to unit range or variance, or to zero mean). Transformations needing the complete matrix of values are here called 'adjustments'. The only one used here is Swan's (1970) method for assessing 'degree of absence'.

The importance of weighting is well known: practical ecologists have usually attached more importance to species presence than to amount, as demonstrated by the traditional cover scales, and there is some evidence (e.g. Smartt, Meacock & Lambert 1976) to support this view. For computing purposes it is convenient to have simple continuous functions with one or two parameters which can be varied to adjust the boost given to rarer species (cf. van der Maarel 1979). Some which have been used, for example log functions (Jensén 1978), are awkward insofar as they are undefined if a datum is zero, and would cause unwanted distortion should it be less than 1 %. But

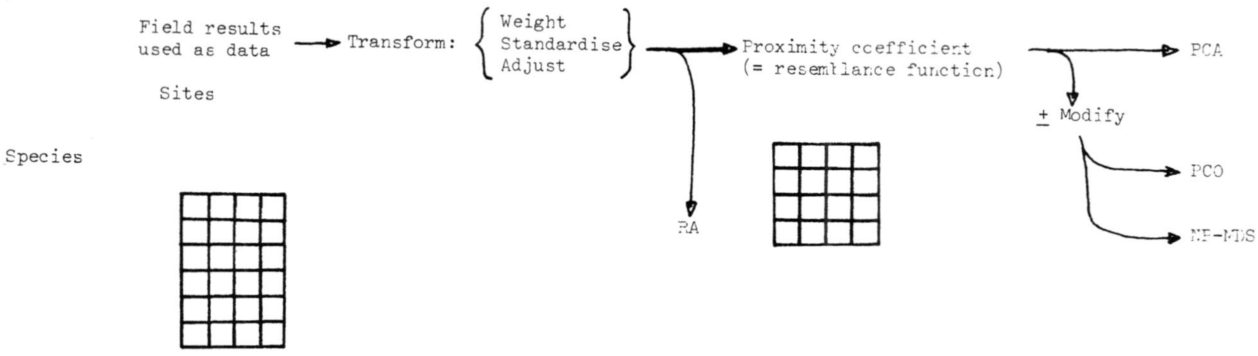

Fig. 4. Stages in an ordination. RA = reciprocal averaging; PCA = principal components analysis; PCO = principal coordinate analysis; NP-MDS = non-parametric multidimensional scaling.

designing such functions is simple. Suppose the original datum, x, can range from 0.0 to 1.0, and the weighted value, $w = f(x)$, is to span the same range. Choose a function, $y = f(x)$, which goes through the point (0,0) and which rises steeply at first but then more slowly. Let $y(1)$ be the value at $x = 1$. Then $w = y/y(1)$ will go through both (0,0) and (1,1). Some functions, such as $w = x^{1/a}$ do this anyway. Table 2 and Fig. 5 show three such single-parameter functions. Given the subjectivity of choice this is probably two more than are necessary, and it is doubtful if there is any good reason for using a two-parameter function. Of the functions in Fig. 5 the root function gives the greatest boost to rare species, the hyperbolic next most, and the exponential least.

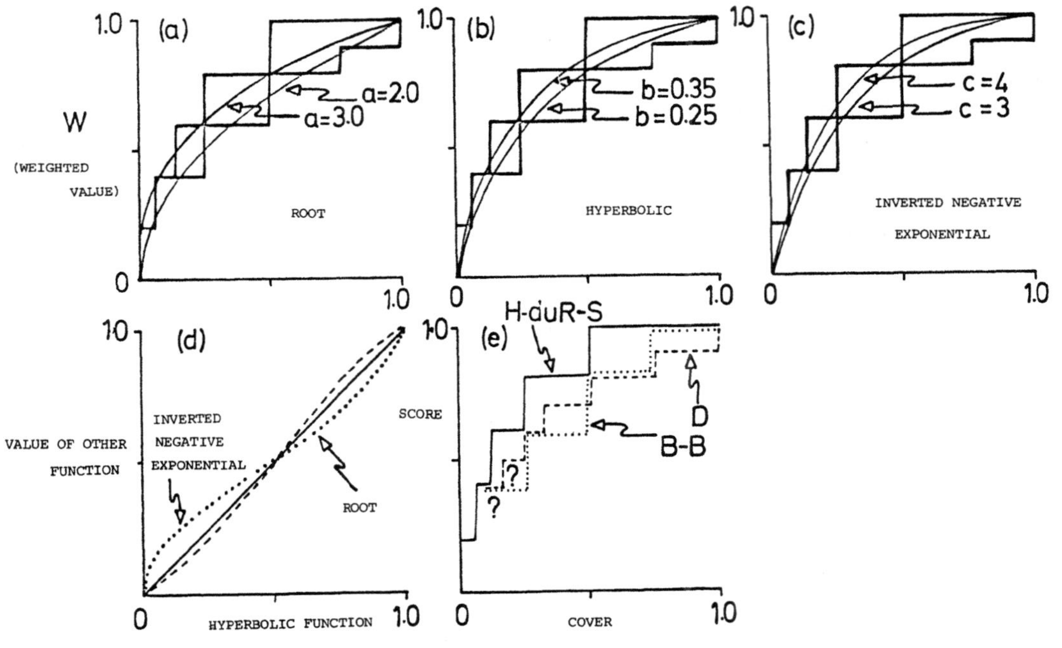

Fig. 5. Three simple weighting functions for two parameter values: (a) $w = x^{1/a}$; (b) $w = x\,(b+1)/(b+x)$; (c) $w = (1-e^{-cx})/(1-e^{-c})$; (d) comparison of these functions with a, b, c such that $x = 0.188$, $w = 0.5$; (e) traditional cover scales (Hult-Du Rietz-Sernander, Domin, Braun-Blanquet). The lower ends of the D and B-B scales are not unambiguous. The envelope formed by these scales is reproduced in (a), (b), (c).

Table 2. Weighting functions, $w = f(x)$; The range of w and x is 0.0 to 1.0.

Name	$y =$	$w =$	$w = 0.5$	for (0.1875, 0.5)
1. Root	$x^{1/a}$	$x^{1/a}$	$a = -\ln(x)/0.693$	$a \simeq 2.416$
2. Hyperbolic	$\dfrac{1}{1+b/x}$	$\dfrac{x(b+1)}{x+b}$	$b = \dfrac{x}{1-2x}$	$b = 0.30$
3. Inverted negative exponential	$1-e^{-cx}$	$\dfrac{1-e^{-cx}}{1-e^{-c}}$	No simple solution	$c \simeq 3.545$

The Hummell Knowe vegetation records contained an unusually large number of species with high frequency (Fig. 3), and many of low frequency. This pattern, if unchanged, would produce analyses dominated by the common species, so the hyperbolic weighting was used, with $b = 0.3$, giving species with frequency 0.19 a weighted value of 0.5. The effect of this on the distribution of weighted frequency is shown in Fig. 3.

The value of standardisation is less clear. Some proximity coefficients (for example the correlation coefficient) include standardising to mean = 0 and variance = 1 and have large effects in some types of analysis – PCA for example. It seems clear however that there is no single criterion which can be applied (Noy-Meir 1971, 1973, Noy Meir, Walker & Williams 1975) to determine what standardisation to use, and that in many cases uncentred data are preferable to centred. Some standardisations increase the importance of rare species, which seems desirable (as already argued), but transformations do this more flexibly. Our interest was chiefly in site-site relations, which might argue for a standardisation by site, but the data as a whole showed unusually little difference in species richness between sites (Fig. 3), so in the end it seemed best to use no overt standardisation.

Swan (1970) devised an adjustment for assessing 'degree of absence' of a species. This is certainly a problem with any metric proximity measure. Swan's adjustment seems to have been applied to field data only rarely, so we tested it on the Hummell Knowe results.

Ordination

Four types of ordination were used (Fig. 4): principal components analysis (PCA), reciprocal averaging (RA), non-parametric multidimensional scaling (NP-MDS) and principal coordinates analysis (PCO).

Reciprocal averaging (Hill 1973) is likely to bend artificial data sets based on vegetation responding to a linear gradient, but it usually keeps the ends apart to give an arch in two dimensions, rather than the convoluted 'horseshoe' of PCA. Comparison of these two methods is often made unfairly, because all axes of an RA plot are usually scaled to the same range, whereas those of a PCA plot are

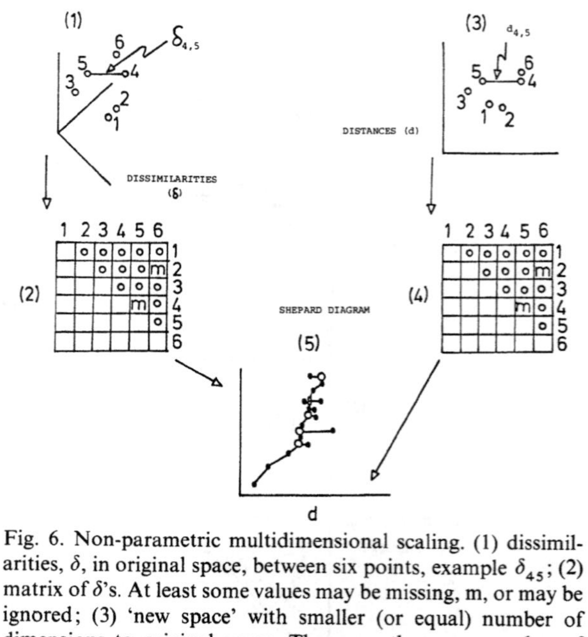

Fig. 6. Non-parametric multidimensional scaling. (1) dissimilarities, δ, in original space, between six points, example δ_{45}; (2) matrix of δ's. At least some values may be missing, m, or may be ignored; (3) 'new space' with smaller (or equal) number of dimensions to original space. There are the same number of distances, d, in new space as dissimilarities in original space; (4) matrix of d's, missing values as in dissimilarity matrix; (5) Shepard diagram with δ plotted against d. The aim is to adjust the points in new space so that the Shepard diagram is monotonic, or, if that is impossible, to minimise the sum of squares of deviations (on the d axis) from a monotonic line. This means that the numerical values of δ are not needed, only their rank order.

commonly scaled in proportion to the square root of the eigenvector, so that curvature is less pronounced. An estimate of the size of the eigenvector in RA is easily got (Hill 1973) from the direct algorithm. The effect of using this scaling is shown in Fig. 12. Reciprocal averaging is relatively cheap to compute, needs little computer store, and may be used on very large sets of data.

Non-metric multidimensional scaling is not yet so well known, but has several advantages, both in theory and in practical tests (Anderson 1971, Dale 1975, Austin 1976, Prentice 1977 and Fasham 1977). It was devised (for use by psychologists) by Kruskal (1964) following earlier work by Shepard and others. Kruskal's account is a model of clarity, and is the basis of Fig. 6.

For simplicity, suppose that six points (sites, individuals, cases etc.) are characterised by three attributes (characters, species, variates etc.). At the top left of Fig. 6 these points are shown (1) in their three dimensional character space. A matrix, usually triangular, of proximity coefficients may be calculated (2). These are commonly dissimilarities, δ. Now suppose there is a 'new space' (3) but with a smaller number of dimensions. The same number of points – six – may be placed at random in this space, as shown at the top right of Fig. 6. Inter-point *distances*, d, may be calculated and arranged in a distance matrix (4). The rule for calculating d may be the same as that for δ, or it may be different. Now plot a graph (5) of δ against d. This graph is called the Shepard diagram. If a line, linking the points, can be drawn that runs always up and to the right (or always down and to the right) then the arrangement of points in the new space is, in this special sense, perfect. The line is monotonic: the *rank order* of d is identical to that of δ. If the rank order is not identical (as it is not in Fig. 6) then measure the imperfection by calculating a 'badness of fit':

$$\sum_{i<j} (d_{ij}-\hat{d}_{ij})^2 \qquad (1)$$

where d_{ij} is the distance in new space between points i and j, and $\hat{d}_{ij}$ is the point with the same δ on the current monotonic line. (The location of $\hat{d}$ is defined later.) The badness of fit may be standardised, so that it is dimensionless and may be compared with the value in other sets of data, by dividing by $\sum (d_{ij}^2)$. This standardised 'badness of fit' measure is a squared quantity (c.f. variance). Taking the square root gives a standard-deviation-like measure of badness of fit called the stress, S:

$$S = [\sum_{i<j} (d_{ij}-\hat{d}_{ij})^2 / \sum_{i<j} (d_{ij})^2]^{\frac{1}{2}} \qquad (2)$$

The positions of $\hat{d}$ are defined as being those which, for a given position of points in new space, are monotonic and minimise S.

One or more points may now be moved in the new space, and a new S calculated. The process is repeated, and the positions of points in new space which best represent the points in the original higher dimensional space are those with the smallest stress S.

The 'best' arrangement must be found by an iterative method, though the work may be reduced by choosing starting positions in new space which are close to those for minimum stress. A more serious difficulty is that there is no guarantee that the solution is the absolute best: it may be a local rather than the global minimum. The results shown here are based on cases where at least three attempts from different starting positions have located very similar final positions in new space.

Three technical choices of important details were made:

(1) Primary treatment of ties was used: if $\delta_1 = \delta_2$ then the rank order of d_1 and d_2 does not matter.

(2) The standardisation used was that shown above, rather than
$\sum_{i<j} (d_{ij}-\bar{d})^2$,
where $\bar{d}$ is the mean of all the d's.

(3) Local rather than global scaling, in the sense used by Sibson (1972a), was preferred. This throws the emphasis on individual site relations.

(4) The distances in the new space were assumed to be Euclidean.

An essential feature of this method is that the *values* of δ_{ij} need never be known, only their rank order. In psychology, the results of experiments are often of this form: people record that they prefer one thing to another, so the rank order of results in the matrix of δ's is known, but that is all. Psychologists call this matrix 'data', whereas ecologists usually calculate it from 'field data'. It is the discovery of metric coordinates in new space from non-metric (rank order) information in old space which has caused the method to become known as 'non-metric' MDS. As Orlóci (1978) points out however this may mislead ecologists, who first calculate what may be a metric proximity coefficient, whose metric properties they then ignore. To avoid this confusion it seems preferable to call the method 'non-parametric' or NP-MDS. Because only the rank order of δ's is used, any transformation of δ should be without effect on the positions in new space. Coefficients of similarity may be used as easily as dissimilarities.

Apart from flexibility in choice of proximity coefficient

(which may be a mixed blessing) NP-MDS has three important and useful features:

(1) Some, perhaps many, of the δ's may be 'missing'. This simply reduces the complexity of the Shepard diagram and reduces the amount of computing. Of course, if for any one site there are fewer δ's than dimensions in the new space then the point will float in new space. Ecological data matrices do not usually have missing δ's, but it may be useful to force some δ's to vanish if only to reduce the amount of computing. This makes it practicable to use NP-MDS on larger data sets: present programs are limited to about 2000 δ's, which would restrict one to about 45 sites for a full matrix or about 65 for a triangular matrix. If instead one requires only that there be, say, 15 δ's for each site, then it becomes possible to deal with about 130 sites. An obvious way of removing δ's is to choose first those which have no species in common, because it is jùst these for which the proximity measure will cause most distortion. Thinning can be extended to those sites with only 1, 2 . . . n species in common. If further thinning is necessary then three obvious ways of achieving it are to remove the smallest, largest or random δ's, always preserving a minimum number of values for each site. Only the first stage of thinning of the Hummell Knowe data is illustrated. This again emphasises local relationships.

(2) There is no fixed functional relation between δ and d. As Kruskal (1964) shows NP-MDS may be used to recover the function even though the δ's are nothing more than rank ordered. To the ecologist this lack of a fixed relation is potentially valuable because it makes the choice of proximity coefficient less important, or, more usefully, means that the coefficient may be chosen for reasons other than those dictated by the ordination method. In practice the scatter of points on the Shepard diagram (e.g. Fig. 15) is often close to a simple curve or even nearly a straight line (Fig. 11). in such cases polynomial MDS (PO-MDS) gives nearly the same result as NP-MDS, but with less computation. In PO-MDS however the *values* of the δ's are used.

(3) The axes in new space are not ordered, so that the temptation to interpret one as 'the most important' and to try to equate the *axis* with variation in one particular environmental variable or complex can be more easily resisted. The arrangement of points in new space *can* be rotated to principal components, and this has been done in this work, but only to make comparison of diagrams simpler.

There remains the problem of how to choose the number of dimensions for the 'new space'. Suppose separate analyses are made with 5, 4, . . . 1 dimensions. It may be that as the number of dimensions is decreased there comes a stage where the stress increases sharply. The Shepard diagram may at that point show a clearly branched pattern too. One may then take the number of dimensions needed to keep the stress low as being in some sense the natural dimensionality of the data. The whole matter is complex however (Kruskal & Wish 1977).

The last of the ordination methods used is principal coordinates analysis (Gower 1966, Krzanowski 1969). This will accept most proximity measures, and has sometimes been described as generalised PCA. It is more useful perhaps to consider PCA as a special case of PCO, using Pythagorean distance as the proximity measure. If the Shepard diagram of NP-MDS approximates a straight line then one might expect PCO to give much the same result as NP-MDS, but with much less computing; but whether or not the Shepard diagram *is* nearly straight will not be known until it is computed, though the occurrence in PCO of many or large negative latent roots may warn of non-linearity.

Proximity coefficients

For RA there is no choice. For PCA the choice between using the dispersion matrix or correlation matrix might be considered a choice of coefficient. Most authors who have tested PCA advocate use of the dispersion matrix so we used that.

For PCO the choice is wider, but as Williamson (1978) points out, incidence (presence-absence) data require the use of the Manhattan (or city-block) metric, m, which may be imagined as the distance along the edge of a hypercube:

$$m_{jk} = \sum_{i=1} |(x_{ij} - x_{ik})| \tag{4}$$

where m_{jk} is the dissimilarity between individuals j and k, and x_{ij} is the value (0 or 1) of the i-th character (of n) of individual j. The site data at Hummell Knowe, after weighting, approached the character of incidence data, so this coefficient was used. It was tried on unweighted data too.

For NP-MDS the choice of coefficient is very wide. Prentice (1977) argues for the use of one particular coefficient (the sum of minima) because it matches the geometry assumed for distances in new space. But NP-MDS requires only the *rank order* of dissimilarities; and the results often, in practice, depend rather little on the choice of coefficient. To test this point three coefficients were compared:

(1) the Manhattan metric

(2) another Minkowski metric, the 'Euclidean distance'

$$d_{jk} = \Big[\sum_{i=1} (x_{ij}-x_{ik})^2\Big]^{\frac{1}{2}} \qquad (5)$$

(3) the non-metric Mountford coefficient, θ, defined by

$$e^{\theta a}+e^{\theta b} = e^{\theta(a+b-c)}+1$$

where a is the number of species at site i, b is the number at site j, and c is the number of species common to both sites. This coefficient (Mountford 1962) is based on an explicit model of species abundance – the log series (Williams 1964) – and for that model the coefficient is independent of sample size. This property is not particularly useful in the present case, but in other cases where there is need to compare records made from samples of different or unknown area it may be very useful (Holloway & Jardine 1968, Clymo and Régnell unpublished). It is important therefore to know which ordination methods may usefully be used with it. In the Hummell Knowe work θ was converted to a dissimilarity, ϕ, using:

$$\phi = \ln 2-\theta$$

Modification of the proximity coefficients

The proximity matrix may be modified in various ways. If one starts from the assumption that the proximity measure

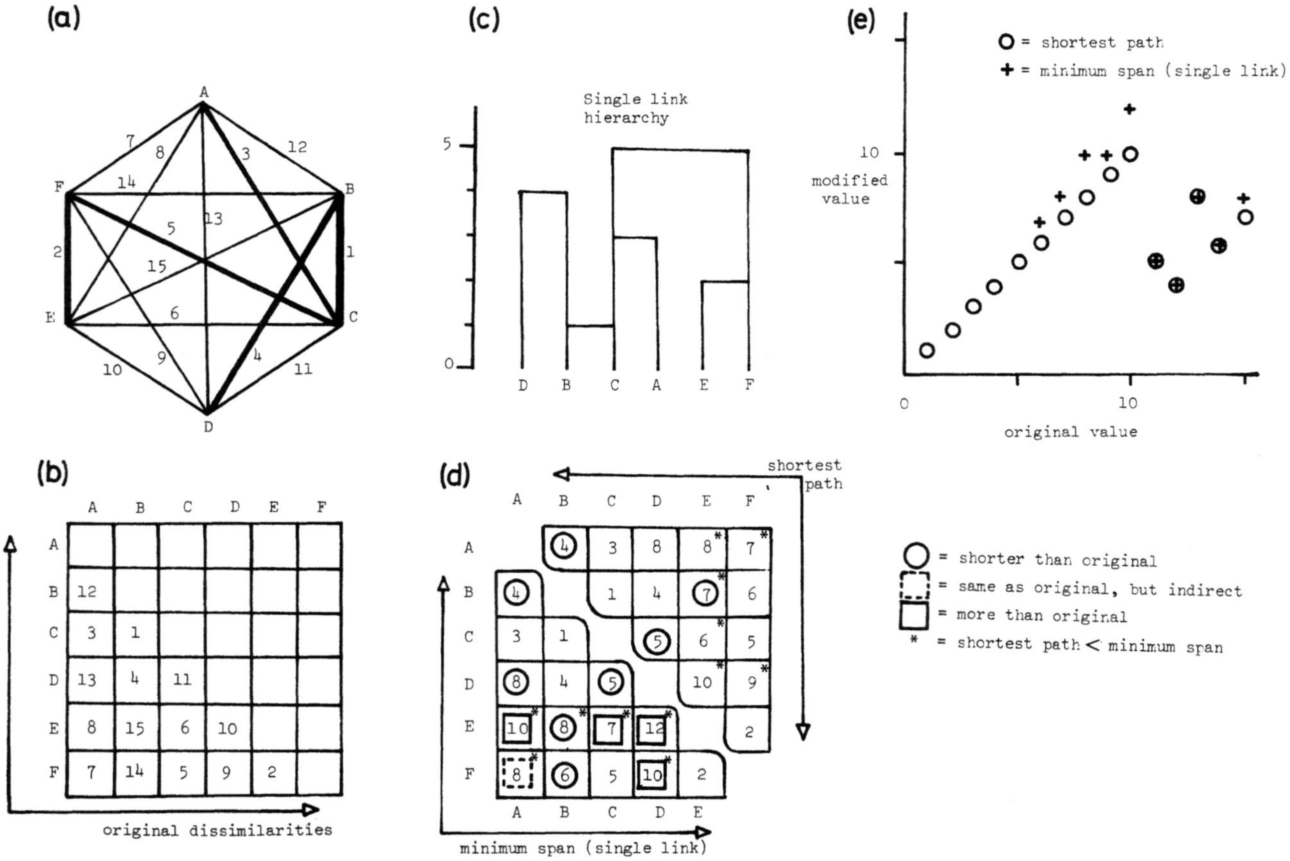

Fig. 7. Modifications to proximity coefficient matrix. (a) graph of network of proximity coefficients linking points A to F; (b) matrix representation of the network; (c) single-link hierarchy, found by using first the shortest link, then the next shortest, and so on; the corresponding structure – the minimum spanning tree – which is the shortest *total* length of links needed to interconnect all points, is shown by thick lines in (a); (d) single link (minimum spanning tree distance) in lower triangle; shortest path by *any* route, not just those of the minimum spanning tree, in upper triangle; (e) original and modified values by minimum spanning tree (single link) and shortest path.

is best known for closely similar sites, it becomes logical to try to replace the larger or dubious or unknown dissimilarities by measures compounded from smaller dissimilarities. One approach is to use the minimum (shortest) spanning tree, which is closely connected to the single-link (nearest neighbour) hierarchy. This is attractive because, there is a highly efficient algorithm, with cost a small multiple of n^2, (Sibson 1972b) for computing single-link structure. Some of the modified elements are smaller than the original, some are unchanged, but some may be larger than the original: Fig. 7 shows an artificially extreme case of this. Some of the paths may have rather a large number of segments – DBCFE in the example. These extremes may be avoided using another approach: the shortest path (Fig. 7). The modified elements are always equal to or smaller than the original ones, and elements which are missing or which have been removed because they have few or no species in common are calculated directly from those which are left – provided there is no disjunction. (Dissimilarities such as the Manhattan and Euclidean distance are already the shortest paths, so this technique is only of use in such cases if some values are to be ignored.) This is the approach used by Williamson (1978). There is a very simple algorithm, shown in Fig. 8, but the computing time is related to n^3. An intermediate approach not used here, is to replace by shortest paths, only those elements which are missing.

Essentially, as Fig. 7 shows, these approaches suppose that larger dissimilarities are over-estimated, or at least that a large dissimilarity is evidence of unreliability. Shortest path compensates to some extent, the minimum spanning tree to a greater extent.

Computation

Three programs were used for the bulk of the computing – DSD (manipulates and summarises data matrices, calculates RA and proximity coefficient matrices and modifies these); GENSTAT (a general statistical program, written at Rothamsted Experimental Station, and used here for PCA and PCO); and KYST2 (for NP-MDS and PO-MDS, from Bell Telephone Laboratories).

Results

Structure and stratigraphy

The general shape of the bog and the depth of the peat may be seen in Fig. 9. The surface is a roughly elliptical eccentric dome about 300 m by 600 m, the centre of which is ca 2 m higher than the edges. Around the perimeter the surface slopes more steeply into valleys to east, west and north. The highest measured point was at F4. The bog has formed in a basin, which may run out in a valley to the north. The southern edge seems to be particularly steep sided, with peat depth increasing from ca 3 m, to ca 10 m over not more than 100 m. The rocks of this region are part of the Carboniferous Limestone, and contain interbedded

Matrix **C**, elements C_{ij} for $i, j = 1, n$.
1. $C_{ii} = 0; i = 1, n.$
2. $C_{jj} = \infty$ for missing or invalid elements
3. $k = 0$
4. $k = k+1$
   for all $i \neq k$ such that $C_{ik} \neq \infty$
   for all $j \neq k$ such that $C_{kj} \neq \infty$
   do $C_{ij} = \min [C_{ij}, (C_{ik}+C_{kj})]$
5. Stop if
   any $C_{ii} < 0$: no solution possible
   $k = n$: solved, C_{ij} are the shortest paths
   else go to 4

Fig. 8. Shortest path algorithm (from Christofides 1975, after Floyd 1962).

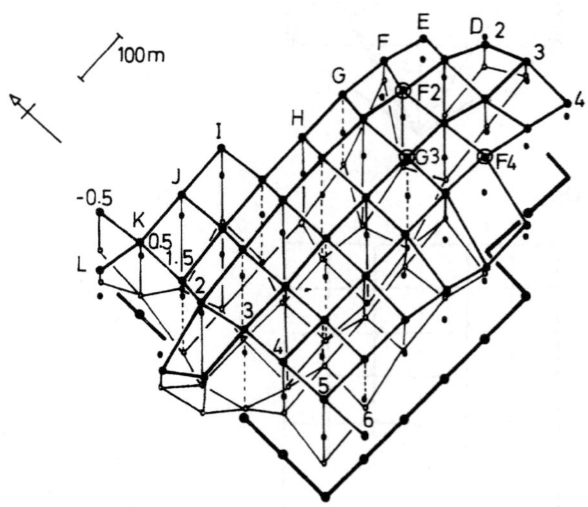

Fig. 9. Surface height and peat depth of Hummell Knowe Moss. The grid interval is 100 paces (except 50 paces between line 1.5 and 2). Surface is shown by large filled circles. Below (except at D2, L2) the small filled circles define a horizontal plane. Peat depth is shown by small unfilled circles – dashed lines indicate interpolated depths. The lower thick line defines the main survey area. Stratigraphy at F2; highest point at F4; greatest peat depth at G3.

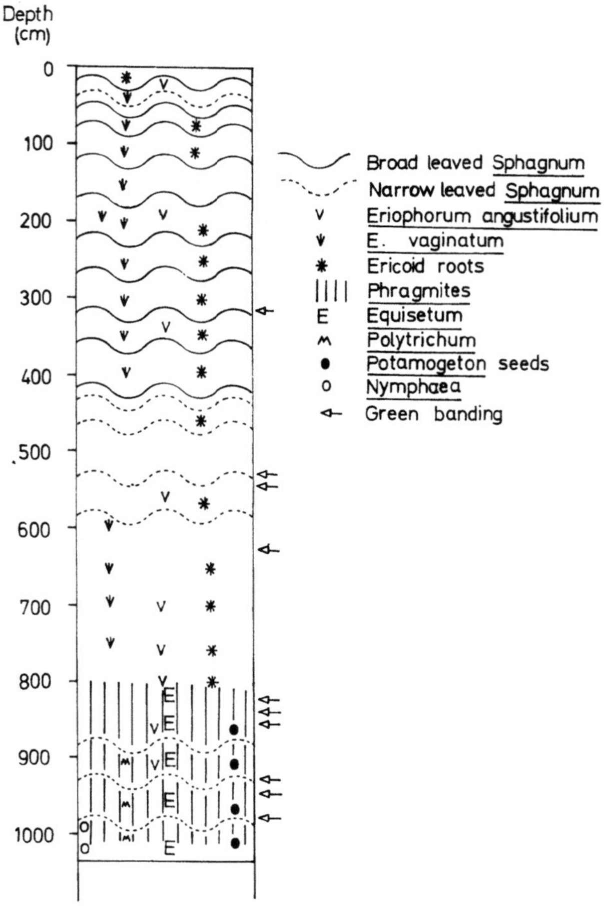

Fig. 10. Stratigraphy at site F2 (Fig. 9).

sandstones, the whole dipping gently to the south. Differential erosion has caused a north-south saw tooth profile, and it is presumably in one of these depressions that the peat has formed.

The deepest peat found was 10.7 m at G3. The stratigraphy was examined in the field at F2, where the peat is 10.4 m deep, and is shown in Fig. 10. Immediately above the basal clay is ca 2.5 m of *Phragmites* peat, with seeds of open water species such as *Nymphaea* and *Potamogeton*, and some remains of *Equisetum* and of a small leaved *Sphagnum* and of *Polytrichum*. Taken together these indicate a shallowing lake, probably tending to be acid. Above, there is a fairly sharp transition to amorphous peat with remains of *Eriophorum* and ericaceous roots. Nearer the surface the structure of *Sphagnum* is well preserved. The profile is remarkably similar to that at Coom Rigg, though no brushwood peat was found overlying the *Phragmites*. Clearly one needs more evidence before concluding anything from the *absence* of a feature: it may have been overlooked. It is clear however that at least part of Hummell Knowe Moss was at one time a lake, and the Moss as a whole is probably therefore partly of a raised bog nature.

Surface-water chemistry

The chemistry of surface waters in September proved rather uniform. In the central area pH ranged from 3.9 to 4.4, and conductivity, after allowing for the contribution from H^+, from 57 to 134 $\mu S\ cm^{-1}$. At the northern sites pH was higher – 5.5 to 6.0 – but conductivity was no greater, with the single exception of a sample from the lagg at the foot of Hummell Knowe itself, where conductivity was 730 $\mu S\ cm^{-1}$.

Vegetation

In general, the species and their abundance on the central bog plain are remarkably similar to those at Coom Rigg (Chapman 1964a). It is convenient to consider one analysis in some detail first, then to examine the effects, in turn, of transformations; of different methods of ordination; of modifications of the proximity coefficient matrix; and of different proximity coefficients. It becomes clear that there is a 'central' group of 32 sites, which are then re-analysed separately. Finally the species composition of sites in various regions of the ordinations is considered.

Before examining the results it is essential to set standards against which the usefulness of the methods may be assessed. This was done *explicitly*, in writing, before making the analyses. The following groups (Fig. 2) were expected:

(1) D2, D3, E2 and perhaps E3, associated with erosion at the eastern side.

(2) E5, F5, G5, I6, associated with the southern margin.

(3) L4, similar to group (2) but with more species characteristic of the central bog plain.

(4) Possibly K4, having some similarity to group (3), but not at all floristically distinct from the central bog plain. The real distinction was in the size of tussocks of *Eriophorum vaginatum*. Any useful method must reveal these patterns *and* it should suggest others.

First analysis

All 41 sites were used with hyperbolic weighting and Manhattan dissimilarity in an NP-MDS, shown in Fig. 11(a).

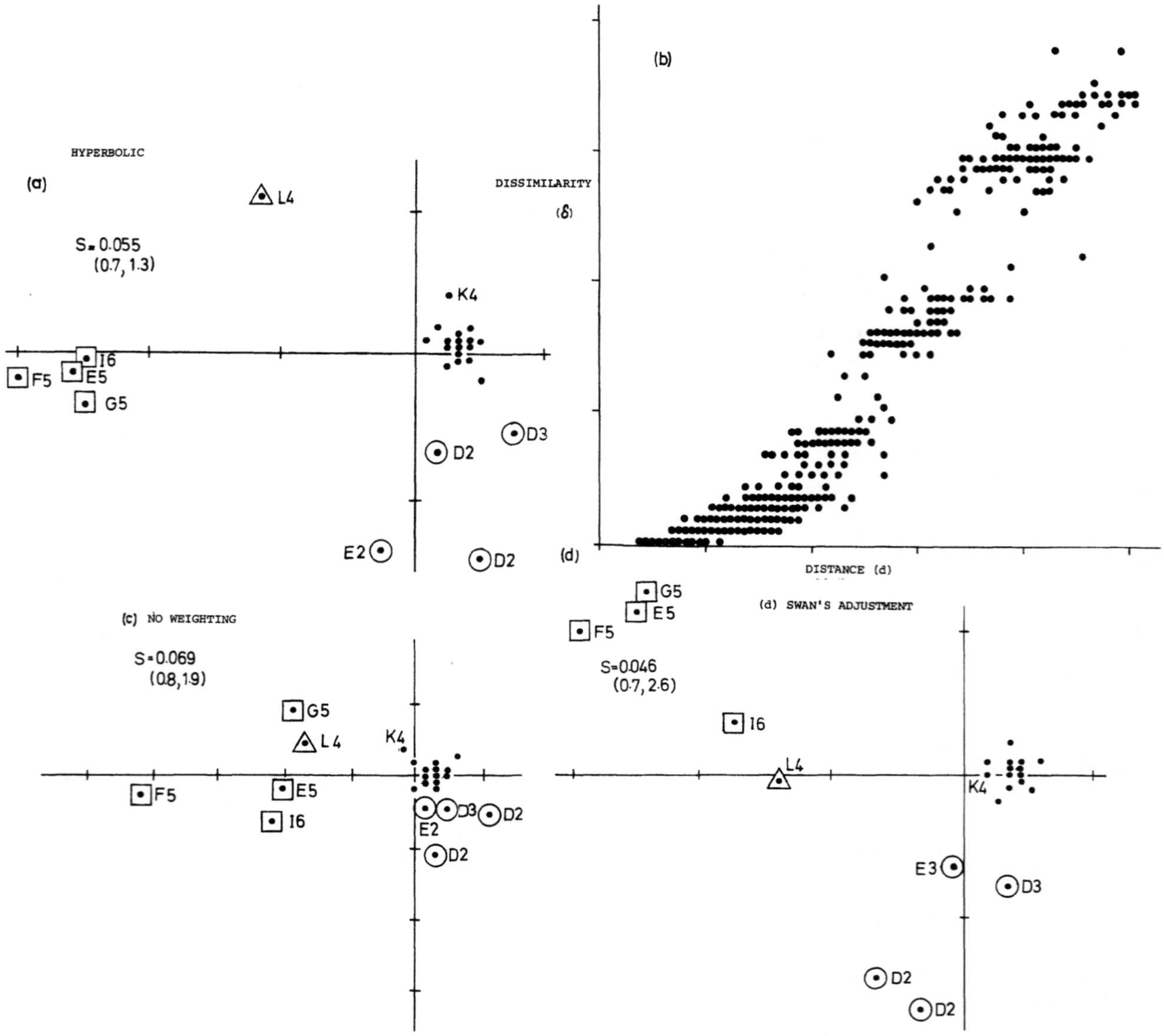

Fig. 11. Ordinations and Shepard diagram of all 41 sites at Hummell Knowe Moss, showing the effect of transformation. All use Manhattan dissimilarity. (a) hyperbolic weighting, NP-MDS; (b) Shepard diagram for (a); (c) no weighting; (d) Swan's (1970) adjustment for degree of absence. S = stress: 0.05 is low. Figures in parentheses are size of stress (relative to that in two dimensions) for three dimensions and for one dimension. In other Figs, with ordinations by RA, PCA and PCO, the values in parentheses are the proportion of variance accounted for by axes 1 to 3. Squares, triangles and circles identify groups of sites referred to in the text and mapped in Figs 2, 17.

The expected groups have appeared with embarrassing clarity: the southern margin group are the most distant from the origin; L4 is clearly separate but between the margin and centre; K4 is rather outside the central group, though were it not expected to be so, the significance of this would be overlooked; and the erosion group is also clearly distinct, though E3 is not separated. The 'replicates' of D2 are as different as any pair in this group, but they do appear in the same general region.

The Shepard diagram, Fig. 11(b), is not far from linear, and the stress in two dimensions at 6 % is low. Adding a third dimension (not illustrated) allows the erosion and marginal groups to differentiate, except for the replicates of D2, but the reduction in stress to 0.7 of that in two

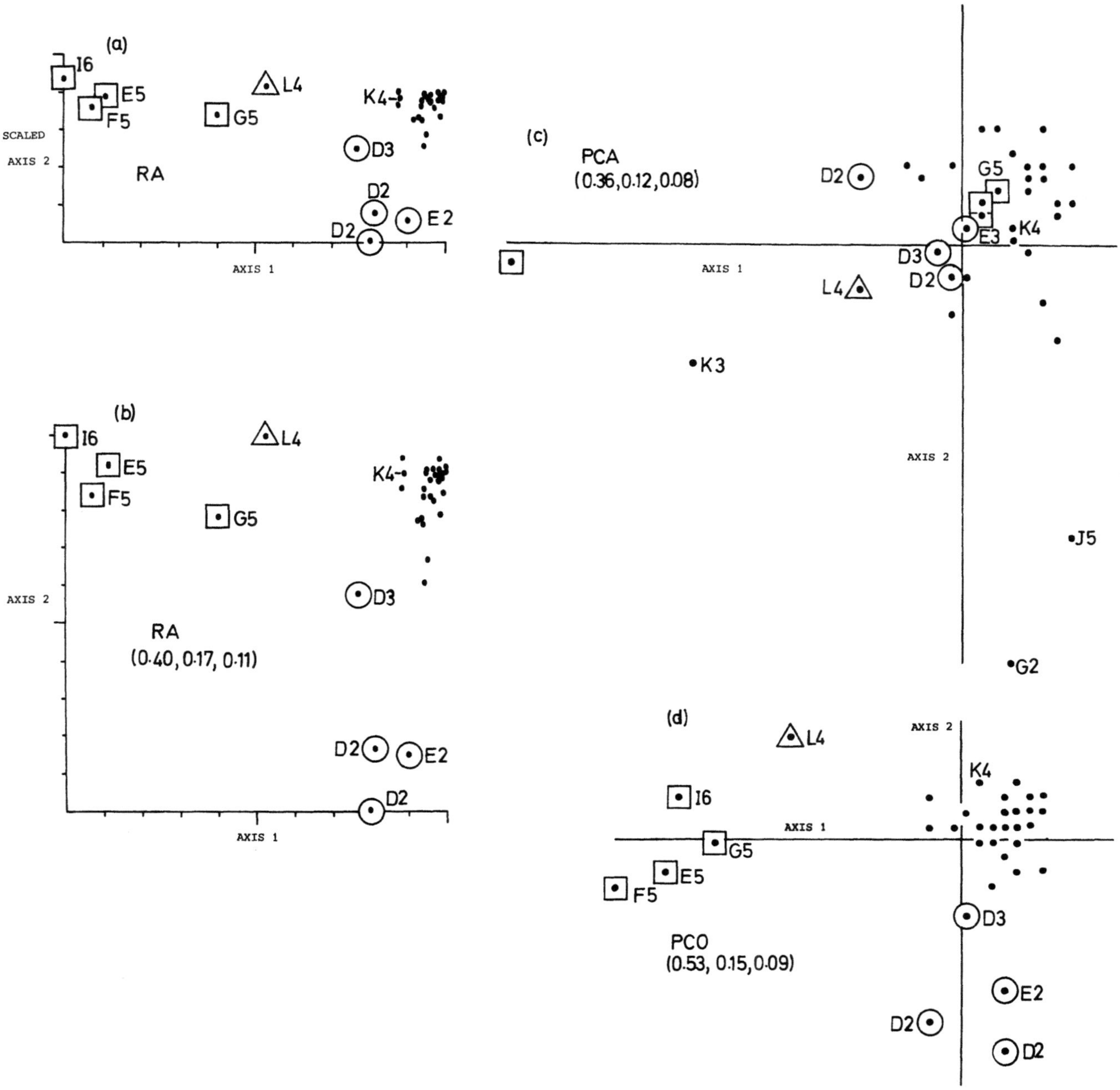

Fig. 12. Ordinations of all 41 sites at Hummell Knowe Moss, showing the effect of type of ordination. All use hyperbolic weighting. (a) RA, second axis scaled in proportion to square root of eigen vector; (b) RA, same results as (a), but second axis unscaled; (c) PCA, dispersion matrix; (d) PCO, Manhattan dissimilarity. Compare with Fig. 11 (a). Other conventions as in Fig. 11.

dimensions is not marked. Nor indeed is the increase in stress in the one dimension analysis.

Transformations

Using no transformation, Fig. 11(c), produces no marked change from Fig. 11(a) in the sense that one can see the resemblance very clearly, but the *differences* are important in that if one had no prior expectation then interpretation would be much more awkward. Probably a single linear gradient would be suggested with F5 at one end and D2 at

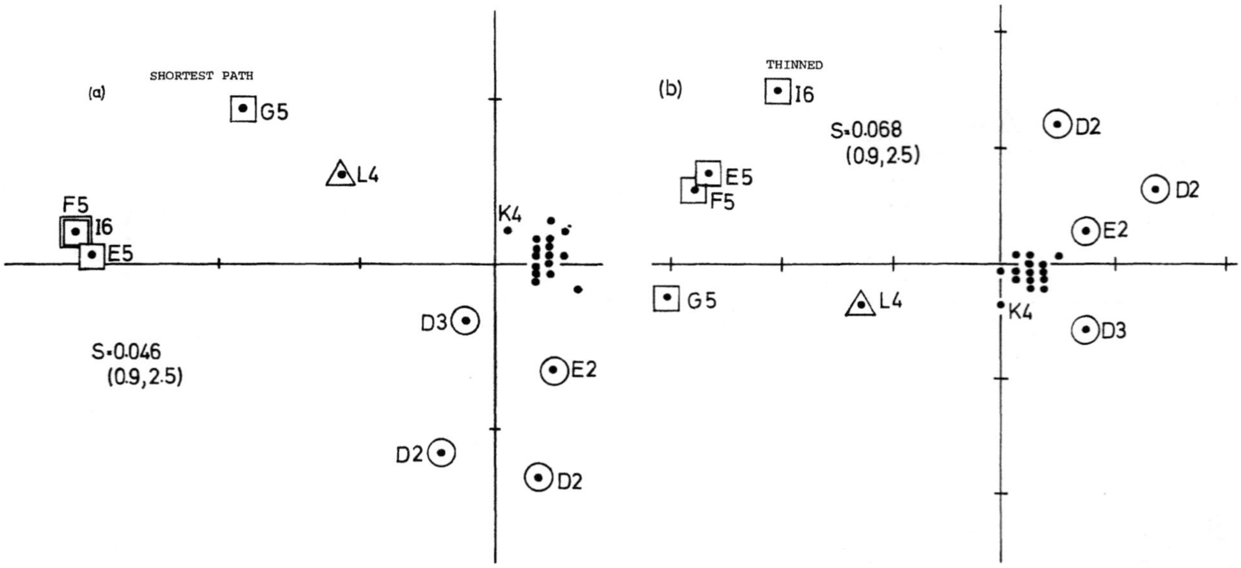

Fig. 13. NP-MDS ordinations of all 41 sites at Hummell Knowe Moss, showing the effect of modifying the dissimilarities. Both ordinations use hyperbolic weighting and Manhattan dissimilarity. (a) dissimilarities between sites with fewer than 8 species in common replaced by shortest paths; (b) data thinned by ignoring dissimilarities between sites with fewer than 8 species in common (this reduces the number of dissimilaraties from 820 to 709). Compare with Fig. 11(a). Other conventions as in Fig. 11.

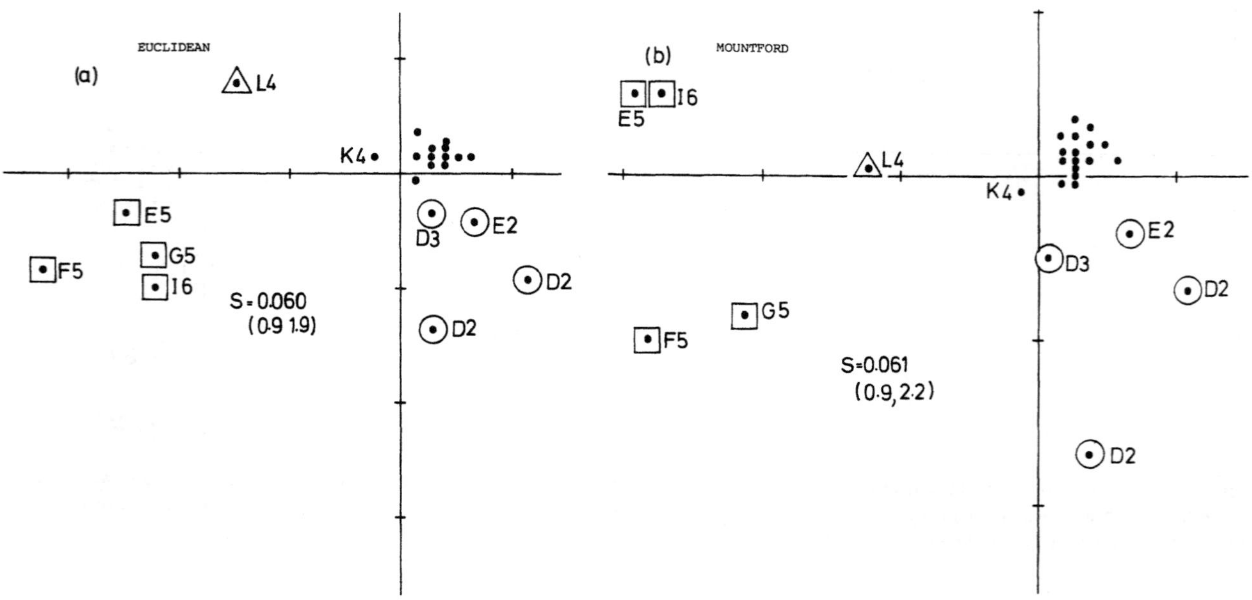

Fig. 14. NP-MDS ordinations of all 41 sites at Hummell Knowe Moss, showing the effect of different dissimilarity coefficients. (a) hyperbolic weighting, Euclidean distance; (b) modified Mountford's coefficient, ϕ (see text). Compare with Fig. 11(a). Other conventions as in Fig. 11.

the other. This is not ecological sense, and results from the small effect of the rarer species. The stress is about 7 %, and it is clear that low stress *per se* is no guide to ecological usefulness (although high stress should always leave one dubious about an interpretation). Swan's adjustment produces a pattern, Fig. 11(d), which is arguably superior to Fig. 11(a) in that the grouping of the D2 'replicates' is closer, and that I6, which is physically between the marginal sites E5, F5, and G5 but closer to the south western lagg, is placed on the ordination diagram close to the intermediate L4 site. (But see the discussion of Fig. 14.) The stress is lower than in Fig. 11(a), and two dimensions give a considerable improvement over one, though three do not produce much further improvement.

Ordination methods

The results of analyses on hyperbolically weighted data are shown in Fig. 11(a) and Fig. 12. Reciprocal averaging produces a pattern similar to NP-MDS, but makes I6 an extreme and places G5 close to L4. The adjacent K4 is not an outlier of the central group. As with Fig. 11(c), the similarity to Fig. 11(a) is obvious, but interpretation of Fig. 12(a), (b) – the scaled and unscaled plots – would be more awkward than that of Fig. 11(a) or 11(d). The PCA, Fig. 12(c), accounts for 48 % of the variance on the first two axes, but is ecologically quite misleading. The PCO Fig. 12(d), was made on the Manhattan dissimilarity, and produces results similar to that of NP-MDS, though K4 is less distinct. The first two axes account for 68 % of the variance, and the analysis, given the proximity coefficient matrix as data, costs about 15 % of the three NP-MDS ones.

Modifications of the proximity matrix

Again, hyperbolically weighted Manhattan dissimilarity was used, with NP-MDS, so the results in Fig. 13 may be compared with those in Fig. 11(a). No pair of sites had fewer than two species in common, so it is unlikely that shortest path calculations will have much effect. For illustration, it was assumed that the dissimilarity for site pairs with fewer than eight species in common was unknown, and this was calculated from the shortest path, Fig. 13(a). This may be compared with the result, Fig. 13(b), got when thinning the data by simply ignoring such cases (about 10 % of the total). The shortest path result is very similar to Fig. 11(a). The thinned result has spread the erosion and marginal groups out, as might be expected, because comparisons are now more dependent on relations between very similar sites.

Proximity coefficient

Again, NP-MDS of Manhattan dissimilarity, Fig. 11(a), may be compared with Euclidean distance, Fig. 14(a), and the modified Mountford coefficient, Fig. 14(b). Euclidean dissimilarity gives a very similar result to the Manhattan dissimilarity, as might be expected. The Mountford dissimilarity gives the now familiar pattern but it separates E5 and I6 from F5 and G5 which lie between them. This unexpected result proves to be caused by the occurrence of a few species such as *Potentilla erecta* and *Galium saxatile* in E5 and I6 and a few (different) species such as *Carex panicea* in the other pair. More interestingly, it draws attention to the greater abundance of *Polytrichum commune* in E5 and I6, and of *Sphagnum recurvum* in F5 and G5, a feature previously overlooked. These differences have obvious ecological interpretations.

The use of NP-MDS and PCO with the non-metric Mountford's dissimilarity are explored further in Fig. 15. The shortest path (for all data, whether or not they have species in common) is shown in Fig. 15(a). The separation of E5 and I6 from F5 and G5 is not obvious, though the order is still there. The site D3 is now less conspicuously separated: it is on the edge of the eroded area and still contains most of the central bog species, though they are infrequent. The Shepard diagram is markedly curved, and the stress is only 3 %, with little change in three dimensions, but a big increase in one dimension. The effect of thinning, by ignoring all pairs for which the species list of one is identical to or a subset of the other is shown in Fig. 15(c). The 60 % remaining pairs, now emphasising *differences* between sites, still give the familiar pattern, though as with the thinning in Fig. 13(b), the erosion sites are now less distinctly separated as a group.

The result with PCO, containing as it does two large negative latent roots, is, as expected, ecologically misleading. All the extreme sites, of all kinds are grouped together, but apparently equally extreme are Fl.5, F2, Gl.5, Hl. 5 at one side and H5, K2 at the other. That these groups are floristically similar is reasonable – the first group are adjacent, and the second group are in the same subjective vegetation zone (Fig. 2). The relation between these groups and the 'extreme' erosion and marginal groups is not simple however. The whole diagram seems, by comparison with all the others (except the unhelpful PCA) to be warped and differentially stretched.

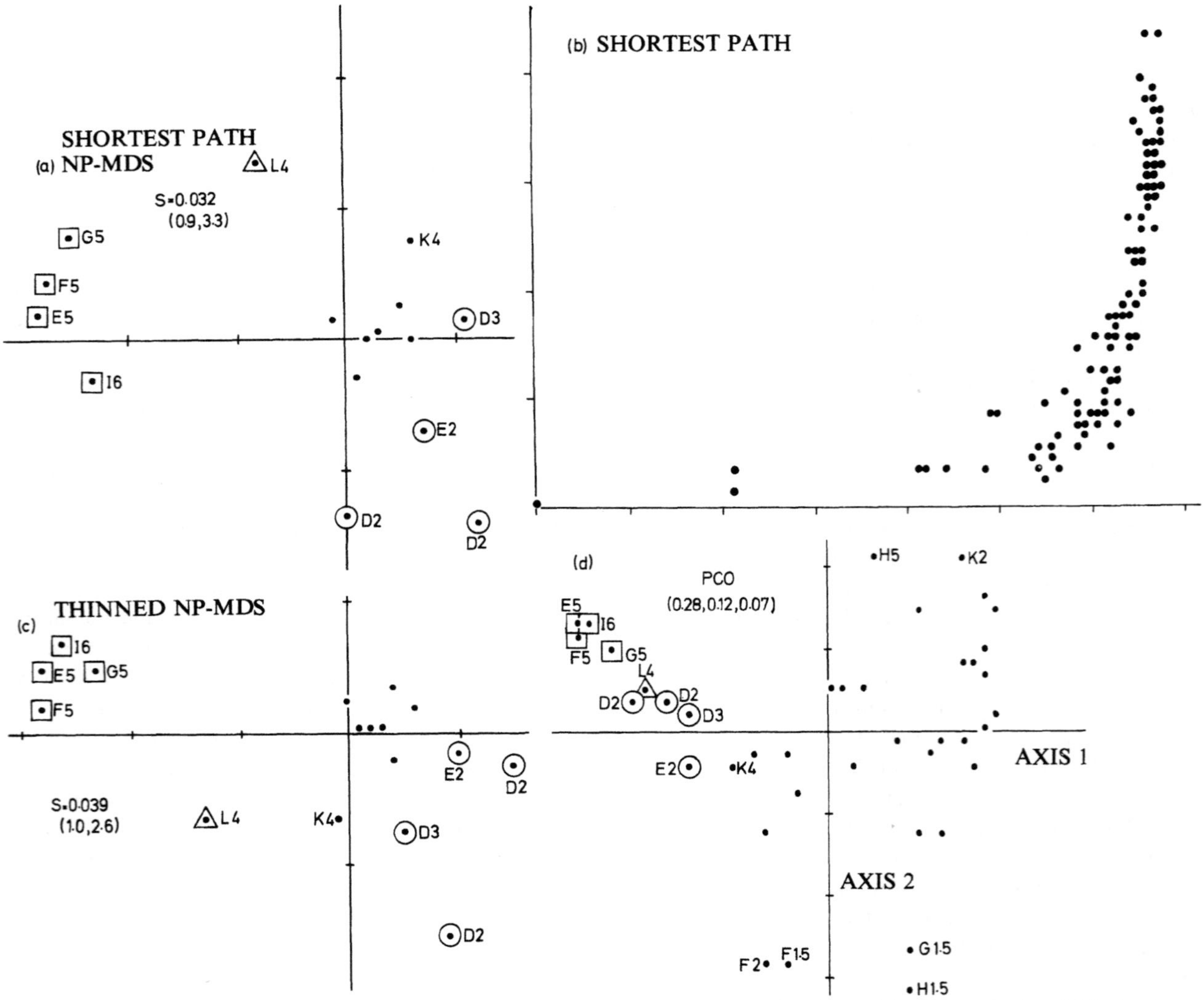

Fig. 15. Ordinations and Shepard diagram of all 41 sites at Hummell Knowe Moss, using the modified Mountford coefficient, ϕ (see text). (a) NP-MDS dissimilarities replaced by shortest paths, whether with species in common or not; (b) Shepard diagram for (a); (c) NP-MDS, ignoring dissimilarities between sites which have identical species lists, or between sites one of which is a subset of the other. (This reduces the number of dissimilarities from 820 to 495); (d) PCO. Compare with Fig. 11(a). Other conventions as in Fig. 11.

Analyses of the 32 'central' sites

So far the methods have in the main been tested for their ability to reveal the ecologically obvious. The next stage was to look for further unobvious patterns. To do this, the outlier erosion and marginal sites were removed (but leaving K4 which should therefore appear as an outlier in the new analyses) and the remaining 32 sites were re-analysed. The results of analyses of hyperbolically weighted data are shown in Fig. 16. One must expect random 'noise' to be more prominent once outliers have been removed, and the stress of 17 % for the NP-MDS (Manhattan dissimlarity) reflects this. Nevertheless the NP-MDS and PCO (same dissimilarity) gave a generally similar pattern, with the exception perhaps of J0.5, which had a relatively impoverished flora, though neither its stress (NP-MDS) nor residual (PCO) were notably large. The RA is considerably different, though the same outliers appear, and when plotted on a map of the bog surface, c.f. Fig. 17, produced no obvious patterns. Finally PCO on the shortest path replacements for pairs of sites with fewer than 12 species in common (emphasising comparisons of

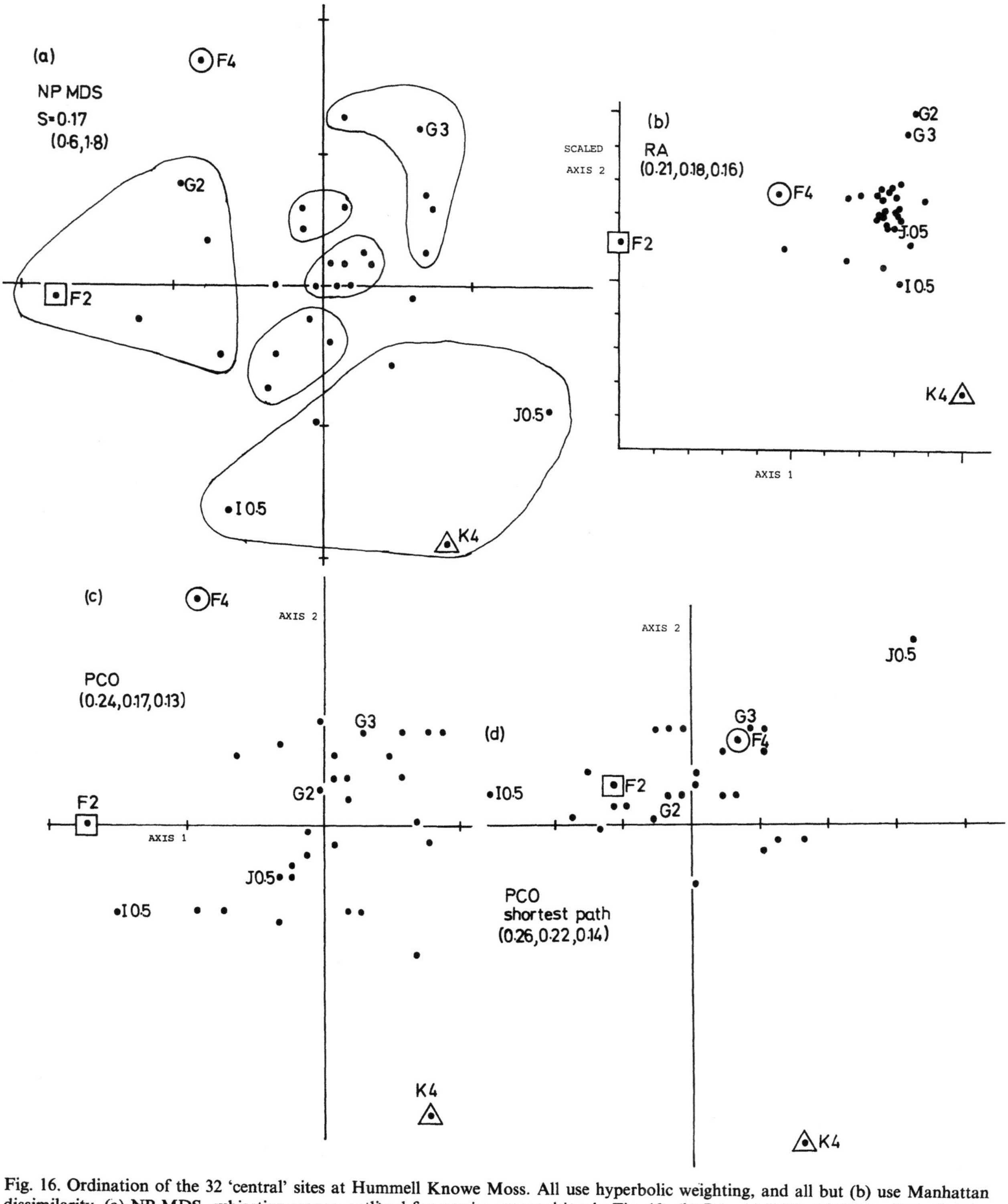

Fig. 16. Ordination of the 32 'central' sites at Hummell Knowe Moss. All use hyperbolic weighting, and all but (b) use Manhattan dissimilarity. (a) NP-MDS, subjective groups outlined for species composition in Fig. 18; (b) RA; (c) PCO; (d) PCO, dissimilarities between sites with fewer than 12 species in common replaced by shortest paths. Large symbols identify sites, or groups of sites, referred to in the text.

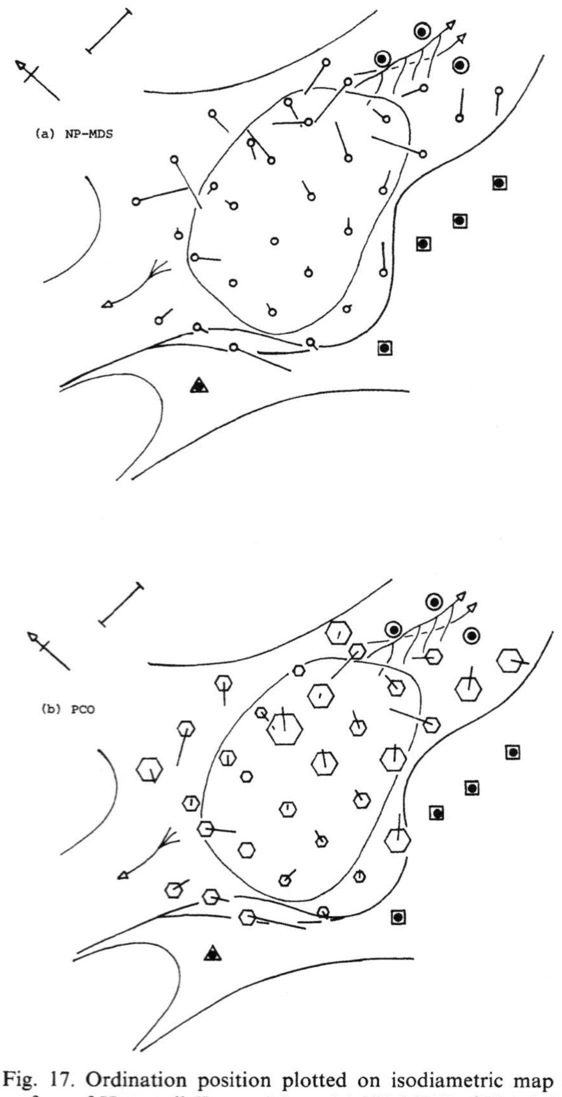

Fig. 17. Ordination position plotted on isodiametric map of surface of Hummell Knowe Moss. (a) NP-MDS of Fig. 16(a); (b) PCO of Fig. 16(c). The length and direction of line indicates the distance and direction from the origin of the ordination. Circles, squares and triangles as in Fig. 2. Hexagons in (b) show residual at that point: larger hexagon = larger residual.

closely similar sites) confirms the uncertainty about site J0.5, and pushes G2 and F2 toward the centre, indicating that their peripheral position in other analyses is owed to superficial resemblance to the peripheral sites.

The positions of the sites in the NP-MDS and PCO ordinations are plotted in Fig. 17 on a map of the bog surface. Tentatively one may identify for further investigation:

(1) An 'erosion' component pointing 'westerly' on the ordinations, and including D4, E3, Fl.5, F2 and, surprisingly, G2. It is characterised by lower frequency, or loss, of species compared to the 'central' area, and is carried further in the 'erosion' group D2, D3, E2 already removed from the analysis. The species composition is shown in Fig. 18. Particularly noticeable are the reduction and eventual loss of *Sphagnum magellanicum*, *Drosera rotundifolia*, *Eriophorum angustifolium* and *Lepidozia setacea* (but not *Mylia* or *Odontoschisma sphagni*), and the gain or increase in *Sphagnum capillifolium* and lichens.

(2) A 'marginal' component pointing 'southerly' or 'south-easterly' on the ordinations, including site K4, and characterised by reduction in frequency of *Narthecium ossifragum*, *Sphagnum tenellum*, *Drosera rotundifolia*, *Trichophorum cespitosum* and *Lepidozia setacea*. The extreme form of this is seen in the marginal sites E5, F5, G5, I6 removed from the analysis, and having none of the first three species, but high frequency (Fig. 18) of *Sphagnum recurvum*, *Polytrichum commune*, *Molinia caerulea* and *Deschampsia flexuosa*.

(3) A 'northerly' or 'north-easterly' component including sites E4, F3, G3, G4, H2, H3, H5 and I2. This is characterised by reduced frequency of *Sphagnum tenellum* ('north-easterly') or *Calluna vulgaris* ('northerly'), small changes in frequency of leafy liverworts, and small increases in *Sphagnum capillifolium* and in lichens. This component may be a slightly drier variant of the 'central' bog type. It may be significant, however, that this component clashes with the 'southerly' one along the line of the fence, so differences in management may be involved.

Species ordinations

These produced two polar groups. One contained the species of the central bog plain, most of high frequency. The other contained species of the margins and erosion, of lower frequency. These ordinations revealed nothing new, so are not shown here.

Discussion

1. NP-MDS, PCO, and to a smaller extent RA, produce patterns which seem ecologically sensible.
2. The PCO and NP-MDS analyses of the 'central' 32 sites have produced interpretable patterns, and not simply random 'noise'.
3. These patterns were not obvious either on the spot, or in the collected results. Two of the three components seem

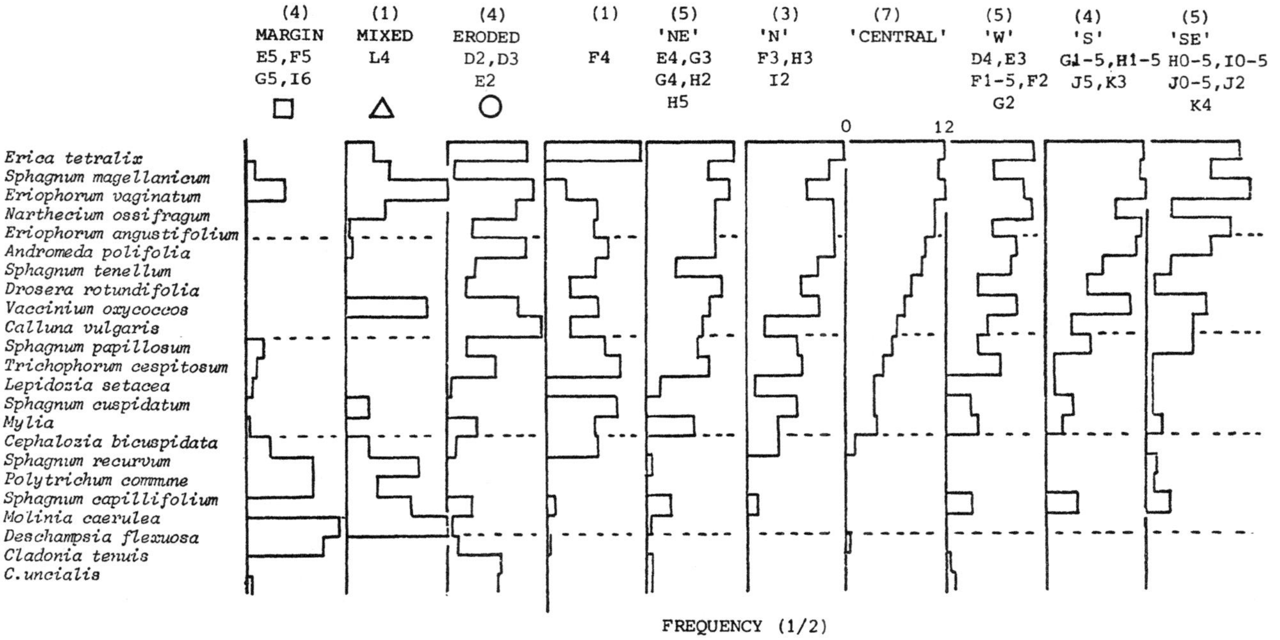

Fig. 18. Frequency of species at subjectively grouped sites. Figures in parentheses are number of sites in group. Leftmost three groups were separated in analysis of all 41 sites (e.g. Fig. 11 (a)). Others are shown in Fig. 16(a). Directions – 'NE', 'N' etc. – refer to position on the ordination diagram, not necessarily on the bog. Species are ordered by frequency in the 'central' group, except the last seven which are arranged to show associations in the first three groups. A complete species list is shown in Table 1.

to be less extreme variants of ones already obvious, but the third is new.

4. The differences between these groups are subtle and in some cases seem to point to particular species in an unexpected way.

5. It is obvious from Fig. 17 that the subjective boundary of the central bog plain is not at all well identified: the whole should be moved to the south-west and reduced at the north and east sides.

There is sufficient here to direct plans, and to that extent the work has succeeded.

Summary

(1) Hummell Knowe Moss is one of several fine and relatively undamaged peat bogs in northern England, close to the border with Scotland.

(2) The central, eccentrically domed, mass of peat is ca 300 × 600 m, and much of this is 7 to 10 m deep.

(3) A single profile has 8 m of bog peat overlying 2.5 m of *Phragmites* peat with seeds of *Potamogeton* and *Nymphaea*. The bog as a whole is probably part 'raised' and part 'blanket', as are others in the area.

(4) Much of the present surface is wet, with *Sphagnum magellanicum*, *Eriophorum* spp., *Andromeda polifolia* etc. There are some eroded areas with less *Sphagnum* and more abundant fruticose lichens (*Cladonia* spp.), and marginal areas with *Molinia caerulea*, *Deschampsia flexuosa*, *Sphagnum recurvum* and *Polytrichum commune*.

(5) Numerical analyses show the importance of data transformation. Of the ordination methods tried, PCA produced unhelpful results, but RA, PCO and NP-MDS were all adequate. PCO can take a variety of dissimilarity measures, but not all produce useful results. NP-MDS is more tolerant, and can also be used to adjust the relative importance of inter-pair dissimilarities in a more flexible way than can PCO.

(6) Further analyses by RA after removal of outliers were not of great use, but those by PCO and NP-MDS revealed clear patterns. Two of the groups of sites were less clearly developed versions of the erosion and marginal vegetation types. The third type may be developed in slightly drier conditions, or may in some places result from differences in management.

(7) The usefulness of the methods in this survey was NP-MDS > PCO > RA > PCA. Computing cost of

NP-MDS, PCO and RA parallels usefulness: more useful, greater cost.

References

Anderson, A.J.B., 1971. Ordination methods in ecology. J. Ecol. 59: 713–726.

Austin, M.P., 1976. Performance of four ordination techniques assuming three different non-linear species response models. Vegetatio 33: 43–49.

Chapman, S.B., 1964a. The ecology of Coom Rigg Moss, Northumberland. I. Stratigraphy and present vegetation. J. Ecol. 52: 299–313.

Chapman, S.B., 1964b. The ecology of Coom Rigg Moss, Northumberland. II. The chemistry of peat profiles and the development of the bog system. J. Ecol. 52: 315–351.

Chapman, S.B., 1965. The ecology of Coom Rigg Moss, Northumberland. III. Some water relations of the bog system. J. Ecol. 53: 371–384.

Christofides, N., 1975. Graph theory, an algorithmic approach. Academic Press, New York, 400 pp.

Clapham, A.R., T.G. Tutin & E.F. Warburg, 1962. Flora of the British Isles. 2nd edn. Cambridge University Press. Cambridge, 1591 pp.

Dale, M.B., 1975. On objectives of methods of ordination. Vegetatio 30: 15–32.

Duncan, U.K. with P.W. James, 1970. Introduction to British Lichens. Buncle, Arbroath, 292 pp.

Fasham, M.J.R., 1977. A comparison of nonmetric multidimensional scaling, principal components and reciprocal averaging for the ordination of simulated coenoclines and coenoplanes. Ecology 58: 551–561.

Floyd, R.W., 1962. Algorithm 97 – shortest path. Comm. Assoc. Comput. Mach. 5: 345.

Gower, J.C., 1966. Some distance properties of latent root and vector methods used in multivariate analysis. Biometrika. 53: 325–338.

Hill, M.O., 1973. Reciprocal averaging: an eigenvector method of ordination. J. Ecol. 61: 237–249.

Holloway, J.D. & N. Jardine, 1968. Two approaches to zoogeography: a study based on the distributions of butterflies, birds and bats in the Indo-Australian area. Proc. Linn. Soc. Lond., 179: 153–188.

Jensén, S. 1978. Influences of transformation of cover values on classification and ordination of lake vegetation. Vegetatio 37: 19–31.

Krzanowski, W.J., 1969. The algebraic basis of classical multivariate methods. Statistician. 20 (4): 51–61.

Kruskal, J.B., 1964. Multidimensional scaling by optimizing goodness of fit to a nonmetric hypothesis. Psychometrika. 29: 1–27.

Kruskal, J.B. & M. Wish, 1977. Multidimensional scaling. Sage Publications. Beverley Hills, California.

Maarel, E. van der. 1979. Transformation of cover-abundance values in phytosociology and its effects on community similarity, Vegetatio 39: 97–114.

Mountford, M.D., 1962. An index of similarity and its application to classificatory problems. In: P.W. Murphy (ed): Progress in Soil Zoology, p. 43–50. Butterworths. London.

Noy-Meir, I., 1971. Multivariate analysis of the semi-arid vegetation in south eastern Australia. Proc. Ecol. Soc. Austral. 6: 159–193.

Noy-Meir, I., 1973. Data transformations in ecological ordination. I. Some advantages of non-centering. J. Ecol. 61: 329–341.

Noy-Meir, I., D. Walker & W.T. Williams, 1975. Data transformations in ecological ordinations. II. On the meaning of data standardization. J. Ecol. 63: 779–800.

Orlóci, L., 1978. Multivariate analysis in vegetation research. 2nd edn. Junk, The Hague, VIII + 450 pp.

Pearson, M.C., 1960. Muckle Moss, Northumberland. J. Ecol. 48: 647–666.

Prentice, I.C., 1977. Non-metric ordination methods in ecology. J. Ecol. 65: 85–99.

Ratcliffe, D.A. (ed), 1977. Nature conservation review. Vol. 2. Site Accounts, 320 pp.

Sibson, R., 1972a. Order invariant methods for data analysis. J. Roy. Statist. Soc. B. 34: 311–349.

Sibson, R., 1972b. SLINK: an optimally efficient algorithm for the single-link cluster method. Computer J. 16: 30–34.

Smartt, P.F.M., S.E. Meacock & J.M. Lambert. 1976. Investigations into the properties of quantitative vegetational data. II. Further data type comparisons. J. Ecol. 64: 41–78.

Smith, A.J.E., 1978. The Moss Flora of Britain and Ireland. Cambridge University Press. Cambridge.

Swan, J.M.A., 1970. An examination of some ordination problems by use of simulated vegetational data. Ecol. 51: 89–102.

Watson, E.V., 1968. British Mosses and Liverworts. 2nd edn. University Press. Cambridge.

West, R.G., 1968. Pleistocene geology and biology. Longmans. London.

Williams, C.B. 1964. Patterns in the balance of nature. Academic Press. London & New York.

Williamson, M.H., 1978. The ordination of incidence data. J. Ecol. 66: 911–920.

Accepted 14 December 1979

CLASSIFICATION AND ORDINATION IN THE INDIAN PEAKS AREA, COLORADO ROCKY MOUNTAINS

Věra KOMÁRKOVÁ*

Institute of Arctic and Alpine Research and Department of Environmental, Population, and Organismic Biology, University of Colorado, Boulder, Colorado 80309, USA

Keywords:
Alpine vegetation, Classification, Colorado Rocky Mountains, Ordination, Syntaxonomy

Introduction

The present paper concerns an attempted integration of classification and ordination methods in a study of alpine vegetation of the Indian Peaks area, Colorado Rocky Mountains (Komárková 1979). Classification, including Braun-Blanquet's, and ordination techniques have been combined, for example, by van Groenewoud (1965), Lacoste (1976), Kortekaas, van der Maarel & Beeftink (1976), and Feoli-Chiapella & Feoli (1977). The need for integration of syntaxonomy with numerical phytosociology has been recently emphasized by van der Maarel, Orlóci & Pignatti (1976) and Noy-Meir & Whittaker (1977).

The Indian Peaks, a temperate, mid-latitude mountain range of the Northern Hemisphere, are located between 40°10′25″ and 40°00′N, and 105°44′ and 105°32′30″ W in the easterly positioned Front Range of the Rocky Mountains. Their environment is characterized by high altitude, continental climate, north-south orientation, and substrate that locally shows a high content of calcium. The flora has Arctic and Asiatic connections, and the vegetation has connections with that of mountains in continental climate regions of the Northern Hemisphere.

The alpine environment of the Indian Peaks, including the vegetation, is regarded as a universe defined by its relative environmental and biotic uniformity, and contained by the boundary of the alpine treeline. To reduce and retain the vegetational and environmental complexity, the numerical methods have been applied to preestablished Braun-Blanquet associations. The predetermined syntaxonomy and the implications of numerical analysis for individual syntaxa have been reported in Komárková (1979).

The following hypotheses were examined:
1. Higher-level units of the Braun-Blanquet hierarchy and their diagnostic taxa groups can be derived by numerical methods from an association taxa data matrix.
2. On the basis of this data, numerical methods can demonstrate some of the environmental gradient complexes controlling compositional variation and the response of syntaxa and diagnostic taxa to these gradients. The results will agree with those suggested by the Braun-Blanquet tablework.
3. The compositional distinctiveness and uniformity of syntaxa paralells environmental distinctiveness and uniformity.

This numerical analysis of the Indian Peaks data is preliminary. Additional work is aimed at the understanding of the relationships between the syntaxa, and the environmental response of taxa and syntaxa.

Methods

Sampling

545 relevés were sampled according to the Braun-Blanquet method of vegetation analysis (Westhoff & van der Maarel 1978). 482 relevés and 404 taxa appear in the final syn-

* Most of the figures in this paper are from a book version of a University of Colorado, Department of Environmental, Population, and Organismic Biology, Ph. D. thesis which was produced at the Institute of Arctic and Alpine Research. I would like to thank Dr. P.J. Webber, the thesis advisor, for his support, for suggesting the numerical methods, and for computer programs which were written by W.F. Reid. Vicki Dow, Marilyn Joel, and Karen Sproul drafted the majority of figures. I am grateful to Gwen Archer for editorial help.

taxonomic table. The sampling approach assumed that by sampling all recognizable habitat and vegetation types, the response of plants and plant communities to the environment will be represented. The most accessible areas yielded the highest relevé density. An effort was made to distribute relevés evenly, replicating the same habitat or vegetation type.

Sampling of the environment involved descriptions of soil profiles and topographical site characteristics, and estimates of climatic and surface conditions on subjective gradient scales (details in Komárková 1979).

Data

To conform to the computer facilities available at the time, the relevé matrix was reduced to a matrix of 63 Braun-Blanquet associations (cf. van der Maarel 1969, 1972). Taxa occurring in one or only a few relevés were omitted; 340 taxa were retained. The continuity of the data was thus increased.

The taxa cover-abundance values were transformed after Moore et al. (1970). The mean transformed cover-abundance value for associations was calculated similarly to the 'Deckungswert' (Braun-Blanquet 1964). Because the analysis of this quantitative data was less satisfactory than the analysis of qualitative data, only the latter is presented. The 63 associations/340 taxa matrix was further reduced to include only the 94 most important taxa; the analysis of this matrix was satisfactory for only the association index calculations.

Data analysis

The relevé data were analyzed by standard Braun-Blanquet classification methods (Westhoff & van der Maarel 1978).

With Pearson's product-moment correlation coefficient as the clustering criterion, agglomerative clustering of associations based on the taxa data produced polythetic, hierarchical dendrograms by the average linkage method (Sokal & Sneath 1963). Inverse analysis produced clusters of taxa. Associations were also clustered on the basis of environmental data.

Linkage diagrams for 94 taxa, constructed by the elementary linkage method (Sokal & Sneath 1963), were based on the index of amplitudinal correspondence (Bray 1956). The members of a diagnostic taxa group were placed in the same sequence as in the syntaxonomic table and linked according to association similarity levels. All association similarities above 80 % are plotted. Taxa associated with the diagnostic group members above this similarity are included in the diagrams; associations among these additional taxa are not shown. The clusters were compared according to a proportion of the possible links between the diagnostic group members which have been realized, the similarity level of the links, and the number of diagnostic group-nonmembers entering the cluster (cf. McIntosh 1978).

Polar ordination (Bray & Curtis 1957) determined the relationship of compositional variation to environmental gradients and located taxa in a phytosociological space. This ordination method was found more informative than some other ordination methods by Clark (1977). The effective ordinations arranged groups of associations, i.e., alliances and orders, and taxa groups, i.e., alliance and order diagnostic taxa groups, along axes representing environmental gradients (cf. Orlóci 1966, Noy-Meir & Whittaker 1977).

Product-moment correlations between the three axes of the ordination of associations and selected environmental variables indicate the environmental gradient complexes which are responsible for compositional variation. Partial composition of these complexes is revealed by agglomerative clustering and correlation of all variables. Isolines connect equal values of selected environmental variables on three ordination planes; relative importance of selected taxa on these planes is also shown. This approach has been used by May (1973) and Webber (1978).

Results

Classification

Seven classes, 8 orders, 17 alliances, and 63 associations were recognized in the Braun-Blanquet system. Alliances and orders were characterized by diagnostic taxa groups and newly described: European classes were used provisionally. Komárková (1979) presented the total classification.

A hierarchical dendrogram of associations based on qualitative taxa data reflects the syntaxonomic table of associations (Fig. 1). Orders are formed at an average of 30 % similarity, alliances at 37 %. The definition of suballiances and alliances was usually better in the dendrograms for individual classes and orders.

In Fig. 1, four groups of orders (classes) can be observed at low levels of similarity. *Betulo-Adenostyletea* and *Montio-Cardaminetea*, which are very close and located at the end of the syntaxonomic table, are separated

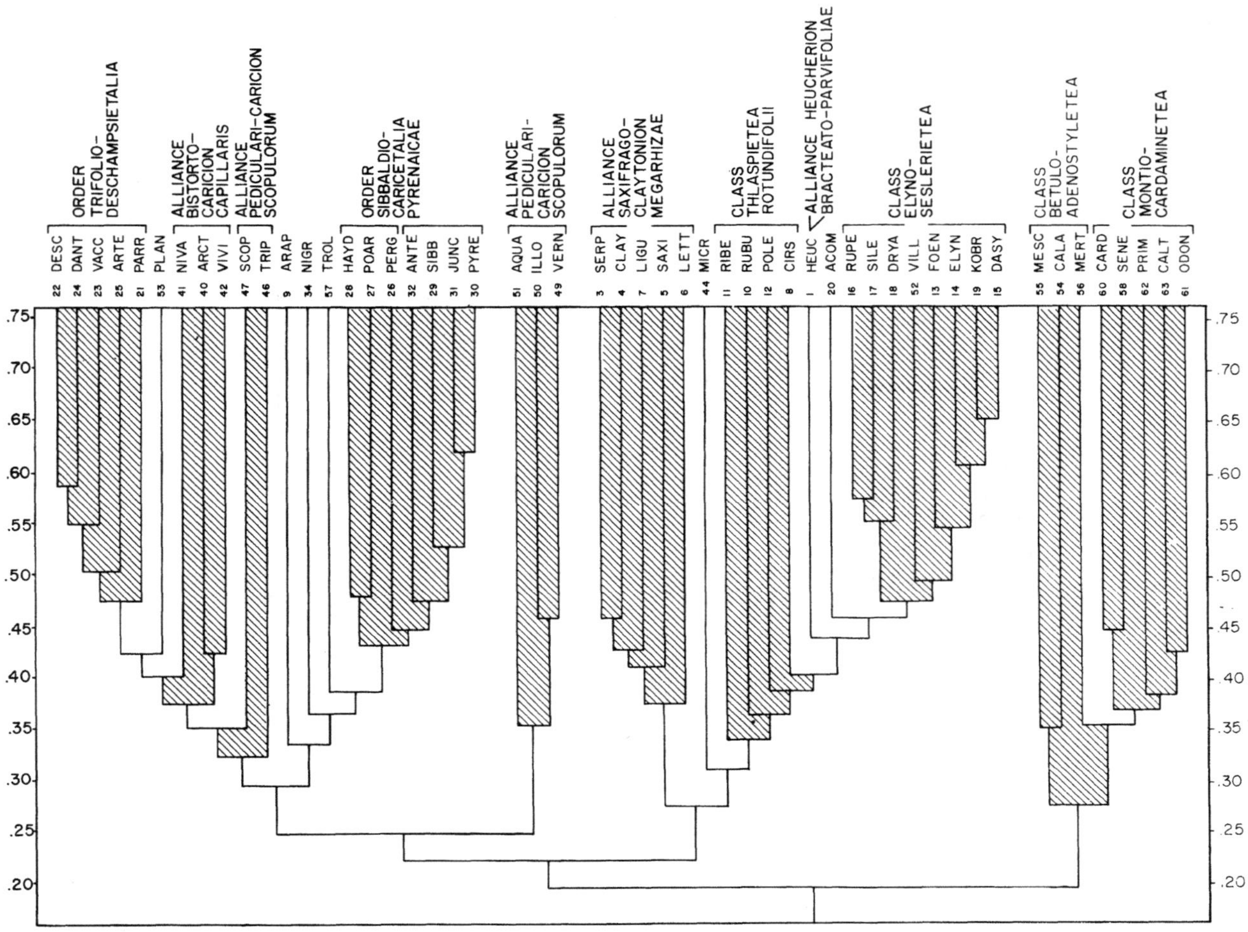

Fig. 1. Hierarchical dendrogram of 52 associations, based on qualitative taxa data and constructed by the average linkage method. Product-moment correlation coefficients label theY axis; numbers and four-letter codes identify associations. Names at the top indicate orders (classes) and alliances, determined earlier by tablework.

Alliances are identified at an average 36 % of similarity, orders(classes) at 30 %. Of the four clusters diverged at 25 %, the third cluster and a part of the first cluster constitute the first successfully ordinated group of associations, and the first, second, and fourth cluster constitute the second group (Fig. 5).

from all other associations at the lowest level 19 %. *Asplenietea rupestria*, *Thlaspietea rotundifolii*, and *Elyno-Seslerietae*, which are located in this order at the beginning of the table, form a cluster separated at 23 % similarity. *Salicetea herbaceae* with orders *Trifolio-Deschampsietalia* and *Sibbaldio-Caricetalia pyrenaicae*, and two alliances of *Scheuchzerio-Caricetea fuscae* form a cluster at 25 % similarity, with only a few associations of *Pediculari-Caricion scopulorum* in a separate cluster at the same level.

The syntaxa, which were easily established during the tablework and which usually have a high number of diagnostic taxa, cluster at the highest similarity levels in the dendrogram, e.g., *Elyno-Seslerietea* at 47 %, *Trifolio-Deschampsietalia* at 47 %, and *Sibbaldio-Caricetalia pyrenaicae* at 43 %. The associations, which were difficult to place and were positioned on the basis of cover-abundance values, are often found among associations of different higher syntaxa in the dendrogram than in the table.

The clusters of associations in the hierarchical dendrogram based on environmental variables (Fig. 2) resemble less the arrangement of associations in the syntaxonomic table than the clusters in Fig. 1. However, they still suggest an environmental distinctiveness of, and uniformity within individual orders (classes). This uniformity also characterizes many alliances, but their clusters formed in only a few instances in Fig. 2; they were usually clearly apparent in the dendrograms for individual classes and orders.

The floristic and environmental resemblance between various syntaxa apparent from the dendrograms often corresponds to that implied during the tablework. For example, *Kobresio-Caricetalia rupestris (Elyno-Seslerietea)* and *Trifolio-Deschampsietalia*, which occur in closely related habitats and which share a group of diagnostic taxa, were placed into two different classes in the table based on the overall floristic differences. These orders are separated at 23 % similarity in Fig. 1, but form a cluster at 67 % in Fig. 2. *Trifolio-Deschampsietalia* appears then as an order intermediate between *Elyno-Seslerietea* and *Salicetea herbaceae*, closer to *Elyno-*

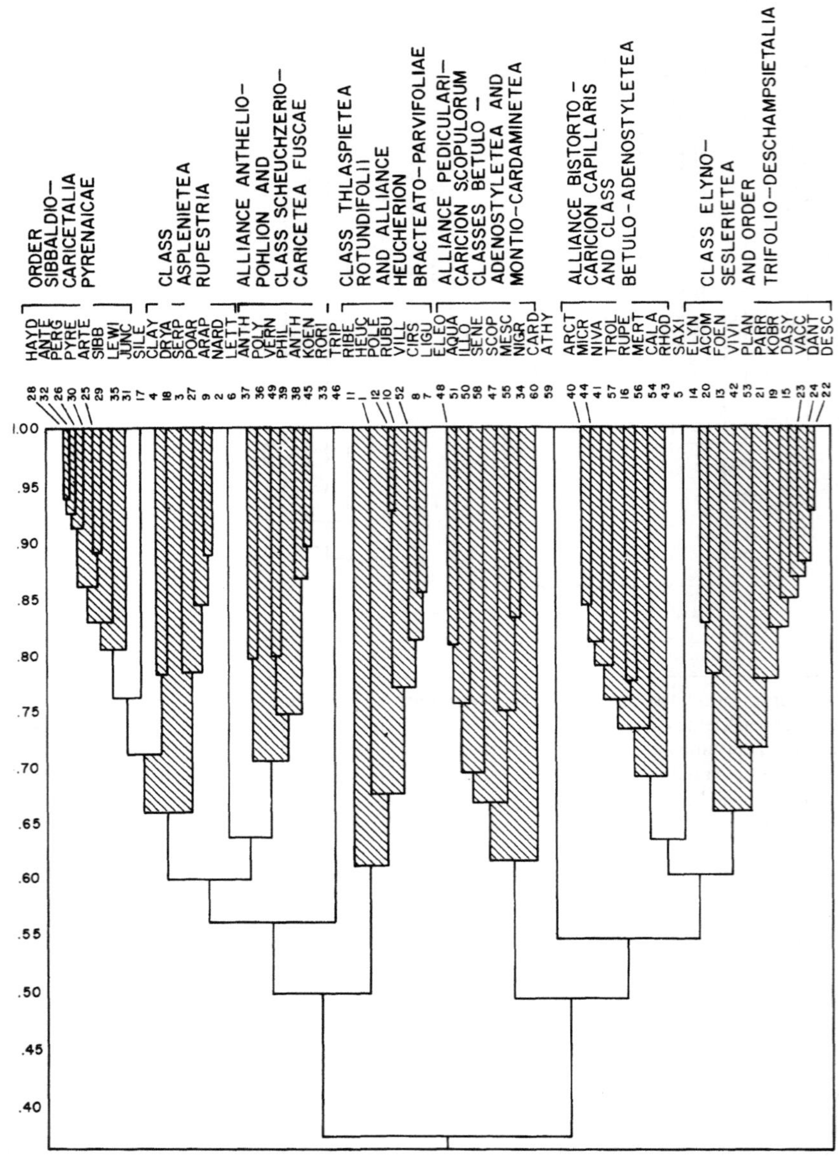

Fig. 2. Hierarchical dendrogram of 60 associations, based on environmental variables and constructed by the average linkage method. Product-moment correlation coefficients label the Y axis; numbers and four-letter codes identify associations. Names at the top indicate orders (classes) and alliances, determined earlier by tablework.
Most orders (classes) appear as separate clusters, which indicates their environmental distinctiveness and uniformity.

Seslerietea environmentally than floristically. The relationships between individual associations and higher syntaxa are also clarified.

Dendrograms of associations based on groups of related environmental variables (not presented) showed that one group of variables was responsible for clustering which was the closest to the Braun-Blanquet hierarchy. This group included estimates of site moisture, duration of snow cover, temperature, and surface age: i.e., factors also underlying the axis along which relevés and associations are recorded in the syntaxonomic tables.

A hierarchical dendrogram based on qualitative association data (Fig. 3) groups taxa into clusters which more or less correspond to the diagnostic taxa groups in the syntaxonomic table of associations. Both the regional taxa group and the group of diagnostic taxa of *Kobresio-Caricetalia rupestris (Elyno-Seslerietea)*, which are among the best-defined taxa groups in the syntaxonomic table,

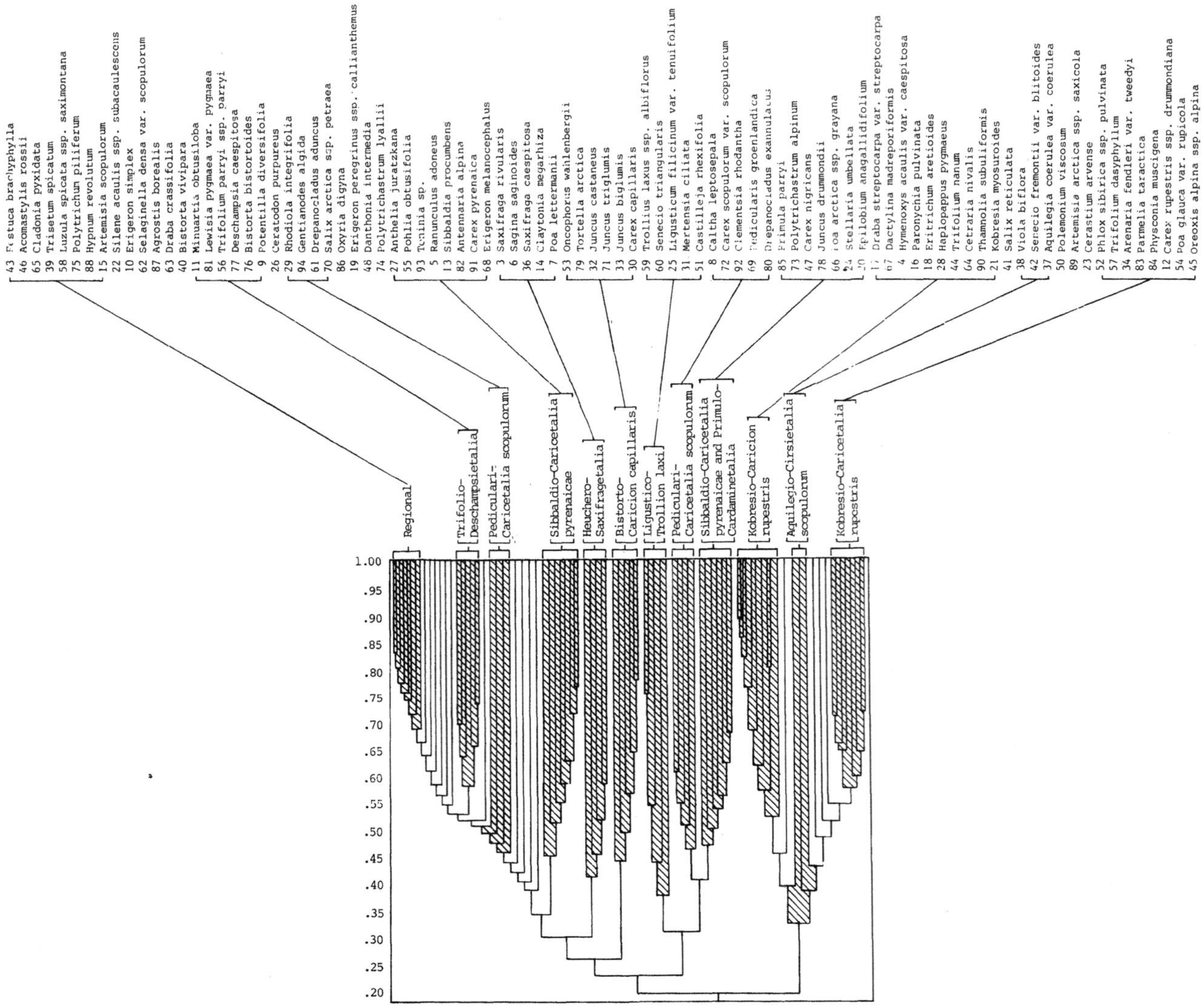

Fig. 3. Hierarchical dendrogram of 94 taxa, based on qualitative association data and constructed by average linkage method. Product-moment correlation coefficients label the Y axis. The syntaxon name indicates the diagnostic taxa group, determined earlier by tablework; the group members are identified by their number and name. Some taxa are listed under a higher- or lower-level syntaxon rather than under their own.
The correspondence between the composition of the clusters and the diagnostic taxa groups is fairly good Except for regional taxa,the similarity level at which the clusters are formed does not appear to be directly related to the position of the syntaxon in the hierarchy.

form clusters at high levels of similarity. The two clusters diverging at 22 % similarity contain taxa of subxerophilous to mesophilous and mesophilous to subhygrophilous associations, respectively. The dendrogram demonstrates the relationship of regional taxa to mesophilous *Trifolio-Deschampsietalia*, in which most of these taxa have one of their optima.

Like some associations in Fig. 1, some taxa appear in different groups in the dendrogram than in the table because the quantitative data was disregarded in the dendrogram. For instance, *Caltha leptosepala* joins the diagnostic taxa group of *Pediculari-Caricetalia scopulorum (Scheuchzerio-Caricetea fuscae)* in the dendrogram. In the table, *Caltha leptosepala* belongs among the taxa of *Primulo-Cardaminetalia (Monte-Cardaminetea)* in which it reaches its highest cover-abundance values.

Fig. 1 and Fig. 3 correspond to some degree. In the table and both dendrograms, clusters at high similarity levels include *Elyno-Seslerietea (Kobresio-Caricetalia rupestris)*, *Trifolio-Deschampsietalia*, and *Sibbaldio-Caricetalia pyrenaicae*. Close relationships are also maintained between *Thlaspietea rotundifolii (Aquilegio-Cirsietalia scopulorum)* and *Elyno-Seslietea*, and *Betulo-Adenostyletea (Salici-Trollietalia)* and *Montio-Cardaminetea (Primulo-Cardaminetalia)* resp. The results of agglomerative clustering appear to show a greater correspondence to the syntaxonomic table in the direction of associations than in the direction of taxa.

Fig. 4 is an example of a linkage diagram based on association index values for a diagnostic taxa group. These diagrams corresponded well to the results of tablework and clarified relationships among individual taxa. The taxa identified as regional by the tablework were associated with at least 30 % of all taxa in at least 87 % of occurrences and, therefore, not included in the diagrams. Among the orders, *Kobresio-Caricetalia rupestris* has the best-defined cluster (shown) which reflects the size and clear definition of its diagnostic taxa group in the syntaxonomic table, and the high similarity at which this cluster appears in Fig. 3. The respective diagnostic taxa groups of *Kobresio-Caricetalia rupestris* and *Kobresio-Caricion rupestris* are not as clearly defined as in Fig. 3. Within the diagnostic taxa group for both *Kobresio-Caricetalia rupestris* and *Trifolio-Deschampsietalia* are fewer high-similarity links than within the diagnostic taxa groups belonging only to *Kobresio-Caricetalia rupestris*.

Ordination

Polar ordination of the data set consisting of all associations was not successful. The data set was then divided by splitting the syntaxonomic table, traditionally arranged approximately along the moisture gradient, into two halves along the axis ordering the associations. This division was made on the basis of only the syntaxonomic table; later, similar division appeared in most dendrograms. The partial data sets with narrower environmental and compositional ranges produced two ordinations (Fig. 5).

Well-defined clusters of associations corresponding to predetermined syntaxa were formed within both ordinations. Most of the syntaxa are clearly spaced along the ordination axes. In general, syntaxa which commonly occur in average, zonal environmental conditions *(Kobresio-Caricion rupestris, Trifolio-Deschampsietalia)* take up less ordination space than syntaxa of marginal, nonzonal habitats (e.g., *Aquilegio-Cirsion scopulorum, Saxifrago-Claytonion megarhizae)*.

The X axis in both ordinations is significantly correlated with a complex environmental gradient represented by site moisture, duration of snow cover, and site temperature estimates. This complex environmental gradient also underlies the syntaxonomic table; the division of the syntaxonomic table into two halves approximately separated subxerophilous to mesophilous associations from mesophilous to subhygrophilous associations. Different variables join this primary complex gradient in each ordination. The Y and Z axes are also significantly correlated with some of the X axis variables. Site temperature and site moisture estimates were frequently correlated with ordination axes also in ordinations of individual orders (not presented).

Dendrograms of environmental variables (Fig. 6) and correlation matrices partially explained the composition of environmental gradient complexes correlated with ordination axes. Site moisture, duration of snow cover, site temperature, and surface age occur in different clusters, each of which may represent one of the environmental gradient complexes controlling the compositional variation.

The environmental interpretation of the position of syntaxa in the ordination agreed with the mean values of respective environmental variables. Isolines (Fig. 7) illustrate the complex relationship between environmental variables, the ordination axes, and syntaxa. Isolines

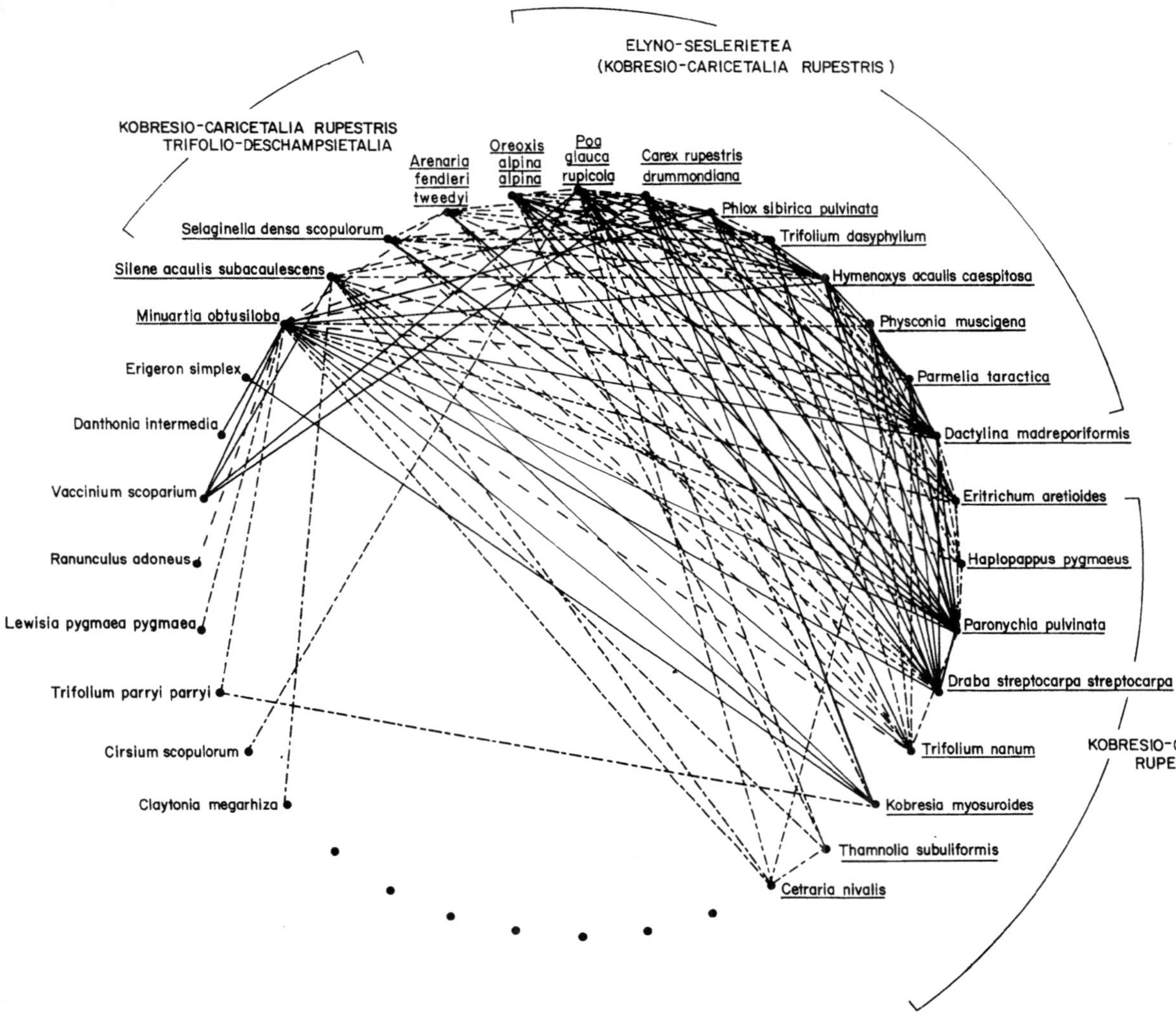

Fig. 4. Linkage diagram of the diagnostic taxa group of *Kobresio-Caricetalia rupestris* /*Elyno-Seslerietea*/. The diagram is based on association indices calculated from qualitative data for 94 taxa. All associations above 80 % similarity are plotted. Members of the diagnostic taxa groups are underlined; associations among nonmember taxa and regional taxa are not shown.

Among clusters for individual order diagnostic taxa groups, this one has the highest number of high-similarity-level links among its members, and the lowest number of nonmember taxa entering the group. The taxa of the second alliance of this order, *Caricion foeneo-elynoidis*, were not included in the analysis.

have also shown environmental trends not detected by correlation with ordination axes.

A comparison between ordinations of associations (Fig. 5) and plots of relative taxa importance within these ordinations (Fig. 8) implies that syntaxa and their diagnostic taxa have similar ranges along the controlling environmental gradient complexes.

These plots complement the polar ordination of taxa (Fig. 9), for which the relationship of the ordination axes with environmental variables could not be quantified. Clusters of taxa corresponding to predetermined diagnostic taxa groups can be found within this ordination, but all the clusters overlap along at least one ordination axis. The same environmental gradient complexes as in the ordination of associations (Fig. 5) may be related to the ordination axes.

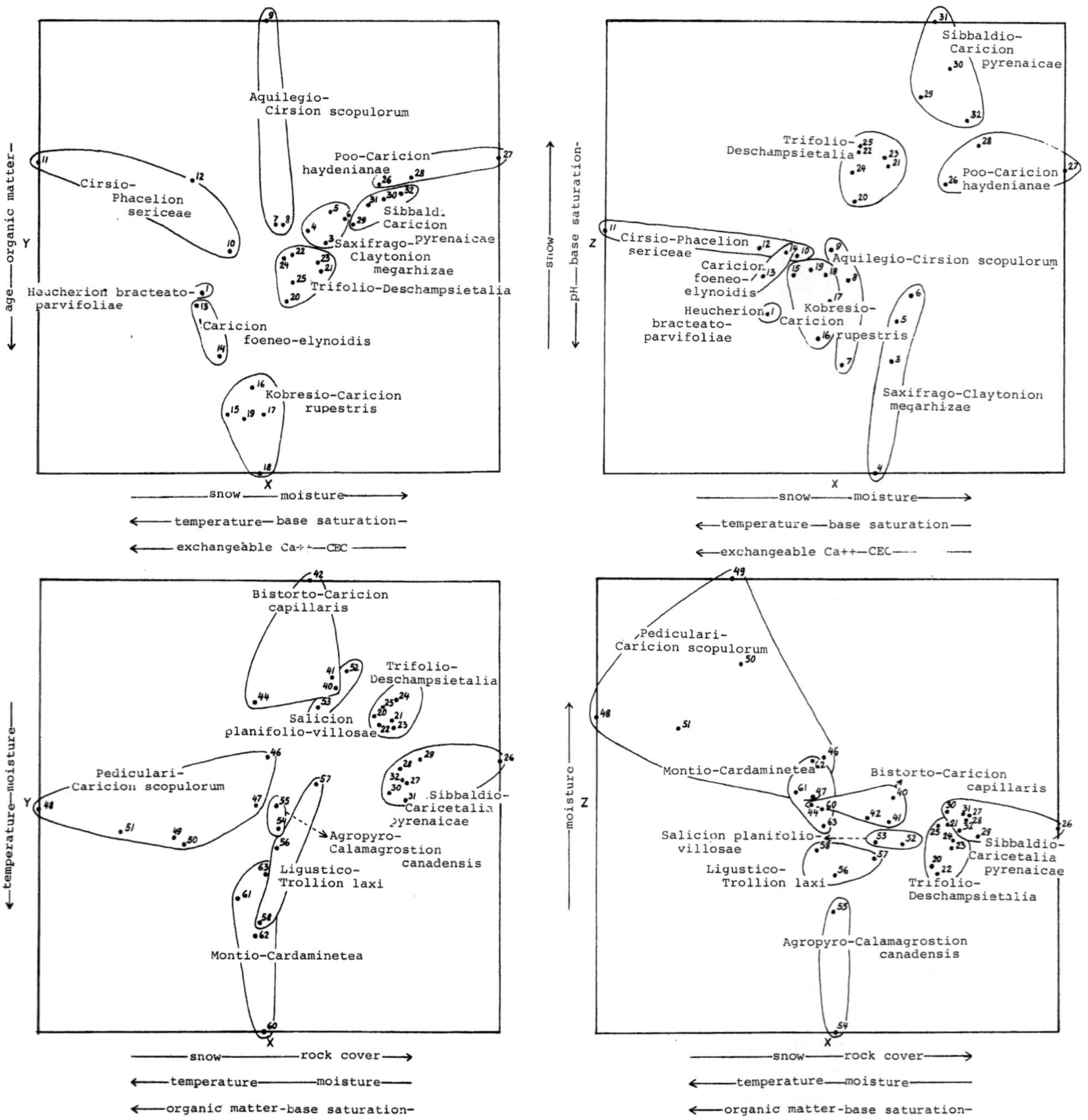

Fig. 5. Polar ordinations of qualitative taxa data for associations. Syntaxa, determined earlier by tablework, are named and outlined; numbers identify the associations. Environmental variables significantly or highly correlated with the ordination axes are indicated. An arrow shows the direction from low to high factor level.

The upper ordination includes 31 (out of total 39) subxerophilous to mesophilous associations of *Heuchero-Saxifragetalia* (*Asplenietea rupestria*), *Aquilegio-Cirsietalia* (*Thlaspietea rotundifolii*), *Kobresio-Caricetalia rupestris* (*Elyno-Seslerietea*), and *Salicetea herbaceae*. The lower ordination includes 34 (out of total 44) mesophilous to subhygrophilous associations of *Salicetea herbaceae*, *Pediculari-Caricetalia scopulorum* (*Scheuchzerio-Caricetea fuscae*), *Salici-Trollietalia* (*Betulo-Adenostyletea*), and *Primulo-Cardaminetalia* (*Montio-Cardaminetea*). *Salicetea herbaceae* appear in both groups for comparison.

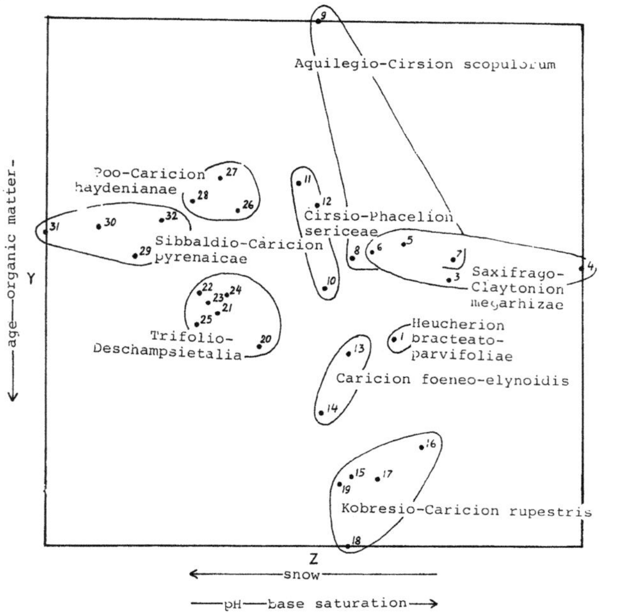

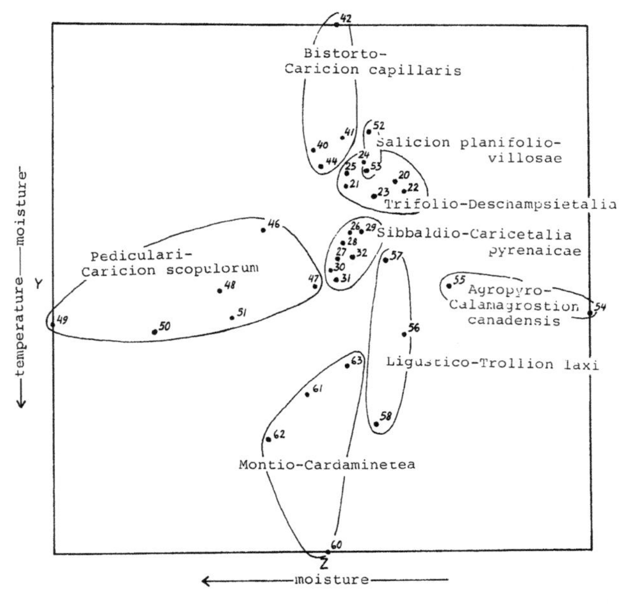

Discussion

Hypotheses

1. The alliances and orders of the Braun-Blanquet hierarchy correspond to groups of associations produced by agglomerative clustering. The average similarity level at which these groups were formed in the present data set agree well with the similarities reported from elsewhere (Westhoff & van der Maarel 1978) for both orders and alliances. At lower levels of the hierarchy, Kortekaas et al. (1976) detected associations in a matrix of individual relevés at 40 to 60 %, subassociations at 60 to 70 %, and variants at 70 to 80 %. One part of the hypothesis remains untested: classes, which cannot be derived on the basis of local data.

Clusters of taxa formed by agglomerative clustering and association analysis correspond relatively well to the diagnostic taxa groups of syntaxa. This has also been reported. For instance, Feoli-Chiapella & Feoli (1977) found a high correlation between taxa groups derived by single linkage clustering on the basis of the product-moment correlation coefficient and syntaxonomy.

The similarity between the syntaxonomic table and the results of numerical classification appears to be greater in the direction of syntaxa than in the direction of taxa. The data discontinuity appears to be greater in the same direction. This could reflect the considerably smaller proportion of the data analyzed to group taxa (63 associations and 94 taxa) than to group associations (63 associations and 340 taxa; see the next section). However, it it could also be a result of sampling and initial tablework which were oriented toward establishment of discontinuous vegetation types.

Although the overall discontinuity of the data appears to be considerably greater than its continuity, continuity can be observed in the syntaxonomic table, where groups of taxa diagnose two orders at once, each from a different class *(Kobresio-Caricetalia rupestris* and *Trifolio-Deschampsietalia)* or two alliances, each from a different order and class *(Cirsio-Phacelion sericeae* and *Caricion foeneo-elynoidis)*.

2. The environmental variables, significantly correlating with the association ordination axes and representing the environmental gradient complexes which control the compositional variation, generally corresponded to the environmental variables which order the syntaxonomic table. The response of syntaxa and taxa plotted on the ordination planes to these complex gradients was obvious, but the clarification of the composition of the complexes was only partly successful. The response of taxa to similar environmental gradient complexes was suggested by the taxa ordination. The response of taxa, syntaxa, and diagnostic taxa groups to individual environmental variables in the data set remains to be shown. Current work is being directed toward this end.

3. Numerical methods more or less demonstrated the parallels between the environmental and compositional

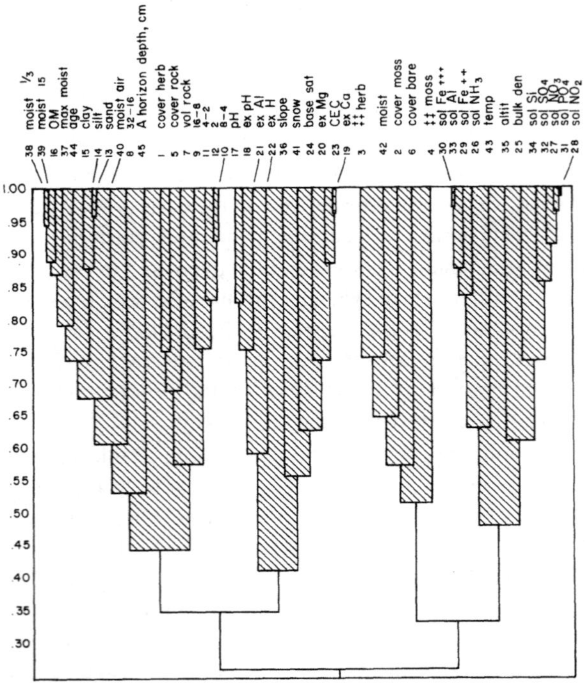

Fig. 6. Hierarchical dendrogram of environmental variables for 63 associations, constructed by average linkage method. Product-moment correlation coefficients label the Y axis, numbers and codes identify variables.

Clusters reflect the partial composition of environmental gradient complexes correlated with ordination axes in Fig. 5.

uniformity of the syntaxa. Ordinations suggested corresponding environmental response of syntaxa and their diagnostic taxa groups, and their members. The indicated environmental continuity of syntaxa is low, as the syntaxa are clearly spaced along the ordination axes. The two halves of the syntaxonomic table correspond to the two largest successful ordinations; the orders (classes) and alliances correspond to the major habitat categories (e.g., McVean & Ratcliffe 1962). Clusters produced by agglomerative clustering based on environmental data suggested correspondence between environmental and compositional variation. Other studies have produced similar results. For example, Feoli-Chiapella & Feoli (1977) found a high correlation between syntaxonomy and the ecological conditions in the vegetation as revealed by Q-PCA ordination based on the product-moment correlation between the relevés on the basis of their scores in the log-similarity ratio × 100 matrix.

Data reduction

Limited reduction of data appeared to facilitate the syntaxonomic and environmental interpretability of the numerical results, while further reduction appeared to have the opposite effect. The point of change probably depends, among other things, on the mode of analysis, vegetation type, the breadth of compositional and environmental variation, and size of the data set.

The compression of the relevé/taxa matrix by syntaxonomic tablework produced a more manageable association/taxa matrix. In this new matrix, reduction of the number of both associations and taxa had an effect on the interpretability of the results.

The original number of taxa was somewhat reduced initially, while the number of associations remained essentially the original one: 63. In this matrix, agglomerative clustering of associations produced interpretable results for both the entire data set and individual classes (orders).

Ordination of associations based on the almost complete association/taxa matrix was not successful until the environmental and compositional variation were approximately halved. Further reduction to individual orders (classes) produced better spacing of syntaxa along ordination axes, but fewer significant correlations between ordination axes and environmental variables. This lower number of significant correlations probably reflects the small sample size or perhaps an environmental uniformity within the sampled universe (cf. Webber 1978). However, in ordinations of situations of narrow vegetational and environmental variation, the assumption of linear variation should be more acceptable (e.g., Austin & Noy-Meir 1971).

Based on the association/data matrix in which the number of taxa was reduced to 94, agglomerative clustering did not produce clusters of associations representing higher syntaxa, and only a few correlations between ordination axes and environmental variables were significant. This loss of information appeared to increase with the decreasing size and distinctiveness of the diagnostic taxa group of the syntaxon in question. The similarity between environmental correlations in these ordinations and in those based on a matrix with 340 taxa decreased in the order of X, Y, and Z axis.

Based on the 63 associations/94 taxa matrix, both agglomerative clustering and association index analysis produced groups of taxa comparable to the diagnostic taxa groups.

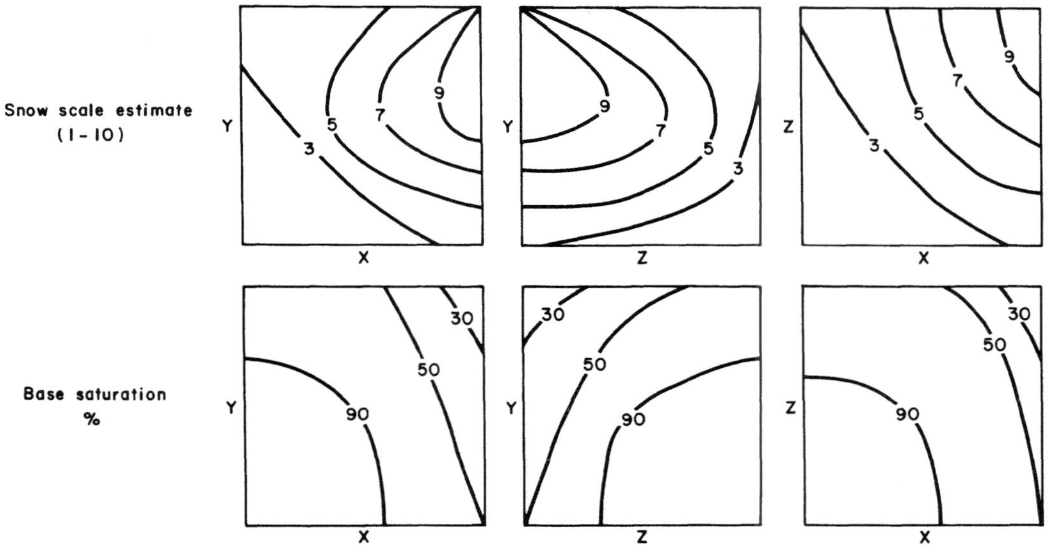

Fig. 7. Isolines connect equal values of two environmental variables on the planes of the ordination of subxerophilous to mesophilous associations.

Values of estimated duration of snow cover and measured percentage base saturation are inversely related. *Salicetea herbaceae* (Fig. 5) are located in the area of the longest-lasting snow cover and the lowest percentage base saturation.

Similarity: – 100 %; – – – 90–99 %; –.–.– 80–89 %.

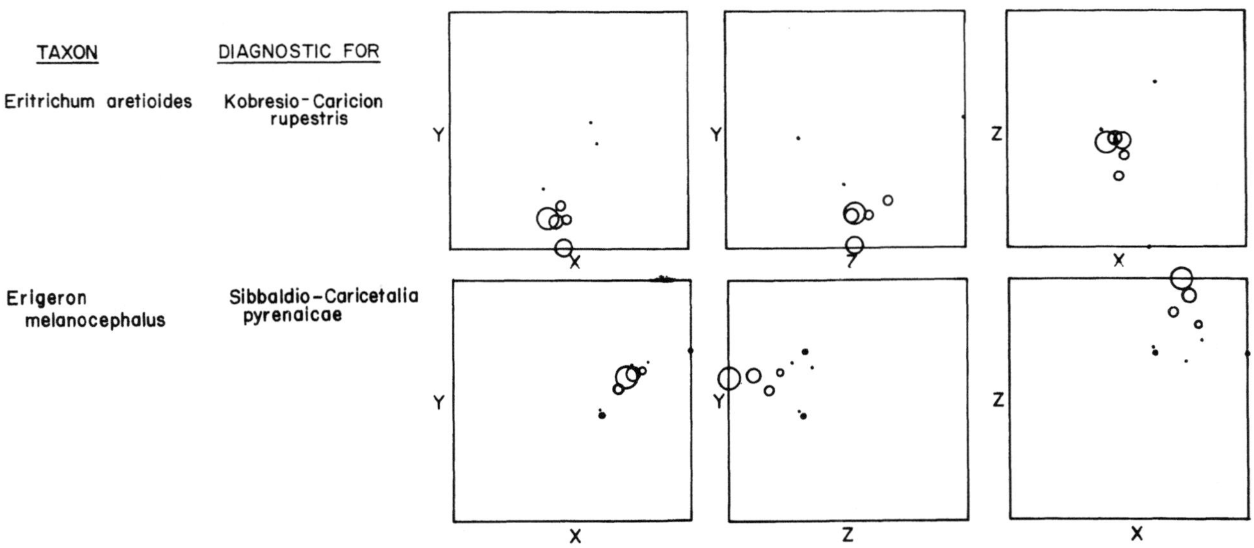

Fig. 8. Relative importance of two diagnostic taxa on the planes of ordination of subxerophilous to mesophilous associations. The largest circle indicates the association with the maximum cover of the taxon within the ordination, smaller circles are proportional to its cover in other associations. The associations are located in circle centers.

Syntaxa (Fig. 5) and their diagnostic taxa have similar relationship to the controlling environmental gradient complexes.

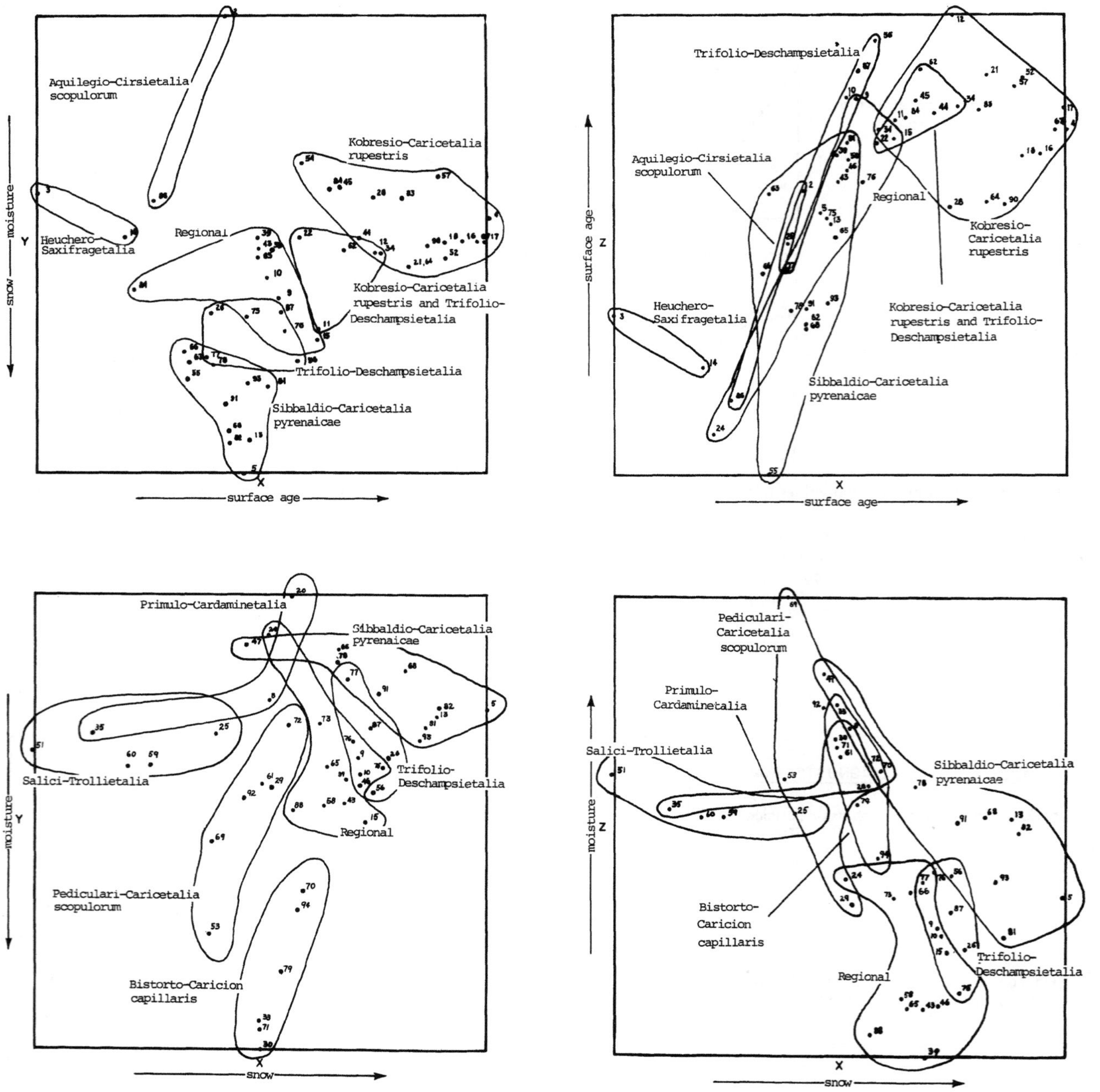

Fig. 9. Polar ordinations of qualitative association data for 94 taxa. Diagnostic taxa groups, determined earlier by tablework, are named and outlines. Numbers identify the taxa. Suggested complex environmental gradients are indicated along the ordination axes. An arrow shows the direction from low to high factor level.

The upper ordination includes subxerophilous to mesophilous taxa of *Heuchero-Saxifragetalia (Asplenietea rupestria), Aquilegio-Cirsietalia scopulorum (Thlaspietea rotundifolii), Kobresio-Caricetalia rupestris (Elyno-Seslerietea), Salicetea herbaceae*, and regional taxa. The lower ordination includes mesophilous to subhygrophilous taxa of *Salicetea herbaceae, Pediculari-Caricetalia scopulorum (Scheuchzerio-Caricetea fuscae), Salici-Trollietalia (Betulo-Adenostyletea), Primulo-Cardaminetalia (Montio-Cardaminetea)*, and *Salicetea herbaceae.Salicetea herbaceae taxa* appear in both groups for comparison.

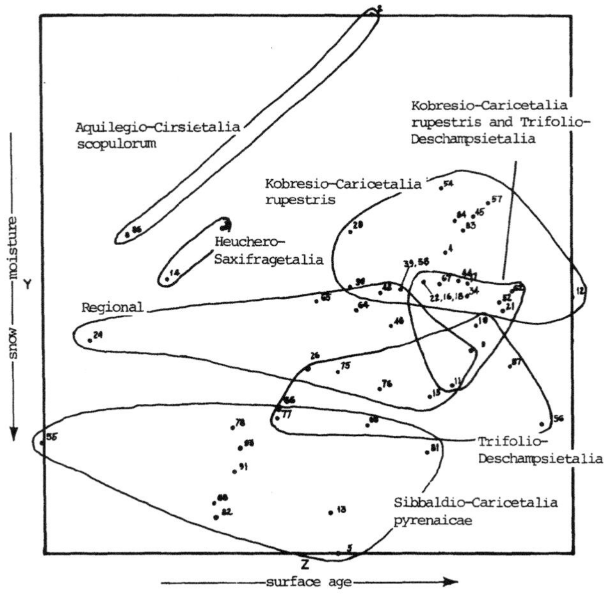

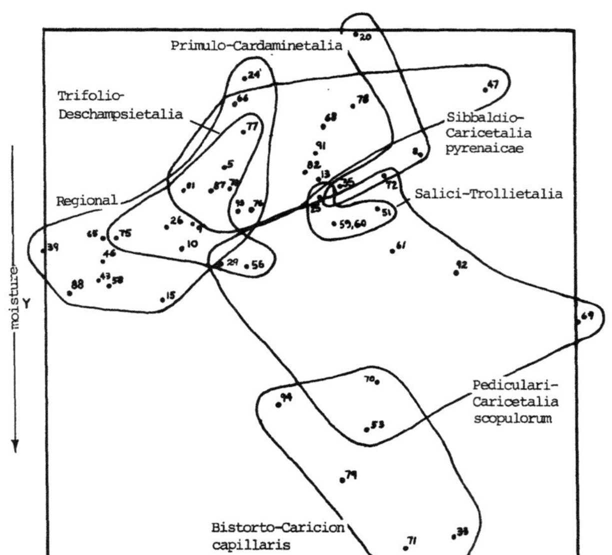

Qualitative and quantitative data

The results of numerical analysis based on qualitative data approximated the syntanonomic results considerably better and transmitted more environmental information than the results based on quantitative data. This probably reflects the primary importance of qualitative data for establishment of alliances, orders, and their diagnostic taxa groups in the investigated alpine vegetation.

Quantitative data could be more important at the association level. In the syntaxonomic table, only very few taxa were identified as diagnostic for associations, which were usually recognized on the basis of quantitative cover data (cf. Krajina 1933). Quantitative data can be even more important elsewhere; according to Kortekaas et al. (1976), in species-poor dominance type communities such as salt marsh vegetation, the quantitative values of the leading taxa entirely determine the two types as distinct associations.

This problem was also discussed by van der Maarel (1979). Noy-Meir & Whittaker (1977) pointed out that if presence-absence data is used in nodal component analysis, overall floristic composition and rare taxa will play major roles in defining noda and these units may then correspond to those of Braun-Blanquet.

Classification and ordination: syntaxonomy and numerical phytosociology

1. Syntaxonomic results can provide a platform for more detailed examination of compositional and environmental variation by numerical methods; the framework of the syntaxonomic results can make the interpretation of numerical results considerably easier. For example, the taxa association index matrix was interpreted with the help of the diagnostic taxa groups in the present work. This reflects syntaxonomy's taking into account a greater part of vegetational variation than any numerical method, including qualitative and quantitative data, and various types of plant response. This flexibility of the traditional classification was stressed by Dale (1977).

Syntaxonomic analysis sections the vegetation data into more manageable, narrower universes, compositionally and environmentally, within which the vegetational response may be linear (Austin & Noy-Meir 1971, Goodall 1978). The ordinations in the present work do not reflect this, however.

2. Numerical analysis can clarify points which cannot be determined by syntaxonomic tablework. Had the numerical analysis been done before the syntaxonomic table was finalized, the placement of syntaxa intermediate between higher-level syntaxa, or the placement of bimodal taxa could have been more informed (cf. Austin 1976a, 1978, van der Maarel 1979).

Both *Phleo-Caricetum nigricantis* (intermediate between *Pediculari-Caricetalia scopulorum* and *Sibbaldio-Carice-*

talia pyrenaicae) and *Stellaria umbellata* (a bimodal taxon with one optimum in *Heuchero-Saxifragetalia* and the other in *Salici-Trollietalia*) were difficult to place in the table. Although *Phleo-Caricetum nigricantis* appears near other associations of *Sibbaldio-Caricetalia pyrenaicae*, among which it was placed in the table, it joins their cluster in the dendrograms at a relatively low similarity level. *Stellaria umbellata*, which was placed among regional taxa in the table, appears with the taxa of *Primulo-Cardaminetalia* and *Sibbaldio-Caricetalia pyrenaicae*, where it possibly belongs, in the dendrogram.

3. The fact that the sampled area was a geographically limited universe, characterized by relative environmental, biotic, and historical distinctiveness and uniformity, probably contributed to the success of this study. In such a universe, the total taxa diversity and the entire response range of most plants and plant communities is, presumably, taken into account (cf. Krajina 1933).

The steep environmental gradients in the sampled area produce a wide range, high diversity, and clear definition of habitat and vegetation dominance types. These attributes of vegetation and environment were probably the major source of the success of this study. Among other such attributes possibly were the resulting relatively low continuity and importance of dominance, a wide range and high enough taxa diversity, degree of maturity, competition, structure development, and environmental stability (cf. van der Maarel 1966, 1975, 1979, Austin 1976a, b), or an appropriate degree of the liquidity of vegetation composition (Dale 1975).

Only classification, both numerical and syntaxonomic, handled the entire Indian Peaks data set effectively. Both classification and ordination worked well at the alliance and order level; according to van der Maarel (1966, 1975, 1979), both these techniques are at their best at these hierarchical levels.

In a vegetation from a different geographical region, which has attributes of different states, or in a partial data set, including one class rather than all, these methods may not have produced interpretable results. Differences between local vegetation models may be such that methods applicable in one local data set are not applicable in a data set from a different region. Local vegetation models should be derived from field data (cf. Austin 1978); because it summarizes and simplifies field data so effectively, syntaxonomy can play an important role in building these local vegetation models.

Summary

A mid-latitude, Northern Hemisphere alpine vegetation in the Colorado Rocky Mountains was suitable for analysis by syntaxonomic and numerical methods, which interpreted the Braun-Blanquet association data, distributed by habitat and vegetation dominance types, correspondingly. Classification and ordination yielded complementary, interpretable results, which combined were more informative than results of either technique alone. Syntaxonomic results can facilitate the interpretation of numerical results; the syntaxonomic relationships can be clarified by the results of numerical analysis.

Alliances and orders of the Braun-Blanquet hierarchy, and their diagnostic taxa groups were derived by numerical methods from an association/taxa matrix. On the basis of this matrix, numerical methods demonstrated some of the environmental gradient complexes controlling the compositional variation. The compositional distinctiveness and uniformity of orders and alliances parallel environmental distictiveness and uniformity.

The following attributes of the sampled vegetation and environment were probably responsible for the successful combination of syntaxonomic and numerical results: 1. The sampled area as a geographically limited universe, characterized by relative environmental, biotic, and historical distinctiveness and uniformity. 2. The wide range, high diversity, and clear definition of habitat and vegetation dominance types. Because it effectively summarizes and simplifies field data, syntaxonomy can play an important role in building local vegetation models.

References

Austin, M.P. 1976a. Performance of four ordination techniques assuming three different non-linear species response models. Vegetatio 33: 43–49.

Austin, M.P. 1976b. On non-linear species response models in ordination. Vegetatio 33: 33–41.

Austin, M.P. 1978. Current prospects in vegetation analysis. In: G.P. Patil, M. Rosenzweig, et al. (eds.), Contemporary quantitative ecology and related ecometrics. 2nd Intern. Ecol. Congress, Satellite Program in Statistical Ecology. Satellite C(S12). Jerusalem, 25 pp.

Austin, M.P. & I. Noy-Meir. 1971. The problem of non-linearity in ordination: experiments with two gradient models. J. Ecol. 59: 763–773.

Braun-Blanquet, J. 1964. Pflanzensoziologie, Grundzüge der Vegetationskunde. 3rd ed. Springer, Berlin, 865 pp.

Bray, J.R. 1956. A study of mutual occurrence of plant species. Ecology 37: 21–28.

Bray, J.R. & J.T. Curtis. 1957. An ordination of the upland forest communities of southern Wisconsin. Ecol. Monogr. 27: 325–349.

Clark, S.J.V. 1977. The vegetation of Rocky Flats, Colorado. M.A. thesis, University of Colorado, Boulder, Colorado. 172 pp.

Dale, M.B. 1975. On objectives of methods of ordination. Vegetatio 30: 15–32.

Dale, M.B. 1977. Planning an adaptive numerical classification. Vegetatio 35: 131–136.

Feoli-Chiapella, L. & E. Feoli. 1977. A numerical phytosociological study of the summits of the Majella Massive (Italy). Vegetatio 34: 21–39.

Goodall, D.W. 1978. Numerical classification. 2nd ed. In: R.H. Whittaker (ed.), Classification of plant communities p. 247–287. Junk, The Hague. 408 pp.

Groenewoud, H. van. 1965. Ordination and classification of Swiss and Canadian coniferous forests by various biometric and other methods. Ber. Geobot. Inst. Rübel, Zürich, 36: 28–102.

Komárková, V. 1979. Alpine vegetation of the Indian Peaks area, Front Range, Colorado Rocky Mountains. Flora et vegetatio mundi 7, R. Tüxen (ed.). 2 vols. Cramer, Vaduz. 591 pp.

Kortekaas, W., E. van der Maarel & W.G. Beeftink. 1976. A numerical classification of European Spartina communities. Vegetatio 33: 51–60.

Krajina, V. 1933. Pflanzengesellschaften des Mlynica-Tales in den Vysoké Tatry (Hohe Tatra). Beih. Bot. Centralbl. 50: 774–957, 51: 1–224.

Lacoste, A. 1976. Relations floristiques entre les groupements prairiaux du Triseto-Polygonion et les megaphorbiaies (Adenostylion) dans les Alpes occidentales. Vegetatio 31: 161–176.

Maarel, E. van der. 1966. On vegetational structures, relations, and systems, with special reference to the dune grasslands of Voorne, The Netherlands. Thesis Utrecht, 170 pp.

Maarel, E. van der. 1969. On the use of ordination models in phytosociology. Vegetatio 19: 21–46.

Maarel, E. van der. 1972. Ordination of plant communities on the basis of their plant genus, family and order relationships. In: E. van der Maarel & R. Tüxen (eds), Grundfragen und Methoden in der Pflanzensoziologie. Ber. Int. Symp. Rinteln, p. 183–192, Junk, Den Haag.

Maarel, E. van der. 1975. The Braun-Blanquet approach in perspective. Vegetatio 30: 213–219.

Maarel, E. van der. 1979. Multivariate methods in phytosociology, with reference to the Netherlands. In: M.J.A. Werger (ed.), The study of vegetation, p. 163–225. Junk, The Hague. 316 pp.

Maarel, E. van der, L. Orlóci & S. Pignatti. 1976. Data-processing in phytosociology, retrospect and anticipation. Vegetatio 32: 65–72.

May, D.C.E. 1973. Models for predicting composition and production of alpine tundra vegetation from Niwot Ridge, Colorado. M.A. thesis, University of Colorado, Boulder, Colorado. 99 pp.

McIntosh, R.P. 1978. Matrix and plexus techniques. 2nd. ed. In: R.H. Whittaker (ed.), Ordination of plant communities, p. 151–185. Junk, The Hague. 388 pp.

McVean, D.N. & D.A. Ratcliffe. 1962. Plant communities of the Scottish Highlands. Monographs of Nature Conservancy 1, Her Majesty's Stationery Office, London. 445 pp.

Moore, J.J., P. Fitzsimmons, E. Lambe & J. White. 1970. A comparison and evaluation of some phytosociological techniques. Vegetatio 20: 1–20.

Noy-Meir, I. & R.H. Whittaker. 1977. Continuous multivariate methods in community analysis: some problems and developments. Vegetatio 33: 79–98.

Orlóci, L. 1966. Geometric models in ecology. I. The theory and application of some ordination methods. J. Ecol. 54: 193–215.

Sokal, R.R. & P.H.A. Sneath. 1963. Principles of numerical taxonomy. Freeman, San Francisco. 359 pp.

Webber, P.J. 1978. Spatial and temporal variation of the vegetation and its production, Barrow, Alaska. In: L.L. Tieszen (ed.), Plant ecology of the Alaskan arctic tundra. Springer, Berlin. (in press).

Westhoff, V. & E. van der Maarel. 1978. The Braun-Blanquet approach. 2nd ed. In: R.H. Whittaker (ed.), Classification of plant communities, p. 287–399. Junk, The Hague. 408 pp.

Accepted 20 December 1979

MULTIVARIATE ANALYSIS OF MULTISPECTRAL REMOTE SENSING DATA ON GRASSLANDS FROM DIFFERENT SOIL TYPES*

P. VAN HECKE[1], I. IMPENS[1], R. GOOSSENS[2] & F. HEBRANT[3]

[1] Departement Biologie, Universitaire Instelling Antwerpen, Universiteitsplein 1, B-2610 Wilrijk, Belgium
[2] Onderzoekscentrum voor Bosbouw, Bosbedrijfsvoering en Bospolitiek, Geraardsbergsesteenweg 123, B-9231 Gontrode, Belgium
[3] Instituut voor Tropische Geneeskunde, Nationalestraat 155, B-2000 Antwerpen, Belgium

Keywords:
Canonical correlation analysis, Canonical variate analysis, Densitometry, 'False colour' infrared transparencies, Grassland, Minimum variance Clustering, Principal component analysis, Soil type, 'True colour' transparencies

Introduction

In the last few decennia, it has become customary to explore vast regions of the earth, with the help of aerial photographs (see e.g. Barrett & Curtis 1974, 1977). This study forms part of a research project under the supervision of the third author, which started in 1976, and is concerned with inventory and evaluation of vegetation, using teledetection tools such as aerial photographs on true colour and false colour infrared emulsions. The principal aim of this project is to establish a key of density values, which will enable to detect (1) different phases of a plant disease in single specimens or in a stand of vegetation and (2) different vegetation patterns. Coniferous and deciduous woodland, poplar plantations, arable crops, natural and permanent grasslands are especially envisaged.

The present contribution deals with the second aspect and concentrates on grasslands. By applying classification and ordination techniques (Armstrong & Clayton 1977, Owen-Jones 1977) an attempt is made to recognize possible grassland types by means of densitometric measurements, and to find a relation with soil types.

Study area and methods

A rectangle of about 60 sq. km has been chosen in the province of East-Flanders (Belgium), mainly covering the area of the municipalities Zelzate, Wachtebeke (c. 51°09'N, 3°54'E) and Eksaarde.

A series of stereoscopic aerial photographs (23 cm × 23 cm) were taken of the entire region on June 6th 1976, distributed over thirteen parallel flight-strips, at a height of about 600 m and a scale of 1 : 2000. Two emulsions are used: Kodak Aerochrome Multispectral 2448 (MS) and Kodak Aerochrome Infrared 2443 (IR).

As it is impossible to include all soil types and subdivisions involved, 13 units have been selected from soil maps at a scale 1 : 20 000. They include sand, sand with loam, loam with sand, clay, peat, peat-clay and marl.

A total number of 58 grasslands (mainly permanent ones) has been selected on ordnance-maps at a scale 1 : 10 000. The distribution over the selected soil types was taken into account.

Species were recorded along a double meter line, which was laid down twice in each grassland (Daget & Poissonet 1971). A total of 51 taxa was recorded.

In the laboratory the grasslands were easily localized on both sets of transparencies. Within each pasture ten points were arbitrarily chosen on which densities were measured with a Macbeth TD 504 using Wratten-filters nrs. 92 (red, R), 93, (green, G) and 94 (blue, B) and a spot size of 2 mm which corresponded to a sampling area of 12.6 sq. m. For each emulsion and each filter the ten density values were averaged by a Wang 600 calculator; consequently six means were obtained for each grassland. Two corrections on the computation of the density means were included: one concerns the great overlap between the

* Nomenclature follows De Langhe e.a. (1973).

three filters resulting in a broad band density measurement, the second the radial decrease in light intensity on the photographs as well as the effects due to the variation in lens vignetting, sun angle and view-angle shadow.

For multivariate analyses, Minimum Variance Clustering (MVC; Orlóci 1967), Principal Component Analysis (PCA), Canonical Correlation Analysis (CCA) and Canonical Variate Analysis (CVA) were applied.

Results

Classification and ordination by means of density values

As a first step scatter plots were constructed using actual density values (Table 1: nrs. 1–6) in different combinations; these values were not normalized (Carter & Gardner 1977, Owen-Jones 1977). Group structure in the grasslands could not be correlated with environmental discontinuities, nor with a structure in the density values. Scatter diagrams of grasslands dimensions formed by different ratios of density values either from the same emulsion or from both emulsions reflected a continuum – like variation.

Neither with a centred PCA on the first six densitometric variables (Table 1) nor with a S-matrix, high loadings were found on component I, in contrast to some on component II: MS B (0.54), MS G (0.57) and IR R (0.52). The first two eigenvalues explained 60.4 % resp. 19.0 % of the total variance.

Table 1. Densitometric variables.

01	MS B	12	$\frac{\text{MS G}}{\text{IR R}} \times \frac{\text{IR B}}{\text{IR R}}$
02	MS G	13	$\frac{\text{MS B} - \text{IR B}}{\text{MS B} + \text{IR B}}$
03	MS R	14	$\frac{\text{MS B} - \text{IR R}}{\text{MS B} + \text{IR R}}$
04	IR B	15	$\frac{\text{MS G} - \text{IR R}}{\text{MS G} + \text{IR R}}$
05	IR G	16	$\frac{\text{IR B} - \text{IR R}}{\text{IR B} + \text{IR R}}$
06	IR R	17	$\frac{\text{MS B}}{\text{IR R}} + \text{MS G}$
07	MS B/IR R	18	$\frac{\text{MS B}}{\text{IR R}} + \text{IR B}$
08	MS G/IR R	19	$\frac{\text{MS G}}{\text{IR R}} + \text{MS B}$
09	IR B/IR R	20	$\frac{\text{MS G}}{\text{IR R}} + \text{IR B}$
10	$\frac{\text{MS B}}{\text{IR R}} \times \frac{\text{MS G}}{\text{IR R}}$	21	$\frac{\text{IR B}}{\text{IR R}} + \text{MS B}$
11	$\frac{\text{MS B}}{\text{IR R}} \times \frac{\text{IR B}}{\text{IR R}}$	22	$\frac{\text{IR B}}{\text{IR R}} + \text{MS G}$

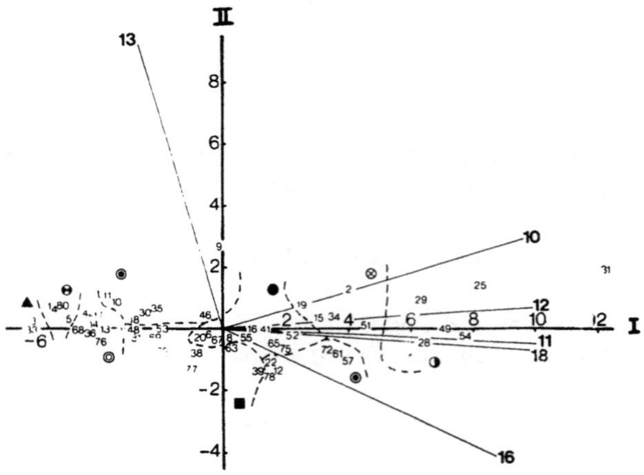

Fig. 1. Principal Component Analysis. Scatter diagram of 58 grasslands, grouped in 9 clusters: ⊖ = cluster 1, ◎ = cluster 2, ▲ = cluster 3, ◉ = cluster 4, ⊗ = cluster 5, ● = cluster 6, ⊛ = cluster 7, ■ = cluster 8, ◑ = cluster 9.
Correlation between 16 densitometric variables (DV) and the first two components. Only DV nrs. 10, 11, 12, 13, 16 and 18 are shown, the remaining ones are situated between nrs. 10 and 12 respectively nrs. 16 and 18.

The number of variables was then extended in order to find combinations of two or more actual density values which might be related to separate groups of grasslands. A PCA was applied to the densitometric variables 7–16. Computing a S-matrix and solving the determinantal equation gave a first eigenvalue λ_1 explaining 96.5 % of the total variance, with high loadings for variables nr. 11 (component I: 0.74), nr. 10 (component II: 0.76) and nr. 12 (component III: 0.76). A calculation, based on the variables nrs. 7–22, produced a first eigenvalue λ_1, extracting 95.3 % of the total variance. Fig. 1 presents the results of a PCA of the grasslands. It appears that var. nr. 11 is highly correlated with component I (0.996), and so are nrs. 12 and 18: 0.990 resp. 0.979. The highest correlation with the second component is obtained by var. nr. 13: 0,916. Secondly, the cluster analysis MVC was carried out on the variables nrs. 7–22 (numbers 13, 15 and 16 were excluded because they sometimes give negative values). This resulted in the formation of 9 clusters of grassland, which could be easily detected in the scatter diagram. However there is still much overlap: cf Duncan's New Multiple Range Test: clusters 3 1 2 4 8 6 7 5 9. Furthermore, the R-matrix has been considered in addition to the S-matrix: λ_1 and λ_2 accounted for 85,7 % and 11.6 % resp. of the total variance. High correlations were found between var. nrs. 18, 12, 11 and the first axis (viz. 0.993, 0.980 and 0.979) and between var. nr. 13 and the second axis (0.941).

The 9 MVC-clusters are easily discovered in the scatter diagram, and they are less dense. Both matrix-analyses agree with each other in that there is apparently no relation between the grassland clusters and the soil types.

Classification and ordination by means of species

Assuming that the density values are mainly determined by the shortwave reflexion of the most important grassland species, i.e. those with high frequency/cover percentage, we may reduce the number of taxa from 51 to 29 (Table 2), which is equivalent to eliminating those taxa never reaching a cover of 25 %. Once more, a PCA has been applied, producing a S- and a R-matrix. The first five eigenvalues of the S-matrix account for nearly 75 % of the total variance (Table 2).

Moreover different taxa are highly correlated to different components (Fig. 2): *Poa trivialis* to axis I (0.758), *Lolium perenne* to axis II (0.828), *Holcus lanatus* to axis III (0.829), *Poa annua* to axis IV (0.911) and *Dactylis glomerata* to axis V (0.740). Regarding the R-matrix, much less of the total variance is accounted for by the first five eigenvalues, viz. 42.1 %, with λ_1 = 10.8 % and λ_2 = 8.5 %. Taxa, giving high correlations with the first five components, occur with low cover percentages or at a very low presence degree. The scatter diagram of the first two axes reflects three peripheral grasslands and a concentration of the others around the origin of both axes. Elimination of the three peripheral grasslands, would reduce the number of taxa from 29 to 24. The importance of the taxa with higher presence and higher cover degrees is increasing: their correlation coefficients are still negative on components I and II (*Lolium perenne* and *Agropyron repens*) but are becoming positive from component III onwards (*Poa trivialis*, *Bromus mollis*, *Holcus lanatus*).

In order to investigate the possible relationship between

Table 2. Principal Component Analysis: 58 grasslands and 29 taxa. Orthonormal eigenvectors and the corresponding eigenvalues based on the S-matrix.

	Components				
Taxa	I	II	III	IV	V
1. Poa trivialis	0.686	0.543	-0.339	0.085	-0.038
2. Lolium perenne	-0.479	0.804	0.199	-0.043	0.023
3. Stellaria media	-0.017	-0.044	-0.016	-0.039	0.259
4. Poa annua	-0.158	-0.050	-0.124	0.908	0.245
5. Alopecurus geniculatus	-0.069	-0.003	-0.014	-0.062	-0.177
6. Trifolium repens	-0.048	0.011	-0.017	0.066	0.075
7. Taraxacum spec.	0.003	-0.002	-0.042	0.061	0.010
8. Agrostis tenuis	-0.002	-0.068	-0.037	-0.031	-0.094
9. Anthoxanthum odoratum	0.018	-0.015	0.019	0.006	-0.007
10. Poa pratensis	0.002	-0.040	-0.012	-0.058	0.016
11. Rumex acetosella	-0.001	-0.035	-0.018	-0.014	-0.040
12. Holcus lanatus	0.432	-0.007	0.846	0.107	0.208
13. Agropyron repens	-0.039	-0.118	-0.109	-0.267	0.325
14. Festuca rubra	-0.005	-0.083	-0.043	-0.118	0.094
15. Ranunculus spec.	0.007	0.010	0.015	-0.002	-0.007
16. Phleum pratense	0.007	-0.021	-0.023	-0.012	-0.050
17. Bromus mollis	0.241	0.054	-0.260	-0.095	0.346
18. Dactylis glomerata	-0.105	0.025	-0.088	-0.151	0.708
19. Urtica dioica	0.016	0.016	0.015	0.003	-0.002
20. Lolium multiflorum	0.092	-0.093	-0.128	0.129	-0.156
21. Festuca arundinacea	0.005	-0.016	-0.018	-0.009	-0.038
22. Juncus acutiflorus	-0.001	-0.030	-0.017	-0.015	-0.046
23. Iris pseudacorus	-0.001	-0.028	-0.016	-0.020	-0.019
24. Phleum spec.	-0.016	0.007	-0.001	-0.012	-0.021
25. Holcus lanatus + H. mollis	-0.002	-0.072	-0.039	-0.035	-0.110
26. Caltha palustris	0.007	0.011	0.017	-0.002	-0.007
27. Phragmites australis	-0.003	-0.063	-0.036	-0.046	-0.044
28. Carex hirta	0.009	0.014	0.021	-0.002	-0.009
29. Plantago major	-0.016	0.007	-0.001	-0.013	-0.021
eigenvalues	1824,8	1606,6	1145,3	670,6	477,2
eigenvalues as of total variance	23,8	21,0	15,0	8,8	6,2
eigenvalues as cumulative	23,8	44,8	59,8	68,6	74,8

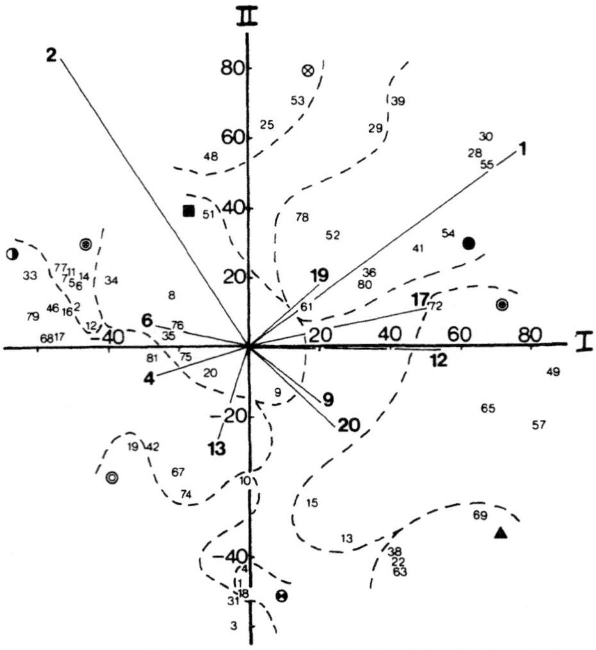

Fig. 2. Principal Component Analysis: (see Fig. 1) Correlation between 29 taxa variables (TV) and the first two components. The TV nrs. not represented are situated in the four quarters but are giving very low correlations.

the most important taxa and the groups of grasslands, a MVC was carried out on the same 29 species (Table 3). These MVC-clusters could easily be separated in the scatter diagram based on the S-matrix (Fig. 2), but this was nearly impossible in both R-matrix analyses.

Canonical analyses

With Canonical Correlation Analysis we might have a second way to judge if some of the most important taxa are highly correlated with some of the densitometric variables. According to the results obtained by the different applications of PCA, the variables nrs. 9, 10, 11, 13 and 18 are withdrawn and so are 25 taxa out of the 29 mentioned above. The differences between the five highest canonical correlations (c.c.) are rather small: I = 0.83, II = 0.76, III = 0.70, IV = 0.62 and V = 0.54. Beyond the third c.c. no single taxon is showing high values. Regarding the first c.c. high values are achieved for the variables nrs. 10 and 18 (0.56 resp. – 0.80), but very low ones (and most probably with no statistical significance) for the 'best' taxa *Caltha palustris* (0,15), *Festuca rubra* (0,13) and *Ranunculus spec.* (−0.14). Regarding the second c.c., rather high values are obtained by the variables nrs. 11 (0.68) and 9

Table 3. Minimum Variance Clustering: 58 grasslands and 29 taxa.

Cluster	Quadrats	n	Description
1	1, 4, 18	3	Poa trivialis absent, all taxa with very low degree of presence.
2	3, 10, 19, 31	4	Poa trivialis absent, moderate to high cover of Agropyron repens.
3	22, 38, 63, 69	4	Poa trivialis absent, very high to complete cover of Holcus lanatus.
4	13, 15, 29, 49, 57, 65, 72	7	Moderate to very high cover of Poa trivialis, usually moderate to complete cover of Bromus mollis.
5	25, 48, 53	3	High to very high cover of Poa trivialis, complete cover of Lolium perenne.
6	28, 30, 36, 39, 41, 52, 54, 55, 61, 78, 80	11	Moderate to complete cover of Poa trivialis, moderate to very high cover of Lolium perenne and Holcus lanatus.
7	2, 5, 6, 7, 11, 12, 14, 77	8	Poa trivialis absent, very high to complete cover of Lolium perenne.
8	8, 9, 20, 34, 35, 51, 75, 76	8	Poa trivialis absent, high to complete cover of Lolium perenne, moderate to high cover of Holcus lanatus.
9	16, 17, 33, 42, 46, 67, 68, 74, 79, 81	10	Poa trivialis absent, moderate to complete cover of Lolium perenne, either moderate to high cover of Poa annua or moderate to complete cover of Dactylis glomerata.

Table 4. Canonical Variate Analysis (= Multiple Discriminant Analysis): 9 grassland-clusters and 5 densitometric variables. Canonical Mean Vectors (rows = MVC – clusters from Tab. 3).

MVC-	densitometric variable nr.				
cluster	10	11	12	13	18
1	-1.985	-1.425	-0.593	0.176	-0.123
2	-1.455	1.018	0.662	-0.282	0.219
3	1.152	0.166	-0.391	-0.411	0.026
4	0.380	-0.513	0.768	0.076	-0.157
5	-0.317	0.985	0.357	-0.315	-0.378
6	0.291	-0.644	0.251	-0.085	0.141
7	0.100	0.232	-0.440	0.007	-0.038
8	0.194	0.550	-0.046	0.671	0.042
9	-0.010	0.101	-0.462	-0.184	0.004

(−0.55) as well as by one species (*Festuca rubra*: −0.50) but only lower ones by *Poa pratensis* (0.24), *Agrostis tenuis* (−0.23), *Rumex acetosella* (−0.22) and *Lolium perenne* (−0.21). All those taxa just mentioned have either a very low degree of presence or (*Lolium perenne*) are very common in nearly each of the 9 MVC-clusters (Table 3).

Due to the complete lack of taxa, having high canonical correlation values with one or more densitometric variables as well as being representative for any detectable grassland-cluster, we can finally apply a canonical variate analysis to examine if clusters can be related directly to some densitometric variable. Therefore, the 9 MVC-groups from Tab. 3 were dealt with together with the densitometric variables nrs. 10, 11, 12, 13 and 18 (nr. 9 used in the canonical correlation analysis has been replaced here by nr. 12). It was found that the first three canonical variates account for 90,6 %, whereas their absolute values were: $\lambda_1 = 28.224$, $\lambda_2 = 22.614$ and $\lambda_3 = 12.322$. From Table 4 it is obvious that with variable nr. 10 it should be possible to distinguish cluster 3 from clusters 1 and 2 and with variable nr. 11 cluster 1 from clusters 2 and 5. Unfortunately with the remaining variables no further separation could be produced. Taking into account a confidence interval of P = 0.90, only cluster 1 could safely be distinguished.

Discussion and conclusion

From this alternative approach we may summarize: (1) There is no relation between soil-types, as derived from soil maps, and MVC-clusters formed by means of either densitometric variables or taxa variables. (2) Both sets of MVC-clusters, traced in the new hyperspaces after the application of PCA, offer a high degree of overlap. (3) Both sets of MVC-clusters do not show a constructional relationship.

Furthermore, the highest canonical correlations are ecologically uninterpretable and CVA is giving a low degree of effectiveness: only 2 or 3 clusters can be well defined on the basis of two densitometric variables.

Gauch & Wentworth (1976) give a survey of the small number of ecological studies where CCA was applied. In only 2 out of the 7 studies CCA appeared successful. In 4 cases PCA gave better results. Austin 1968, Barkham & Norris 1970 and Gauch & Wentworth 1976 have emphasized the unwelcome fact that CCA was required to detect simultaneous linear relationships between two sets of variables. Thus we can agree with the statement that PCA is more informative while CCA is adding little new information. In some cases CVA seems to be more promising.

In spite of the application of different sophisticated multivariate techniques, the results obtained are disappointing. Undoubtedly the explanation lies in the high complexity of the grasslands involved. Owen-Jones (1977: 112) suggested that 'different varieties of the same crop type, variations in sowing time, soil moisture, ... all contribute to a broadening of the spectral distribution, with a consequent increased probability of misclassification'. As our permanent 'crops' are much more complex, we are of the opinion that a lot of other factors will contribute, like the seed mixture, the frequency of mowing, grazing and manuring, the nature of the manure and the kind of cattle.

Summary

From an area of 60 sq. km in East-Flanders (Belgium), two sets of stereoscopic aerial photographs were taken. After selection of 13 soil types, 58 grasslands (mainly permanent ones) were localized. Data on species composition (51 taxa) were quantitatively sampled with a 'double meter'.

Density values were obtained from both emulsions and using three filters. Using the densitometric variables (extended to 22) or taxa variables, Minimum Variance Clustering (MVC) and Principal Component Analysis (PCA) were applied. The recognition of different grasslandtypes and their relative position in both new sample point spaces was discussed as well as their relation to soil type. The results obtained by both Canonical Analyses (CCA and CVA) were examined and where necessary compared to the scarce applications in vegetation ecology.

Finally, the possible reasons for the meagre results were briefly discussed.

References

Armstrong, A.C. & K.M. Clayton. 1977. Objective generalization of Landsat images. In E.C. Barrett & L.F. Curtis. (eds.) 'Environmental remote sensing: practices and problems'. 2, p. 163–189.

Austin, M.P. 1968. An ordination study of a chalk grassland community. J. Ecol. 56: 739–757.

Barkham, J.P. & J.M. Norris. 1970. Multivariate procedures in an investigation of vegetation and soil relations of two beech woodlands, Cotswold Hills, England. Ecology 51: 630–639.

Barrett, E.C. & L.F. Curtis (eds.). 1974. Environmental remote sensing; 1: Applications and achievements. Arnold, London.

Barrett, E.C. & L.F. Curtis (eds.). 1977. Environmental remote sensing; 2: Practices and problems. Arnold, London.

Carter, P. & W.E. Gardner. 1977. An image-processing system applied to earth-resource imagery. In E.C. Barrett & L.F. Curtis. (eds.) 'Environmental remote sensing: practices and problems'. 2, p. 143–162.

Daget, P. & J. Poissonet. 1971. Principes d'une technique d'analyse quantitative de la végétation des formations herbacées. In P. Daget (ed.) 'Méthodes d'inventaire phyto-écologique et agronomique des prairies permanentes'. Document no. 56: 85–100. C.N.R.S. Montpellier.

De Langhe, J.E., L. Delvosalle, J. Duvigneaud, J. Lambinon & C. Vanden Berghen. 1973. Nouvelle Flore de la Belgique, du Grand-Duché de Luxembourg, du Nord de la France et des régions voisines. Edit. Patrimoine Jard. Bot. Nat. Belgique, Bruxelles.

Gauch, H.G. & T.R. Wentworth. 1976. Canonical correlation analysis as an ordination technique. Vegetatio 33: 17–22.

Orlóci, L. 1967. An agglomerative method for classification of plant communities. J. Ecol. 55: 193–205.

Owen-Jones, E.S. 1977. Densitometric methods of processing remote sensing data, with special reference to crop-type and terrain studies. In E.C. Barrett & L.F. Curtis (eds.) 'Environmental remote sensing: practices and problems'. 2, p. 101–124.

Weber, E. 1972. Grundriss der Biologischen Statistik. Fischer, Jena.

Accepted 8 December 1979

ORDINATION AS A TOOL FOR ANALYZING COMPLEX DATA SETS

Robert K. PEET*

Department of Botany, University of North Carolina, Chapel Hill, North Carolina 27514, USA

Keywords:
Forests, Gradient analysis, North Carolina, Ordination, Reciprocal averaging, Vegetation

Introduction

Ordination, or indirect gradient analysis, has been widely used in plant ecology as a tool for examining relationships between environment and vegetation. Since the classic work of Bray & Curtis (1957) which popularized such applications, the mathematical models underlying ordination methodology have become increasingly sophisticated. Principal Components Analysis provided increased conceptual rigor and objectivity. The realization that the assumptions of linearity implicit in PCA were not consistent with representation of nonlinear species responses along environmental gradients (Beals 1973, Jeglum, Wehrhahn & Swan 1971, Whittaker & Gauch 1973) led to introduction of a series of nonlinear methods including Reciprocal Averaging, RA, (Hill 1973, 1974), Gaussian Ordination, GO, (Gauch, Chase & Whittaker 1974, Ihm & Groenewoud 1975) and Nonmetric Multidimensional Scaling, NMDS, (Fasham 1977, Prentice 1977). Subsequent comparative studies (e.g., Fasham 1977, Gauch, Whittaker & Singer 1980) have suggested these nonlinear methods to consistently and accurately recover the structure of simulated data.

Simple, unidimensional coenoclines or compositional gradients appear to be readily recoverable using available nonlinear methods. Gauch et al. (1980) have also demonstrated reasonable success in recovery of second axes in simulated data, where the axes were roughly orthogonal and equal in length. What is rarely stated in such evaluations is that as the complexity of a data set increases, the efficiency of information recovery declines. This is in part a consequence of the underlying environmental factors not conforming to a consistent, geometric model. Multidimensional rectangular niche structures are theoretically attractive but rarely encountered. Rather, nonlinear interactions of factors are the rule; factors vary in importance with changes in other factors. The result is that beyond 2 or 3 dimensions, ordinations are rarely interpretable in terms of underlying environmental factors. In short, ordination can work well for identification of the most important and conspicuous environmental trends in a data set, but these would have been deduced by a competent field ecologist while collecting the data. Ordination has thus far proven much less effective at resolving environmental relations in complex data sets, precisely those data sets where much could be learned if effective ordination methods could be found.

As part of a study of forest succession on the North Carolina piedmont, I sampled, in collaboration with Dr. Norman Christensen (see Peet & Christensen 1979, 1980), 88 hardwood forest stands. One of our objectives was to characterize potential forest vegetation and to examine the environmental basis for its compositional variation. This characterization could then provide a framework for subsequent studies. Oosting's 1942 monograph had previously been the standard reference on the vegetation of this region, but it was based on only minimal sampling and recognized only three upland forest types: preclimax, climax and postclimax. The known edaphic and geologic complexity of the piedmont suggested to us that traditional gradient analysis methods wherein one or two master environmental complexes are recognized would be inappropriate. An ordination methodology was needed which would allow recognition of complex,

* The author gratefully acknowledges the continuing collaboration of Dr. Norman L. Christensen of Duke University. This research was supported by National Science Foundation grants DEB-7708743 and DEB-7804043 to R.K.P. and DEB-7707532 and DEB-7804041 to N.L.C.

multidimensional patterns which were not obvious at the time the data were collected.

The intent of the present contribution is to illustrate a possible solution to the difficulty of applying ordination methods to complex data sets. Although the approach was specifically developed for interpretation of North Carolina piedmont forests, it should be applicable to many complex data sets.

The basic methodology is one of progressive fragmentation, or removal of groups of stands. After initial ordination in 2 or 3 dimensions using a method such as RA or NMDS, the resulting pattern is examined for environmental correlations. Distinctive groups are removed (or a continuum is partitioned) and the remaining stands are reordinated, the impact of the initially observed factor having been greatly reduced. In effect, this resembles a subjective form of Hill et al.'s (1975) Character Species Analysis. The objective, however, is not simply to classify, but to understand the environmental relationships between groups of similar stands.

Methods

Samples were 0.1 ha quadrats, 20 × 50 m. Cover and frequency were recorded for all photosynthetic area below 1 m using a transect of 25 contiguous 0.5 × 2.0 m subquadrates located along the center of plot. All species occurring within the 0.1 ha were recorded and all trees (≧ 1.5 cm) were recorded by diameter at breast height. Importance values were calculated for understory (< 1 m) species as the average of relative cover and relative frequency. Species present in the 0.1 ha sample but absent from the subquadrates were assigned importance values of 0.01 (yielding a possible scale of 0.01 to 100). All 88 stands sampled were dominated by hardwoods, and had received little or no disturbance for over 50 years. Disturbance history was ascertained from records of the Duke Forest and Umstead State Park wherein all samples were collected. Five soil pits were dug in each plot and the upper 10 cm of the A_1 horizon were analyzed for Ca, Mg, K, PO_4, organic matter, pH, and available water capacity (see Peet & Christensen 1979 for further details).

RA was used for all ordinations. Species with less than four occurrences were consistently omitted from the data matrices. After removal of rare species the initial 88 stand matrix contained 202 species. The double relativization method of Bray & Curtis (1957) was consistently used. The tree data were not used in this analysis, the herbs (i.e., leafavea below 1 m) being considered more responsive to minor site differences and having fewer residual effects of past disturbances. Rather, the tree data were used in interpretation of the results by plotting dominant species on the ordinations.

Results

The first ordination revealed a pronounced gradient from alluvial bottomlands to dry, upland sites. The second axis was merely a polynomial version of the first, a result often encountered where the first axis is especially pronounced (Gauch, Whittaker & Wentworth 1976, Hill 1973). A plot of the first and third axes (Fig. 1) showed a distinct group of mesic, nutrient-rich coves. There is some suggestion of a third group characterized by the presence of *Quercus marilandica* and typically occurring on montmorillonitic clays, often derived from diabase dikes.

The distinctive alluvial (5) and cove forest (7) stands were removed from the data set leaving a heterogeneous set of mixed upland hardwoods. After removal of the rare species and relativization, these data were again ordinated (Fig. 2). Two distinct sets of stands emerged. *Quercus marilandica* stands over montmorillonitic clays dominate the lower right of the ordination while stands from north-facing bluffs with disjunct montane species such as *Kalmia latifolia*, *Rhododendron catawbiense* and *Galax aphylla* dominate the upper left. The remaining

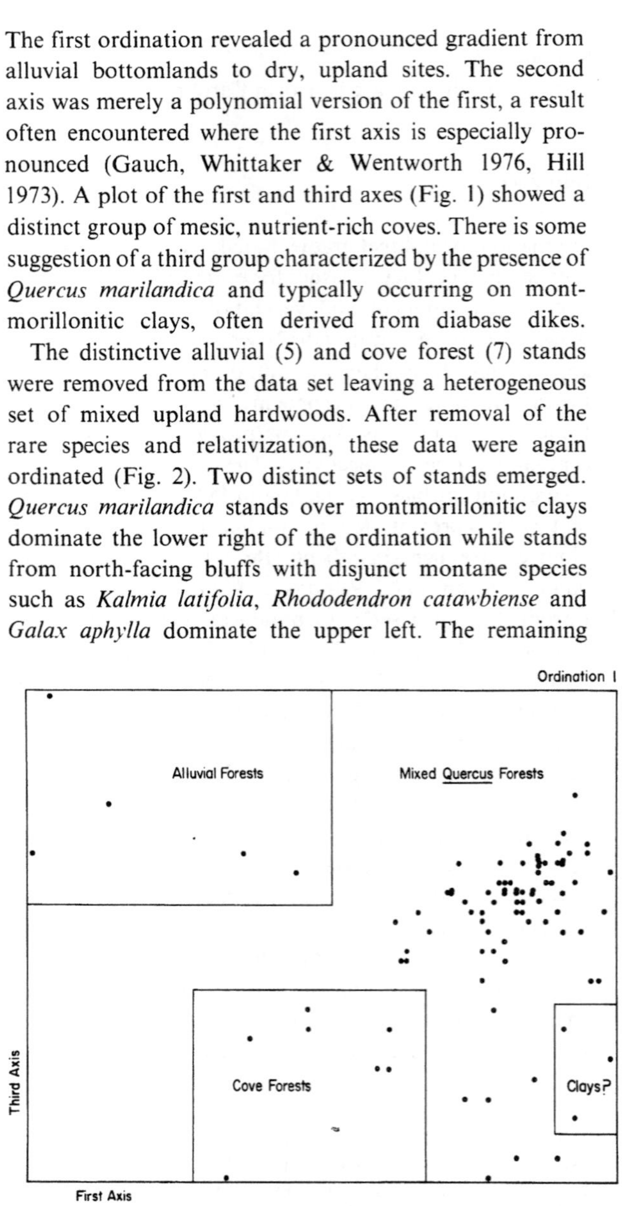

Fig. 1. First ordination of North Carolina piedmont hardwood stands showing segregation of alluvial and cove forest types.

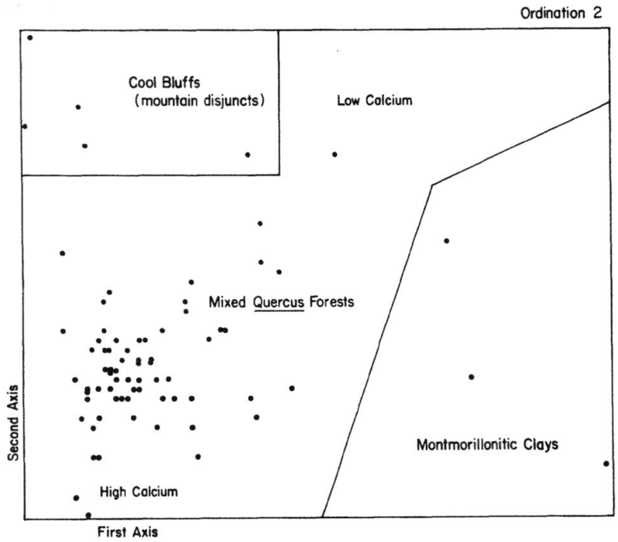

Fig. 2. Second ordination of North Carolina piedmont hardwood stands showing segregation of north-facing bluffs with montane elements, and of sites on montmorillonitic clays, usually with *Quercus marilandica* and *Q. stellata*.

stands again form a heterogeneous group of mixed hardwood stands, but with a suggestion of a gradient in soil pH and cations from base-rich soils in the lower left to more acidic, oligotrophic stands in the upper right.

A third ordination was performed after removal of the north-facing bluff (5) and *Quercus marilandica* (3) sites (not illustrated). The second axis of this ordination revealed a conspicuous group of stands dominated by *Quercus prinus* (3) characteristically occurring on isolated hills which rise up to 100 m above the surrounding landscape. These hills represent Jurassic intrusions of calcium poor volcanic rock. It is reassuring that the stands comprising the *Quercus prinus* group came from locations distant from each other, but similar in tree composition, topographic position and substrate.

The fourth ordination was performed after removal of the *Quercus prinus* stands (Fig. 3). Here pattern appears in what was previously a blurr of mixed hardwood stands. An apparent moisture gradient runs across the diagonal of the ordination from mesic stands dominated by *Fagus*, *Liriodendron* and *Acer rubrum* through mixed *Quercus* and *Carya* stands to xeric and ridge top stands dominated by *Quercus stellata*. Roughly orthogonal to this is a curving gradient in soil acidity and base content, the upper right portion of the gradient containing a group of stands dominated almost exclusively by *Quercus alba* and characterized by soils of extremely low pH ($\simeq$ 3.9) and exchangeable calcium content (< 100 ppm).

A fifth ordination was performed after removal of the oligotrophic *Quercus alba* forests. As illustrated in Fig. 4, this ordination yielded two apparent complex-gradients. The first axis corresponded to a moisture gradient with

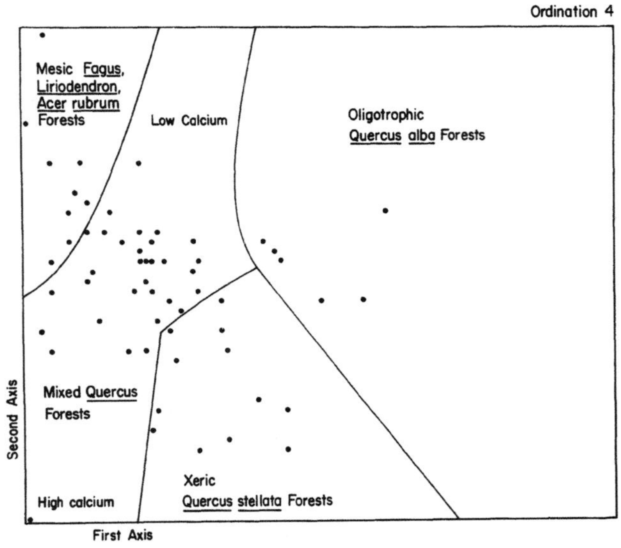

Fig. 3. Fourth ordination of North Carolina piedmont hardwood stands showing a soil pH and calcium gradient from lower left to upper right, and a moisture gradient from the upper left (mesic) to lower right (xeric).

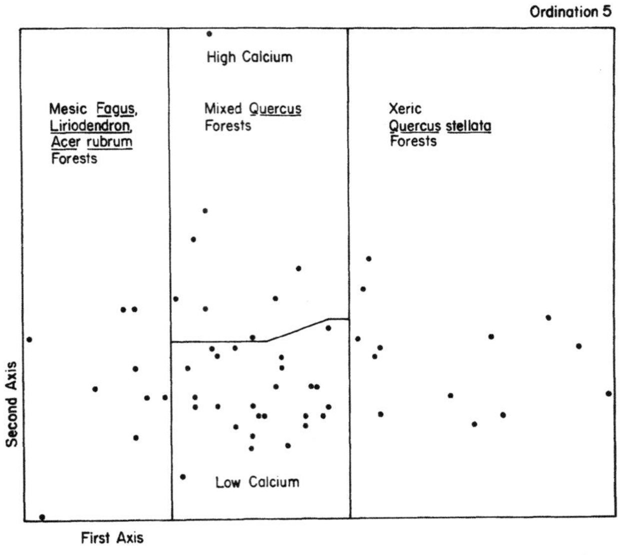

Fig. 4. Fifth ordination of North Carolina piedmont hardwood stands showing a moisture gradient from left (mesic) to right, and a soil calcium gradient from top to bottom.

mesic *Fagus*, *Liriodendron*, and *Acer* stands on the left and xeric *Quercus stellata* stands on the right. The second axis correlated strongly with soil calcium (r = 0.691), magnesium (r = 0.644) and pH (r = 0.515).

The mesic and xeric groups of stands were removed and the remaining set of 31 mixed *Quercus* and *Carya* stands was ordinated. In this sixth ordination (not illustrated) the first axis revealed the expected gradient in soil calcium content. The second axis revealed an axis in soil water storage capacity. This was paralleled by gradients in soil organic content and PO_4 suggesting low available water to correspond with low rates of decomposition.

Discussion and conclusions

The example presented illustrates the potential of a strategy of progressive data set fragmentation for determining environmental relations of vegetation when using ordination. While superficially the approach resembles classification, traditional classification methods do not allow ready examination of environmental relationships. If the stands were first classified into the final groups recog-ized and then ordinated, the environmental relationships would have again been obscured by the correlations, interactions and nonlinearities of the factors.

The approach suggested has an admitted and inescapable element of subjectivity (see Hill et al. 1975 for a more objective analogue), though careful use of correlation coefficients might reduce the severity of the limitation. It must be remembered that no classification scheme can be considered a priori correct; subjectivity will enter into all useful ecological classification schemes, Dale (1980) not withstanding (consistency can and should still be sought). Once the subjective nature of the classification process is recognized and accepted, it should be clear that progressive, subjective removal of stands should not greatly modify the perceived environmental relationships of the stands retained.

One of the major objectives of vegetation science is to understand the environmental patterns underlying vegetation composition. Ordination offers considerable potential as a tool for achieving this objective, if only the difficulties presented by complex data sets can be resolved. In this paper I suggest a simple but potentially useful strategy for coping with these difficulties.

Summary

Ordination has proven to be a useful tool for examining relationships between environment and vegetation in data sets with a simple underlying environmental structure. Complex data sets have proven much less tractable. A strategy is offered for dealing with complex data sets based on progressive removal of sets of stands along identified gradients, and subsequent reordination. This strategy is demonstrated using forests of the North Carolina piedmont.

References

Beals, E.W. 1973. Ordination: Mathematical elegance and ecological naivete. J. Ecol. 61: 23–25.

Bray, J.R. & J.T. Curtis. 1957. An ordination of the upland forest communities of southern Wisconsin. Ecol. Monogr. 27: 325–349.

Dale, M.B. 1980. Syntactic basis for classification. In: E. van der Maarel (ed.), Advances in vegetation science: Classification and Ordination. Vegetatio 42: 111–111.

Fasham, M.J.R. 1977. A comparison of nonmetric multidimensional scaling, principal components and reciprocal averaging for the ordination of simulated coenoclines, and coenoplanes. Ecology 58: 551–561.

Gauch, H.G., G.B. Chase & R.H. Whittaker. 1974. Ordination of vegetation samples by Gaussian species distributions. Ecology 55: 1382–1390.

Gauch, H.G., R.H. Whittaker & S.B. Singer. 1980. A comparative study of nonmetric ordinations. J. Ecol. (in press).

Hill, M.O. 1973. Reciprocal averaging: An eigenvector method of ordination. J. Royal Stat. Soc., Ser. C. 23: 340–354.

Hill, M.O., R.G.H. Bunce & M.W. Shaw. 1975. Indicator species analysis, a divisive polythetic method of classification, and its application to a survey of native pinewoods in Scotland. J. Ecol. 63: 597–614.

Ihm, P. & H. van Groenewoud. 1975. A multivariate ordering of vegetation data based on Gaussian-type gradient response curves. J. Ecol. 63: 767–777.

Jeglum, J.K., C.F. Wehrhahn & J.M.A. Swan. 1971. Comparisons of environmental ordinations with principal component ordinations for sets of data having different degrees of complexity. Can. J. For. Res. 1: 99–112.

Oosting, H.J. 1942. An ecological analysis of the plant communities of piedmont, North Carolina. Amer. Midl. Nat. 28: 1–126.

Peet, R.K. & N.L. Christensen. 1979. Hardwood forest vegetation of the North Carolina piedmont. Veröff. Geobot. Inst. ETH, Stiftung Rübel, 1979 (in press).

Peet, R.K. & N.L. Christensen. 1980. Succession: a population process. In E. van der Maarel (ed.), Advances in vegetation science: Succession. Vegetatio 43: ...–....

Prentice, I.C. 1977. Non-metric ordination methods in ecology. J. Ecol. 65: 85–94.

Whittaker, R.H. & H.G. Gauch. 1973. Evaluation of ordination techniques. In: R.H. Whittaker (ed.), Ordination and classification of plant communities, p. 287–321. Junk, The Hague.

Accepted 10 November 1979

MANAGEMENT AND MULTIVARIATE ANALYSIS OF LARGE DATA SETS IN VEGETATION RESEARCH

Otto WILDI*

Swiss Federal Institute of Forestry Research, CH-8903 Birmensdorf, Switzerland

Keywords:
Data-processing, Large data sets, Multivariate analysis, Vegetation

Introduction

I would like to present a processing system designed for the analysis of vegetational data. It has been developed in 1977/1978 at the Department of Plant Sciences of the University of Western Ontario, London, Canada. An interactive version has been produced for class use in graduate courses of plant ecology and a batch version for processing large data sets (Wildi, Fewster & Orlóci 1978). It is presently integrated into a storage and retrieval system for landscape information, now under development at the Swiss Federal Institute of Forestry Research, Birmensdorf. I would like to thank Dr. L. Orlóci for helpful discussions, and the University of Western Ontario for the facilities to do my research.

Vegetation surveys almost always produce large sets of data, usually in the form of measurements on species performance or of site factors within given stands of a fixed size. The analysis of such data is difficult with the programs commonly available, in that the computer core required is large and the expenses are excessive. Whereas many algorithms are known which can provide for an intensive analysis of vegetation data, (cf. Orlóci 1978), some simpler ones, handling very large data sets, have also been suggested (Lieth & Moore 1971, Ceška & Roemer 1976, Holzner & Stockinger 1973, Spatz & Siegmund 1973, Dale & Quadraccia 1973, Holzner et al. 1978, Van der Maarel et al. 1978, Streng & Schönfelder 1978, Louppen & van der Maarel (1979) and others). These usually do not lead to stable or unique solutions. Most of the existing program packages did not satisfy the types of objectives which I consider important:

1. Sufficient number of options for the method of data analysis.
2. Flexibility in applications under different requirements, given the possibility to interact with the processing system.
3. Help with the simple, but time consuming manipulations of printing tables and graphs, reducing or combining data sets while retaining identification numbers of species and relevés, etc.

Evidently, a single program could not fulfil all these needs since it would occupy too much computer core space. The version which I am describing consists of eleven different programs which are not all needed at the same time.

Different analyses may require different combinations of programs to serve different objectives. A brief outline of the systems concept involved is my first topic, followed by brief descriptions of the operating procedure, and some examples.

The operating system

In any investigation, data processing is just one of several logically related steps. Considering the purpose of the analysis, three basic types of processes are of interest:

1. Data handling. This involves manipulations such as data storage and retrieval, subdividing or combining data sets, reducing the number of species or relevés, condensing data by forming frequency or contingency tables.
2. Data analysis. Numerical analyses are usually based on resemblance matrices. The present processing system allows ranking procedures, multidimensional scaling (to produce coordinates for ordination), classifications

* A detailed introduction with a description of the processing system described in this paper is available from the author at no cost.

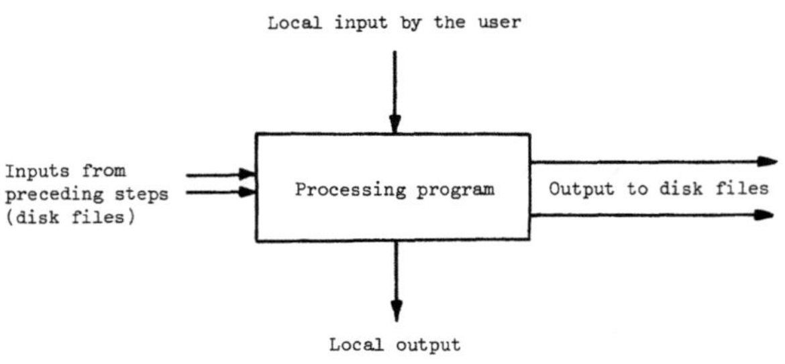

Fig. 1. General structure of the programs.

(to find groups), identifications (to reallocate new observations) and also statistical tests to measure the success of classifications.

3. Response programs. Some of the results need to be presented graphically in order to facilitate the interpretation. Vegetation tables, dendrograms and scatter diagrams (ordinations) are examples.

Since most of these processes need information from previous ones, programs have to be linked by input- and output files. The general structure of a typical program is shown in Fig. 1. All disk files are written in standard format and are read or written automatically. The local input specified by the user allows to influence the analysis by selecting the option which serves the purpose of the investigation. Results are given in the form of a local output. Their interpretation may suggest the programs to be run next. It is, of course, also possible to repeat an analysis with a different option if the results are not satisfying to the investigator.

System structure and operation

Each program of the package (Table 1) performs one step in the analysis. Therefore, a variable number of programs may have to be run in order to get results. Fig. 2 illustrates this; the logically acceptable steps in the analyses are indicated by arrows. The user will probably not need all the algorithms, but a limited subset according to personal preference and the objectives of the analysis. Typical pathways are shown in Fig. 3:

The first, indicated by a dotted line, may be used by investigators preferring subjective classification. The analysis will start with program EDIT, which stores the initial arrangement and also the group membership of relevés and species. Program TABS then prints a vegetation table, a frequency table (Stetigkeitstabelle) and a contingency table. Program CIAC (for a canonical analysis

Table 1. List of programs. See main text for further explanations.

Name	Function	Purpose
1 INIT	Handling	Initializes data files, performs transformations and computes frequencies.
2 EDIT	Handling	Stores classifications from other programs or from the user, reduces data sets.
3 TABS	Handling and presentation	Prints vegetation tables, computes and prints frequency- and contingency tables.
4 RESE	Analysis	Computes resemblance vectors or matrices.
5 IDEN	Analysis	Reallocates new relevés or species to existing groups.
6 RANK	Analysis	Ranks species by orthogonal functions.
7 CLTR	Analysis and presentation	Does cluster analyses, prints dendrograms and forms groups.
8 PCAB	Analysis	Computes component analysis (R and Q) or correspondance analysis (reciprocal ordering).
9 GRID	Analysis	Finds groups in ordinations.
10 ORDB	Presentation	Prints scatter diagrams.
11 CIAC	Analysis	Measures success in tabular sorting.

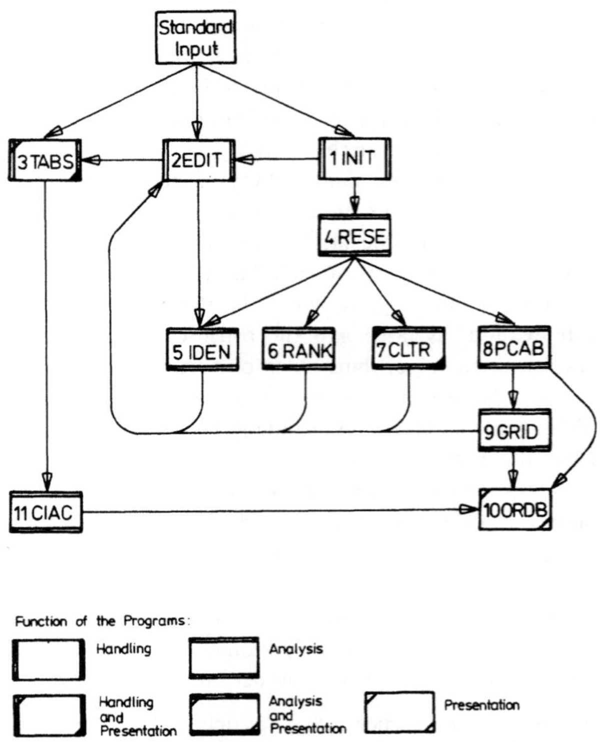

Fig. 2. Flow chart of operations. The arrows indicate logical steps in the analysis.

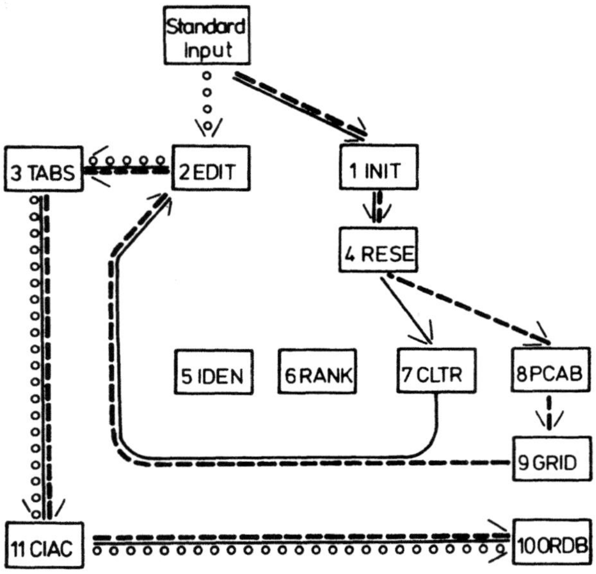

1. o o o o Pathway for subjective classification followed by measurement of success in tabular sorting

2. ▬ ▬ ▬ Pathway for detecting groups in an ordination by grid analysis

3. ——— Pathway for detecting groups by agglomerative clustering

Fig. 3. Three typical pathways in a vegetation analysis.

```
SPECIFY FILE TO BE USED
OPT1  ORIG.DAT
OPT2  TAST.DAT
OPT3  RECR.DAT
OPT4  CO.DAT
TYPE OPT:
1

1
OPT1: PRINT RAW TABLE
OPT2: PRINT ORDERED TABLE
OPT3: COMBINE 2 TABLES
TYPE OPT:
1

1
ENTER A NEW FORMAT, OTHERWISE TYPE NO:
NO

NO
  TABLE NO.      1
  TEST DATA SET 5/79
  ARTIFICIAL DATA

+------------------------------------------+----------
I                                          .
I                                          .
I                                          .
I      RELEVE NO.                          . 1 2 3 4 5
+------------------------------------------+----------
I   1 SPECIES 1                            . R   +
I   2 SPECIES 2                            . 1 1 2   1
I   3 SPECIES 3                            .       5
I   4 SPECIES 4                            . 3   4
I   5 SPECIES 5                            .   R R   +
+------------------------------------------+----------
```

Fig. 4. Output from program TABS (unordered vegetation table).

```
SPECIFY FILE TO BE USED
OPT1  ORIG.DAT
OPT2  TAST.DAT
OPT3  RECR.DAT
OPT4  CO.DAT
TYPE OPT:
1

1
TYPE "YES" FOR SIGNUM TRANSFORMATION:
NO

NO

CODE REPLACEMENT IS ACTIVE!

NEXT TRANSFORMATION:
OPT0: NO TRANSFORMATION
OPT1: RELEVE VECTORS TO UNIT LENGTH
OPT2: SPECIES VECTORS TO UNIT LENGTH
OPT3: TO DEVIATION FROM EXPECTATION
TYPE OPT:
3

3
TABLE NO.            1
NUMBER OF RELEVES       5
NUMBER OF SPECIES       5
DATE OF SAMPLING    4-79  TO  5-79
PLOT SIZE       0 SQUARE METER

TEST DATA SET 5/79
ARTIFICIAL DATA

CODE:       R      +      1      2      3      4      5
VALUE:   3.0    5.0    6.0    7.0    8.0    9.0    9.0    0.0    0.0    0.

RELEVE          1    2    3    4    5
DIVERSITY       3    2    4    1    2

SPECIES         2    5    1    4    3
FREQUENCY       4    3    2    2    1

OUTPUT OF DATA COVERS   16 LINES.

TYPE "YES" FOR OUTPUT OF DATA:
YES

YES

SPECIES NO.:
        1         2         3         4         5

RELEVE NO.    1:
    0.015   -0.001   -0.031    0.055   -0.036

RELEVE NO.    2:
   -0.015    0.040   -0.017   -0.031    0.023

RELEVE NO.    3:
    0.032   -0.022   -0.044    0.045   -0.011

RELEVE NO.    4:
   -0.015   -0.046    0.112   -0.031   -0.020

RELEVE NO.    5:
   -0.018    0.030   -0.020   -0.038    0.047
```

Fig. 5. Output from program INIT.

of the contingency table) can be applied to measure the success of tabular sorting. The results of CIAC are presented graphically by program ORDB which prints an ordination of relevés and species groups.

In the second case (dashed line in Fig. 3), program INIT initializes the computation. Here, a code, such as Braun-Blanquet, is replaced by numerical values and some transformations may be applied (cf. van der Maarel 1979). Program RESE computes a resemblance matrix and program PCAB computes ordination coordinates for both, relevés and species in Q- and R- type component analysis or reciprocal ordering. GRID analysis finds noda (clusters) in the ordinations, which are picked up and stored by running program EDIT. A vegetation table is printed and tested in the same manner as described in the first case.

In the third case (light line in Fig. 3), the analysis begins with the program INIT. The resemblance matrix (program RESE) is subjected to cluster analysis (program CLTR). This yields a dendrogram, which, after inspection, is subdivided into groups.Program EDIT stores the classifications, TABS prints the vegetation and summary tables and CIAC measures the success of the procedure.

An example

In the following part, I am presenting some computer outputs from the analysis of an artificial data set (5 relevés and 5 species). The pathway followed in the example is similar to the second case in Fig. 3. The data are easiest to inspect when printed with program TABS (Fig. 4). In all cases program INIT is the first in the computations producing general information on the data and replacing a code by numbers (Fig. 5). Frequencies are computed and the raw data is given as output on request. Program

```
TYPE 'R' FOR COMPARING RELEVES, 'S' FOR SPECIES:
R

R


SPECIFY RESEMBLANCE FUNCTION
OPT1  CROSS PRODUCT
OPT2  COVARIANCE
OPT3  CORRELATION COEFFICIENT
OPT4  EUCLIDEAN DISTANCE
OPT5  CHORD DISTANCE
OPT6  OCHIAI COEFFICIENT
OPT7  VAN DER MAAREL'S COEFFICIENT
TYPE OPT:
1

1
TYPE 'YES' FOR COMPARING 1 CHARACTER WITH ALL OTHERS
NO

NO
RESEMBLANCE MATRIX OF RELEVES IS NOW AVAILABLE IN FILE RES.DAT
OPT 1 HAS BEEN USED

OUTPUT OF THE MATRIX COVERS    17 LINES!
TYPE "YES" FOR OUTPUT OF THE MATRIX:
YES

YES

NO.:
      1         2         3         4         5

NO.   1
   0.006   -0.002    0.005   -0.005   -0.004

NO.   2
  -0.002    0.004   -0.002   -0.003    0.004

NO.   3
   0.005   -0.002    0.006   -0.006   -0.003

NO.   4
  -0.005   -0.003   -0.006    0.016   -0.002

NO.   5
  -0.004    0.004   -0.003   -0.003    0.005
```

Fig. 6. Output from program RESE.

```
INFORMATION IS AVAILABLE FOR OPT:
   1
   2

SELECT COORDINATES:
OPT1: RELEVES
OPT2: SPECIES
OPT3: RELEVES AND SPECIES GROUPS
TYPE OPT:
2

2

SPECIFY NUMBER OF DIMENSIONS OF GRAPH (2 OR 3):
2

2
SPECIFY COORDINATE SETS FOR X-,Y-AXES
LENGTH OF X-AXIS IN ROW UNITS (2I1,I3)
12 15

12 15


TEST DATA SET 5/79

PLOT OF SPECIES
+---------------+
I   .           I
I   .           I
I  1.1          I
I   .           I
I   .           I
I...............I
I   .          3I
I  2.           I
I2  .           I
+---------------+

X-AXIS IS 1. COORD. RANGE     -0.0620 TO        0.1245
Y-AXIS IS 2. COORD. RANGE     -0.0683 TO        0.1181
    5 POINTS ARE PLOTTED IN    3  GROUPS

LOCATION OF PLOTS
  -
  -
  2   5
  -
  -
  -
  3
  1
  4
```

Fig. 7. Output from program ORDB.

RESE identifies the rows of data matrix as relevés (Fig. 6). After selecting one out of seven measures available, a resemblance matrix is computed. When deriving coordinates for relevés and species in a component analysis (program PCAB), program GRID may do a nodal analysis. This leads to a classified ordination (Fig. 7, shown for species) printed by program ORDB. The results must be stored in a run of program EDIT. TABS prints an ordered vegetation table, a contingency table and also a frequency table (Stetigkeitstabelle) (Fig. 8). CIAC measures the success in tabular sorting and gives coordinates for relevés and species groups, which are again printed, using program ORDB (Fig. 9).

```
TABLE NO.      1
TEST DATA SET 5/79
ARTIFICIAL DATA

+-----------------------------------------+----------
I                                         .
I     RELEVE GROUP NO.                    . 1 1 2 3 3
+-----------------------------------------+----------
I                                         .
I                                         .
I                                         .
I     RELEVE NO.                          . 3 1 4 2 5
+-----------------------------------------+----------
I  2 SPECIES 2                          1 . 2 1   1 1
I  5 SPECIES 5                          1 . R     R +
I  4 SPECIES 4                          2 . 4 3
I  1 SPECIES 1                          2 . + R
I  3 SPECIES 3                          3 .     5
+-----------------------------------------+----------
TYPE YES FOR FREQUENCY- AND CONTINGENCY TABLES!

YES

YES

CONTINGENCY TABLE

RELEVE GROUP NO.:
     1    2    3

FREQUENCIES IN SPECIES GROUP    1:
     3    0    4

FREQUENCIES IN SPECIES GROUP    2:
     4    0    0

FREQUENCIES IN SPECIES GROUP    3:
     0    1    0

GROUP CENTROIDS ARE IN TAST.

   TABLE NO.   1001
   TEST DATA SET 5/79
   ARTIFICIAL DATA

+-----------------------------------------+------
I                                         .
I                                         .
I                                         .
I     RELEVE GROUP NO.                    . 1 2 3
+-----------------------------------------+------
I  2 SPECIES 2                            . 5   5
I  5 SPECIES 5                            . 3   5
I  4 SPECIES 4                            . 5
I  1 SPECIES 1                            . 5
I  3 SPECIES 3                            .   5
+-----------------------------------------+------
```

Fig. 8. Output from program TABS (ordered vegetation table).

Discussion

It is obvious from what has been shown in the previous chapters that the main difference between this package and most others lies in the great number of programs, data files and options available. A loss of some operational ease is a direct consequence of this, even though manual work, like writing tables, drawing graphs or changing format statements, is made superfluous.

```
INFORMATION IS AVAILABLE FOR OPT:
   1
   2
   3

SELECT COORDINATES:
OPT1: RELEVES
OPT2: SPECIES
OPT3: RELEVES AND SPECIES GROUPS
TYPE OPT:
3

SPECIFY NUMBER OF DIMENSIONS OF GRAPH (2 OR 3):
2

SPECIFY COORDINATE SETS FOR X-,Y-AXES
LENGTH OF X-AXIS IN ROW UNITS (2I1,I3)
12 15



TEST DATA SET 5/79

CANONICAL ANALYSIS OF FREQUENCIES
+---------------+
IS.             I
IR.             I
I .             I
I .             I
I..........*....I
I .             I
I .             I
IS.             I
IR.             I
+---------------+

X-AXIS IS 1. COORD. RANGE     -0.6031 TO      2.4867
Y-AXIS IS 2. COORD. RANGE     -1.5449 TO      1.5449
    6 POINTS ARE PLOTTED IN    0  GROUPS

LOCATION OF PLOTS
 102
   2
   -
   -
   3 103
   -
   -
 101
   1
```

Fig. 9. Output from program ORDB (input is from CIAC).

The user is faced with three major difficulties: 1. He has to observe the systems operational logic (Fig. 2). 2. He must know the objectives of the methods available in order to evaluate his own operational pathway (Fig. 3). 3. The programs, if possible, avoid restricting decisions but leave them open to the user. As an example, the classification, ordination or identification algorithms accept anyone of seven resemblance measures in program RESE, irrespective of their meaningfulness.

The same properties, on the other hand, make up the advantages of the package. Its outcome is based on the idea mentioned by Orlóci (1980), that if multivariate analysis is intended to serve a useful purpose in vegetation studies, the methods must be chosen with a clear understanding of their potential and with a reasoned evaluation of circumstances in the intended application. The diversity of such applications, in fact, is reflected in the number of options and pathways available. For example, 48 different sets of results are possible when clustering relevés or species with program CLTR. This is a direct outcome of the four transformation options in program INIT, six resemblance measures in RESE (number 4 and 5 are redundant!) and the two algorithms available in CLTR.

The system's structure allows the inspection of intermediate results. These permit the user to plan the subsequent steps. In this way, the data become a part of the processing system with direct profits for the user. This reminds us of the phenomenon mentioned by Wedekind (1972), i.e. that in a complex processing system there is hardly a difference between data and programs.

Summary

A processing system consisting of 11 programs, designed for management and analysis of vegetation data is presented. The concept of the system and the operational logic are discussed. Some of the main features are detailed. These include: (1.) availability of many methods, (2.) the possibility for the user to interact with the processing system by altering the analysis through options, and (3.) the computer supported handling of data, such as storing and retrieving data, subdividing or combining data sets, reducing the number of species or relevés, and printing tables and graphs. Differences between this and other program packages are discussed. It is shown that the data, not only the programs, can be made a part of the processing system.

References

Ceška, A. & H. Roemer. 1976 : Ceška-Roemer Program for Identifying Species-Relevé Groups. Manuscript. 14 pp.

Dale, M.B. & L. Quadraccia. 1973. Computer assisted tabular sorting of phytosociological data. Vegetatio 28 : 57–73.

Holzner, W. & F. Stockinger. 1973. Der Einsatz von Elektronenrechnern bei der pflanzensoziologischen Tabellenarbeit. Oesterr. Bot. Z. 121: 303–309.

Holzner, W., M.J.A. Werger & G.A. Ellenbroek. 1978. Automatic classification of phytosociological data on the basis of species groups. Vegetatio 38 : 157–164.

Lieth, H. & G.W. Moore. 1971. Computerized clustering of species in phytosociological tables and its utilization for field work. In: G.P. Patil, E.C. Pielou & W.E. Waters (eds.), Statistical Ecology, Vol. 1, p. 403–422. London, Penn. State Univ. Press.

Louppen, J.M.W. & E. van der Maarel. 1979. Clusla. Vegetatio 40: 107–115.

Maarel, E. van der. 1979. Transformation of cover-abundance data. Vegetatio 39: 97–115.

Maarel, E. van der, J.G.M. Janssen & J.M.W. Louppen. 1978. TABORD, a program for structuring phytosociological tables. Vegetatio 38: 143–156.

Orlóci, L. 1978. Multivariate analysis in vegetation research. Junk, The Hague. 2nd ed., 451 pp.

Orlóci, L. 1980. An algorithm for predictive ordination. In: E. van der Maarel (ed.). Advances in vegetation science: Classification and ordination. Vegetatio 42: 000–000.

Spatz, G. & J. Siegmund. 1973. Eine Methode zur tabellarischen Ordination, Klassifikation und ökologischen Auswertung von pflanzensoziologischen Bestandesaufnahmen durch den Computer. Vegetatio 28: 1–17.

Streng, R. & Schönfelder. 1978. Ein heuristisches Computer-Programm zur Ordnung pflanzensoziologischer Tabellen. Hoppea, Denkschr. Regensb. Bot. Ges. 37: 407–433.

Wedekind, H., 1972: Datenorganisation. Walter de Gruyter, Berlin, New York. 2. Aufl. 271 pp.

Wildi, O., P.H. Fewster & L. Orlóci, 1978: Management and multivariate analysis of large data sets in vegetation research. Manuscript. London, Ont., Canada.

Accepted 5 November 1979

REFLECTIONS ON THE PHYTOSOCIOLOGICAL APPROACH AND THE EPISTEMOLOGICAL BASIS OF VEGETATION SCIENCE*

S. PIGNATTI

Istituto Botanico, cas. Università Via Valerio 30, I-34100 Trieste, Italy

Keywords:
Braun-Blanquet approach, Epistemology, Logic of vegetation science, Phytosociological method, Popper

The conventional approach: phytosociology based on the inductive method

It is generally assumed that phytosociology is a true empirical science. In his fundamental theoretical work Braun-Blanquet (1928) identified the task of phytosociologists with the following words' . . .in der bunten Vielheit der Vegetationsflecke das Uebereinstimmende erkennen und herausarbeiten, das Typische abgrenzen und unter den abstrakten Assoziationsbegriff zusammenfassen . . .': '. . . to recognize and elaborate in the polychrome multiplicity of the vegetation cover the unifying, to delineate the typical and summarize it within the abstract association concept . . .'. Westhoff (1972) summarized the methodological approach of phytosociology in three steps: (a) the comparison between concrete vegetation stands (phytocoenoses) in nature, (b) carrying out of relevés, as samples of the concrete vegetation and (c) reciprocal comparison of the relevés and abstraction of vegetation types (plant associations, phytocoenons). The basic ideas on the essence of plant associations and their diagnostic species are exposed by Westhoff & van der Maarel (1978).

In a more practical sense, the classical working method of phytosociology involves several operations i.e.:

(a) survey in the field	Geländestudien
(b) relevés	Aufnahmen
(c) rough table	Rohtabelle
(d) ordered table	geordnete Tabelle
(e) synoptic table	Uebersichtstabelle
(f) definition of phytocoenons	Definition der Vegetationstypen

From the earliest studies with phytosociological tables in the twenties up to the first tentative applications of computers in phytosociology at the end of the sixties there was little change in this method. Tentative applications of statistics (e.g. by Guinochet, Dagnelie, Raabe, Ellenberg, Pignatti and Mengarda, Barkman etc.) remained isolated examples. On the other hand in fifty years of activity, an enormous amount of information was collected in a remarkable standardized form and entirely available for comparisons: a stock of 200.000 relevés (and probably even more) concerning the European vegetation and a consistent set deriving from other zones of the world.

The epistemological basis of this method can be synthesized in this way:

11. Autoecological approach

111. Each species has an own ecology corresponding to a section of the multidimensional space defined by the different ecological variables (ecological space).

112. There is a probabilistic correspondence between species and ecological variables.

113. The presence of a species in a site indicates that the ecology of the site is included in the ecological space of the corresponding species.

114. First conclusion: The presence of a species in a site gives information on the ecology of that site.

* This paper was written under the auspices of the Programma finalizzato 'Qualità dell'Ambiente' of the C.N.R. (Roma) Subprogetto 5 'Metodologie matematiche e basi di dati'. I thank the following people for helpful comments: E. Feoli, D. Lausi, P. Nimis, E. Pignatti Wikus (all at the Istituto Botanico, Trieste), D. Dal Maso (Istituto di Matematica, Trieste), E. van der Maarel (Botanisch Laboratorium, Nijmegen).

12. Synecological approach

121. Several species a_i, a_j, a_k with a similar ecology (their ecological spaces indicated as S_i, S_j, S_k) define an ecological space $S_i \cap S_j \cap S_k$ which is common to all the given species.
122. The common part $S_i \cap S_j \cap S_k$ is contained in the complete space of the various species involved: $S_i \cap S_j \cap S_k \subset S_i \cup S_k$. Thus it is better precised in terms of ecological variables.
123. The presence of many species in a site indicates, that the ecology of the site is included in the common part of the ecological space defined by the species.

13. Operational deductions

131. Species a_i, a_j, a_k have the common ecological space $S_i \cap S_j \cap S_k$.
132. The combination a_i, a_j, a_k can be recognized in nature and individuate sites with corresponding ecology
133. This observation is generalizable.

Although these statements seem very elementary and obvious, a short discussion is necessary.

The passage from 11. to 12. may be considered from the probabilistic point of view. If the probability to find a site, the ecology of which is included in the ecological space of the species a_i, a_j is indicated with p_i, p_j, then $p_{ij} \leqq p_i$ and $p_{ij} \leqq p_j$

The probability of finding a site with an ecology included in the ecological space common to both species is in general lower than that for the ecological space of a single species. Only if both spaces are entirely coincident or if one space is entirely included in the other is possible to assume equal probability for the event, but these two hypotheses are limit-cases and very rare (may be restricted to some obligate symbionts). As a general case the passage from 11. (Autoecology) to 12. (Synecology) implies a reduction of probability, i.e. an increase of information, which seems to be logically correct.

A similar, but distinct approach starts from the probability of finding species a_i in a site (cf. Yarranton 1969), or species a_j or both a_i, a_j together. If we regard the presence of the species composing the general flora of the region as equi-probable, the probability to find the combination a_i, a_j is lower than that for a_i or for a_j alone, and these probability becomes lower and lower with each new species added to the combination. As theoretical values for the combination between species become rapidly infinitesimal, whereas empirical frequencies are significantly higher, it is possible to observe a situation with low probability, i.e. an event with a high informational value. From these speculations methods were proposed by Cristofolini, Lausi & Pignatti (1970) and more recently by Lausi & Feoli (1979) for weighting the informational value of character-species in vegetation.

In both approaches it seems, that the passage from 11. Autoecology to 12. Synecology is logically correct and consequently also the passage from the individualistic concept of vegetation to the sociological concept.

On the other hand it may be opposed, that the consequent use of the method indicated above as 11. – 12. – 13. always leads to a result, even if the combination a_i, a_j, a_k is accidental or of little information. This happens if the combination concerns widespread species or, on the contrary, if the combination reflects very local conditions. E.g.: The combination between *Quercus petraea* and *Festuca heterophylla* (widespread species) may regularly occur on natural environments, but indicates only, that we are in Europe on limestone-poor soils; or the combination between *Sesleria varia* and *Potentilla aurea* (a very local condition) indicates an acidification on limestone soils in the alpine vegetation belt in the Dolomites near Lienz (Austria), but can indicate quite different ecological conditions in other habitats. It seems that the passage from 132. to 133. (generalization) represents an assumption which cannot become demonstrated. Each combination (interpreted as an assemblage of species) can be recognized, but it is not true, that each combination can be generalized. Consequently, the inductive approach to phytosociology seems somewhat unsatisfactory.

A more detailed analysis of the identification of phytocoenons (this means the passage from 132.to 133.) may be discussed. We compare a typical inductive procedure (a) with the normal procedure (b) in phytosociological work.

(a) *The empirical way* – A number of points are chosen at random in a given territory covered by vegetation. In each point an inventary of the vegetation (relevé) is carried out. Relevés are compared and clustered. Each cluster is identified as a type (phytocoenon).

This procedure requires a large amount of time and the results are often confuse. Each phytocoenon can be individuated only on the basis of several relevées in more or less homotonous plots, i.e. at least 10–20 plots. From large scale phytosociological cartography it is clear, that in a single region under normal conditions some of the vegetation types are largely diffused, while others are very rare, and that mixed vegetation stands, mosaics or ecotones are always present. In consequence, the chance of finding the rarer vegetation types by random sampling is very low and a significant investigation requires a large amount of work. In general rare vegetation types are always insufficiently represented, whereas for the most diffused types the amount of data is too large and makes the elaboration difficult. The procedure can be regarded as really useful only in the testing of a detailed hypothesis derived from earlier typological studies; in this case consequently this is not an example of application of the passage 132. to 133. In conclusion, the precise method of random sampling presents low efficiency and is seldomly used (at least in Europe).

(b) *The well-trained phytosociologist's approach* – A relatively short survey is sufficient to identify vegetation units with a repetitive character. Provisional delimitations of the type (on the basis of species-combination) and verification with relevés in adequate homogeneous stands. Formation of clusters and definition of types.

This is the normal procedure (for a more detailed treatment cf. Westhoff & van der Maarel 1978), with high efficiency; the results are usually more comprehensive as with procedure (a).

Popper's logic and the deductive method in phytosociology

The criticism of the inductive approach in phytosociology exposed as (a) seems in accordance with the concepts exposed by Popper (1959). This author demonstrates that the inductive method is not justified from a logical point of view because it leads to final statements with a metaphysical character.

In this sense, it is now possible to endeavour a quite different interpretation of the phytosociological method. The procedure starts from an axiom and evolves through staments to a conclusion. The problem of corroboration and falsification may be discussed in a further paragraph.

Axiom: Vegetation is organized in communities

Alternative formulations

I) Vegetation is composed by ordered sets of plant species

II) Species cannot be distributed in nature independently of one another.

III) Vegetation reacts to continuously variating ecological gradients with discontinuous variations.

General statements

(a) Communities are recognizable by species combinations
(b) A definite species combination depends on the ecological condition at the present and on long-term floristic evolution.

The long-term floristic evolution leads to specialisation of floras and brings effective differences only between sites belonging to different geographical regions; within a single region there are in general no important differences and then this factor can be neglected. It is a matter of fact, that in some cases the possibility exists, that different types of vegetation occur on the same place (e.g. a pioneer wood, that in some decades evolves in a more mature forest vegetation), but even in these cases it is possible to think, that with the evolution of the vegetation a parallel evolution of the impact of ecological factors takes place. Consequently, the species combination in a given floristic region depends essentially on the ecological condition.

The identification of a phytocoenon occurs along a series of at least 5 steps, that can be exposed in temporal (and logical) order as follows:

21. Basic statement (relevé)

There is the combination of species a_i, a_j, a_k in the space-time region k'

22. Theory

The combination a_i, a_j, a_k corresponds to a given set of ecological factors (the ecological space $S_i \cup S_j \cup S_k$)

23. Falsification I

Make other relevés and test in the field,
if $(a_i, a_j, a_k) \rightarrow S_i \cup S_j \cup S_k$
or if $S_i \cup S_j \cup S_k \rightarrow (a_i, a_j, a_k)$

24. Falsification II

test the falsifiability (Fsb) of other combinations (from others Authors or from other regions), i.e.

if Fsb $(a_i; a_j, a_t) >$ Fsb (a_i, a_j, a_k)

25. Conclusions

251. The combination a_i, a_j, a_k corresponds to the ecological space $S_i \cup S_j \cup S_k$ and is generalizable
252. Species a_i, a_j, a_k characterizes a phytocoenon
253. A phytocoenon exists characterized by a_i, a_j, a_k which occurs in sites included in the ecological space $S_i \cup S_j \cup S_k$.

The problem of falsification

Following Popper (1959) a theory can be corroborated by empirical testing. If the results of the tests are in accordance with the theory, then the theory is verified. But the verification does not exclude the possibility of further tests, which can demonstrate the theory as false. Consequently the central problem is not the verificability of the theory, but the inverse: the falsifiability, i.e. the possibility to decide empirically between the theory and some rival theory.

The theories in phytosociology are falsifiable at two different levels.

The first level of falsification (23.) is carried out in the field by relevés. The basic statement 'combination a_i, a_j, a_k' can by falsified by observing 'combination a_i, a_j, a_t' . . . etc. The falsification by relevés is nevertheless limited to the combinations observable in the field, and consequently it cannot take all possible combinations into consideration. Local conditions or discontinuities in the distribution of the flora can affect the results of this type of falsification. These limitations are removed by 24., i.e. by further investigation.

The second level of falsification (24.) is based on comparisons with data from others authors or from previous own research. This falsification can be carried out at the desk and it is very similar to the central topic of what was called by Braun-Blanquet the study of 'synthetische Gesellschaftsmerkmale'. In general the comparison 24. is sufficient to overcome the limitations indicated under 23.

The analysis of the conventional procedure in phytosociology allows the conclusion, that the theory proposed is entirely falsifiable by anyone; consequently the results can be regarded as being achieved using a method which is in accordance to the logic of scientific discovery. The validity of the Braun-Blanquet's method is thus confirmed.

The use of numerical methods in the classification of relevés does not create a completely different situation. The data processed by the computer are either the relevés carried out under 23. or the data of other sources compared under 24. The falsifiability is not increased by the computer. On the other hand, we must distinguish between potential falsifications and the falsifications effectively tested. With the conventional procedure the scientist does not make use of many of the falsifications possible, because the testing of similar combinations does not generally produce entirely new results, i.e. it is not worth the enormous work required. Consequently, with the conventional procedure a large part of the falsifiability remains at a potential status. On the contrary the computer can test exhaustively each element of the theory and then the amount of falsifications effectively tested increases up to the theoretical maximum.

A further advantage of the use of computers is the possibility of expressing the results of the tests in a quantitative form so that the most satisfactory solution can be easily selected. This cannot be interpreted as if the results of an elaboration by computer were more objective then the results of the conventional procedure. The objectivity is in both cases assumed on the basis of an opinion with a probabilistic basis.

Realizations and perspectives

The consequent use of the phytosociological method procurates the assembly of an enormous amount of data on the European vegetation and its ecology. For Europe the phytosociological analysis may be regarded as reaching the level of saturation (at least for the temperate and cold-temperate zones of the continent); new vegetation types remain to discover only in Southern Europe, in the North etc. The method is valuable for investigations in other parts of the Globe and is used particularly in Japan but also in SW Asia, N Africa, S Africa, in the temperate part of S America etc. But the the most interesting topic in these years is not only the detection of improved plant combinations from extra-european floras.

The axiom indicated as a starting point for phytosociology seems particularly satisfactory in territories with a high vegetational diversity, caused by the synergism of substrate and climate, richness of the flora, glaciations, marine transgressions, man's activity etc. Europe and Japan (the areas most investigated from the phytosociological point of view) show a very favourable condition in this sense. On the contrary it remains to demonstrate whether the axiom can be generalized for the investigation of highly homotonous countries (like the plains in Eastern North America or in the Eurosiberian region), or in equatorial areas, and for these objects the method possibly may undergo some revision.

It must be pointed out, that in general the ecological variables underlying the distribution of species are not expressed as quantitative parameters but only as relations with the ecology of other species. The same is true for vegetation units. For example, the ecology of *Fagus sylvatica* is not defined in absolute terms (optimum growth temperature, light intensity etc.) but as occurring in a cooler habitat than *Quercus pubescens* and in a warmer habitat than *Picea excelsa;* similarly the *Fagetum* is cooler than the *Querco-Lithospermetum* and warmer than the *Piceetum.* In this sense phytosociology is founded within a system of reference determined by species of the continental flora marking the endpoints of basic gradients (energy, moisture). This may appear quite unexact, but is the the easiest way to evaluate environmental factors during a survey in nature and good enough for applications like forestry or environmental planning.

Computer methods applied both to classification and to ordination problems contribute to corroborate and stabilize this reference system. In addition, the computer makes it possible to use methods of numerical analysis which could not be otherwise applied. For these reasons the use of computers will have a positive influence for the progress of phytosociology (cf. van der Maarel 1975).

On the other hand, possible misapplications of the computer or misinterpretations of its output must be avoided (cf. on this point Detering 1971 and Pignatti 1975):

– Computers make the scientist free from fastidous routine-work, but bring him into the danger of reducing elaborations only to this activity and losing the opportunity for applying phantasy and imagination.

– The scientist cannot delegate the computer to decide on the validity of a theory.

– The methodological step 22. (proposition of a theory) must be performed by man and cannot be delegated to the machine.

– Computers can process only a certain type of (quantifiable) data, but this does not allow the neglect of other possibilities

– Wrong data lead to a wrong theory, even if they have been processed by computer.

– The computer cannot save the investigator from this responsibility.

In conclusion the use of numerical methods in phytosociology can be positively judged on the basis of the experimentation done by the Working Group for Data-Processing in Phytosociology in the last years. The phytosociological method becomes confirmed rather than diversified, is brought to a higher degree of efficiency and can be better tested. Anyway, the fundamental step in the procedure is invariate: the theory, which can be proposed only by the researcher on the basis of its experience and phantasy.

Summary

It is tried to identify the logical basis of the phytosociological method. Both the inductive and the deductive approach are discussed. The starting point can be summarized with the axiom 'vegetation is organized in communities'. Some problems related to the use of numerical methods and computers are discussed.

References

Braun-Blanquet. J. 1951. Pflanzensoziologie 2. Aufl. 631 pp. Wien.

Cristofolini, G., D. Lausi & S. Pignatti. 1970. Ueber statistische Eigenschaften der Charakterarten und deren Verwertung zur Aufstellung einer empirischen Systematik der Pflanzengesellschaften. Internat. Symp. Rinteln 1966 pp. 8–25. Den Haag.

Detering, K. 1971. Mensch und Automat. Scheidewege 1 (2): 164–207.

Lausi, D. & E. Feoli. 1979. Hierarchical classification of european salt marsh vegetation based on numerical methods. Vegetatio 39: 171–184.

Maarel, E. van der. 1975. The Braun-Blanquet approach in perspective. Vegetatio 30: 213–219.

Pignatti, S. 1975 . Pflanzensoziologie am Scheideweg. Vegetatio 30: 149–152.

Popper, K.R. 1959. The logic of scientific discovery. London. 480 pp.

Westhoff, V. 1972. Die Stellung der Pflanzensoziologie im Rahmen der biologischen Wissenschaften, in Tüxen R.: Grundfragen und Methoden in der Pflanzensoziologie p. 1–15. Den Haag.

Westhoff, V. & E. van der Maarel. 1978. The Braun-Blanquet approach. 2nd ed. in: R.H. Whittaker (ed.), Classification of plant communities p. 287–399. Den Haag.

Yarranton, G.A. 1969. Plant ecology: a unifying model. J. Ecol. 57: 245–250.

Accepted 20 December 1979

LIST OF PARTICIPANTS

SYMPOSIUM ADVANCES IN VEGETATION SCIENCE, 15–19 MAY 1979

Abdel-Razek, Dr. M., Faculty of Science, Alexandria, Egypt.

Ågren, Dr., G. Swedish Coniferous Forest Project, Uppsala, Sweden.

Austin, Dr. M.P., CSIRO, Division of Land Use Research, Canberra, Australia.

Ayyad, Prof. Dr. M.A., Faculty of Science, Alexandria, Egypt.

Beeftink, Dr. W.G., Delta-Instituut, Yerseke, The Netherlands.

Clymo, Dr. R.S., Department of Botany, Westfield College, London, England.

Dale, Dr. M.B., CSIRO, St. Lucia Qld, Australia.

Debussche, Dr. M., CEPE Louis Emberger, Montpellier, France.

Deknopper, E., Labo voor Beschrijvende Plantkunde, Leuven, Belgium.

Faliński, Dr. J.B., Stacja Geobotanicza UW, Bialowiezá Woj. Bialostok, Poland.

Feoli, Dr. E., Istituto ed Orto Botanico, Trieste, Italy.

Fresco, L.F.M., Lab. v. Plantenoecologie, Haren, The Netherlands.

Garcia Novo, Prof. Dr. F., Departamente de Ecología, Sevilla, Spain.

Glenn-Lewin, Prof. D.C., Dept. Botany-Plant Pathology, Ames, U.S.A.

Grabherr, Dr. G., Institut für Systematik und Geobotanik, Innsbruck, Austria.

Greig-Smith, Prof. P., School of Plant Biology, Bangor, United Kingdom.

Grünig, A., Geobotanisches Institut ETH, Zürich, Switzerland.

Guillerm, Dr. J.-L., CEPE Louis Emberger, Montpellier, France.

Hecke, Dr. P. van, Departement Biologie, Wilrijk, Belgium.

Houssard, Dr. C., CEPE Louis Emberger, Montpellier, France.

Hullu, Drs. E. de, Vakgroep Vegetatiekunde, Utrecht, The Netherlands.

Hulst, Prof. R. van, Dept. Biology, Quebec, Canada.

Jancey, Prof. R.C., Dept. Plant Sciences, London Ont., Canada.

Jensén, Dr. S., Dept. Plant Ecology, Lund, Sweden.

Komárková, Dr. V., Institute of Arctic and Alpine Research, Boulder, Colorado, U.S.A.

Krüsi, B., Geobot. Institut ETH, Zürich, Switzerland.

Lausi, Prof. Dr. D., Istituto ed Orto Botanico, Trieste, Italy.

Lepart, Dr. J., CEPE, Montpellier, France.

Lieth, Prof. Dr. H., Fachbereich Dynamische Systeme, Osnabrück, Germany.

Meyer, M., Institut für Meereskunde, Kiel, Germany.

Moral Prof. R. del, Dept. of Botany, Washington, U.S.A.

Orlóci, Prof. L., Dept. of Plant Sciences, London Ont., Canada.

Peet, Prof. R.K., Dept. of Botany, Chapel Hill, U.S.A.

Persson, Dr. H., Växtbiologiska Institutionen, Uppsala, Sweden.

Persson, S., Dept. of Plant Ecology, Lund, Sweden.

Pletl, Dr. L., Lehrstuhl für Grünlandlehre, Freising-Weihenstephan, Germany.

Prentice, Dr. I.C., Dept. Plant Biology, Newcastle upon Tyne, England.

Regnell, G., Dept. of Plant Ecology, Lund, Sweden.

Romane, Dr. F., CEPE Louis Emberger, Montpellier, France.

Ruthsatz, Dr. B., Institut für Landschaftsökologie, Freising, W. Germany.

Rijmenams, J., Labo voor Beschrijvende Plantkunde, Leuven, Belgium.

Sjörs, Prof. Dr. H., Växtbiologiska Institutionen, Uppsala, Sweden.

Smidt, Dr. J.T. de, Vakgroep Vegetatiekunde, Utrecht, The Netherlands.

Spatz, Dr. G., Institut für Grünlandlehre, München, Germany.

Szöcs, Dr. Z., Research Institute for Botany, Budapest, Hungary.

Tosca, Dr. C., Lab. Ecologie Végétale, Toulouse, France.

Trabaud, Dr. L., CEPE Louis Emberger, Montpellier, France.

Waldemarsson-Jensén, E., Dept. Plant Ecology, Lund, Sweden.

Wiegleb, Dr. G., Fachbereich IV, Universität, Oldenburg, Germany.

Wildi, Dr. O., Eindg. Anstalt für das Förstliche Versuchswesen, Birmensdorf, Switzerland.

Participants and symposium assistents from or related to the Division of Geobotany, Nijmegen.

Henk Butteling (Department librarian).

Rita Dubbers (Department secretary).

Drs. Louis Fliervoet (Ph. D. student).

Drs. Jan van Groenendael (Ph. D. Student, Inst. Ecol. Research – Oostvoorne).

Drs. Rob Hengeveld (guest scientist).

Drs. Wil Kortekaas (Ph. D. student)

Tom Kuyper (student).

Drs. Jo Louppen (staff).

Dr. Eddy van der Maarel (staff).

Drs. Madeleine van Mansfeld (guest scientist)

Dr. Frank van der Meulen (guest scientist).

Drs. Ton Roozen (Ph. D. student)

Marijke Roozen-van Mansfeld (student).

Drs. Marty Rijken (guest scientist).

Dr. Marysia Rijken-Kepczyńska (staff).

Drs. Willem Schenk (staff).

Drs. Matthijs Schouten (staff).

Ing. Harry van de Steeg (staff).

Drs. Karlè Sykora (staff).

Drs. O. van Tongeren (guest scientist).

Dr. Marinus Werger (staff).

Prof. Dr. Victor Westhoff (Head of Department).

GPSR Compliance

The European Union's (EU) General Product Safety Regulation (GPSR) is a set of rules that requires consumer products to be safe and our obligations to ensure this.

If you have any concerns about our products, you can contact us on ProductSafety@springernature.com

In case Publisher is established outside the EU, the EU authorized representative is:

Springer Nature Customer Service Center GmbH
Europaplatz 3
69115 Heidelberg, Germany

Batch number: 10593309

Printed by Printforce, the Netherlands